McCance and Widdowson's

The Composition of Foods

McCance and Widdowson's

The Composition of Foods

Fifth revised and extended edition

B. Holland, A. A. Welch, I. D. Unwin
D. H. Buss, A. A. Paul and D. A. T. Southgate

The Royal Society of Chemistry
and
Ministry of Agriculture, Fisheries and Food

A catalogue record for this book is available from the British Library.

ISBN 0–85186–391–4

© The Royal Society of Chemistry 1991
© The Crown 1991

Reprinted 1992, 1993, 1994, 1995, 1997, 1998

Published by the Royal Society of Chemistry, Cambridge, and the Ministry of Agriculture, Fisheries and Food, London.

Photocomposed by Goodfellow & Egan Phototypesetting Ltd, Cambridge
Printed in the United Kingdom by Redwood Books, Trowbridge, Wiltshire

PREFACE

Following publication of the fourth edition of McCance and Widdowson's *The Composition of Foods* in 1978, the Ministry of Agriculture, Fisheries and Food took on the responsibility for maintaining and updating the official tables of food composition in the United Kingdom. In 1987, the Ministry joined with the Royal Society of Chemistry to begin production of a computerised UK National Nutrient Databank from which a number of detailed supplements and now this fifth edition of *The Composition of Foods* have been produced.

This method of production gives greater accuracy, consistency and flexibility than is possible with manual compilation. It has also allowed the simultaneous production of a computer readable file of the data from this book, which is available from the Royal Society of Chemistry. Similar files of data from the supplements are also available, and a range of additional machine-readable products and printed tables will be prepared as the Databank is extended.

Epidemiological studies involving more than one country have heightened interest in international collaboration on the content of food composition tables. Much work has been done on food coding which seeks to describe food in detail and faciliate more precise computer retrieval, as well as to improve the comparability of nutrient data within and between countries. The RSC and MAFF will monitor and, where appropriate, participate in such developments so that the benefits can be available to users of these Tables.

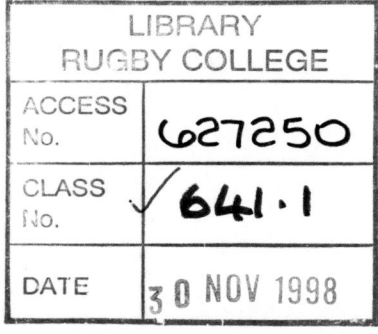

CONTENTS

3. Appendices

FOREWORD

by R. A. McCance and E. M. Widdowson

In 1926 I (R. A. McC) was a medical student at King's College Hospital, London. Dr. R. D. Lawrence, himself a diabetic, was in charge of the diabetic patients, and he was writing a book 'The Diabetic Life'. He wanted to include some values for the carbohydrate content of fruits and vegetables, which were then an important part of diabetic diets, but there were problems with this. First, the values that were being used were derived from Atwater and Bryant's tables published in America in 1906, and these were nearly all obtained 'by difference' that is, water, fat, nitrogen and ash were determined, nitrogen was multiplied by 6.25 to obtain protein, the percentages of these were added together and the sum subtracted from 100 to give the percentage of carbohydrate. Carbohydrate calculated in this way contained not only sugar and starch which were important to the diabetic, but also the 'unavailable carbohydrate' or dietary fibre. Another problem in using the American tables was that most of the analyses had been made on raw materials, whereas people eat most of their vegetables cooked and their composition is altered by cooking. So a grant of £30 a year was obtained from the Medical Research Council for me to analyse raw and cooked fruits and vegetables for total 'available carbohydrate', that is sugars plus starch, which was the value needed for calculating diabetic diets. I analysed 109 different plant materials, each on six separate occasions, in the time I had to spare from my medical studies, and the results were published in 1929 as a Medical Research Council Special Report No. 35 'The Carbohydrate Content of Foods' by R. A. McCance and R. D. Lawrence.

When Professor Cathcart, Professor of Physiology at Glasgow University, read the report he suggested that the work should be extended, and that protein and fat should be determined in meat and fish. The Medical Research Council agreed to provide a grant to cover the salaries of a chemist, H. L. Shipp and a technician, Alec Haynes, and a study of meat and fish began. Sixty-two varieties of fish were analysed, all except oysters cooked, 26 different cuts of meats, 9 varieties of poultry and game and 9 different kinds of 'offal', all cooked in standard ways. Besides total nitrogen, purine N, amino-N and extractive-N were determined and the analyses included fat, carbohydrate when present and minerals Na, K, Ca, Mg, Fe, P and Cl. We also investigated the losses of various constituents when meat and fish were cooked in various ways. Shrinkage caused most of the losses from meat, but not from fish. All the results were published in 1933 as a second Medical Research Council Special Report No. 187 'The Chemistry of Flesh Foods and their Losses on Cooking' by R. A. McCance and H. L. Shipp.

At the end of this study H. L. Shipp left and was replaced by L. R. B. Shackleton, and it was at this point that I (E. M. W.) joined the team. We four started again on fruits, vegetables and nuts. The analyses included 56 varieties of fruit, 9 of nuts, 28 of raw vegetables and 44 of vegetables after cooking. We analysed them for water, total nitrogen, glucose, fructose, sucrose and starch, and for 'unavailable carbohydrate'. The same minerals were determined as in the meat and fish. Losses of sugars, nitrogen and minerals from vegetables while being boiled were also investigated. These results made a third Medical Research Council Special Report, No. 213, published in 1936 'The Nutritive Value of Fruits, Vegetables and Nuts' by R. A. McCance, E. M. Widdowson and L. R. B. Shackleton. The stock of all these reports was destroyed in a fire resulting from an air raid on London during World War II and they have been out of print ever since.

In 1938 we moved to Cambridge. L. R. B. Shackleton left but Alec Haynes came with us. We finished the analyses we had begun in London of cereals, dairy products, beverages and preserves and we put the results of all our analytical work together to make the first edition of 'The Chemical Composition of Foods' by R. A. McCance and E. M. Widdowson. This was published in 1940 as a fourth Medical Research Council Special Report No. 235. The working notebooks containing the details of all the analyses have been deposited with the Wellcome Institute for the History of Medicine.

Since one of the uses of the tables was likely to be the calculation of the composition of diets, and diets generally include cooked dishes we gave some information about their composition. Most of the recipes were taken from standard cookery books, and 90 are to be found in that first edition.

A second edition appeared in 1946, which included values for wartime foods, Household milk, dried eggs and National wheatmeal flour and bread made from it. Values for the composition of about 20 'economical' dishes were included.

In the 1950s we began to work on a third edition. By then many new foods had become available, and those introduced in wartime had disappeared from the market. Alec Haynes had left, and Dr. D. A. T. Southgate joined us. He, with the help of a technician, Janet Adams, was responsible for analysing more than 100 new foods for the same constituents as we had previously done.

By the 1950s methods for the determination of vitamins had improved, and many foods had been analysed for one or more of them. We decided to depart from our original principle of including only the results of our own analyses in the tables, and to use values taken from the literature. Dr. W. I. M. Holman, who knew a great deal about the determination of vitamins in foods, undertook the task of reading every paper he could find on the vitamin content of foods published in the past 15 or 20 years. This involved abstracting well over 1000 papers. He selected those reporting results which he believed to be reliable, and then he left us and his abstracts to take up a post in South Africa. Miss I. M. Barrett joined us, and she constructed the tables of the vitamin content of foods from the information Dr. Holman had collected together.

Values for the amino-acid content of the main protein-containing foods, cereals, meat, fish, eggs, milk and its products, and of some nuts and vegetables were also included in the third edition. These were partly taken from the literature and partly from analyses made by Dr. B. P. Hughes who was working with us at the time. The third edition was published in 1960, with a change of title to 'The Composition of Foods'. As time had gone on some cookery experts had been rather critical of our original recipes, so the whole of the section on the composition of cooked dishes was revised with the help of members of the cookery department of King's College of Household and Social Science.

Up to the third edition we had the ultimate responsibility for the tables. I (R. A. McC) retired in 1966 and it became clear that a decision had to be made about the future of 'The Composition of Foods'. Tables such as these must be revised from time to time or they become obsolete and therefore useless. In the late sixties I (E. M. W) raised the matter at a meeting of the Interdepartmental Committee on Food Composition. It was unanimously agreed that the tables must not be allowed to die. The Interdepartmental Committee on Food Composition accepted responsibility for the revision of the tables, and appointed a Steering Panel under the chairmanship of Dr. D. A. T. Southgate to advise those responsible for the revision, leading to the fourth edition. In the event meats were completely reanalysed. The conformation of farm animals had altered and methods of butchering had changed since the 1930s when the original samples were collected. Cereals, milk and milk products were also extensively revised, but most other foods were not re-analysed, and about a third of the values in the fourth edition, published in 1978, were our original figures, obtained by what are regarded nowadays as very primitive methods 40 years before. Those methods were no less accurate than the modern automated ones, but they took a much longer time.

Since 1978 several supplements to the tables have been published covering the composition of different groups of foodstuffs as these have been revised, and tables showing the composition of foods used by immigrants in the United Kingdom have been made available. Now the fifth edition of 'The Composition of Foods' has been prepared. This represents the work of many people including those who were responsible for making the analyses as we had done half a century ago. We are happy that we are still part of it.

July 1991

ACKNOWLEDGEMENTS

A large number of people have helped at each stage of this book.

Most of the values in this book are based on those in the supplements to 'The Composition of Foods' and, for those food groups where revisions to the fourth edition have yet to be undertaken, to the data in that work. We are therefore indebted to all the people who have helped with this series of books.

The majority of the new analyses of foods included in this edition were conducted at the Laboratory of the Government Chemist and the Reading and Norwich laboratories of the AFRC Institute of Food Research. The analytical teams were headed by Mr I Lumley and Mrs G Holcombe at the LGC, Mr K J Scott and Mr E Florence at the IFR Reading, and Mr R M Faulks at the IFR Norwich. Most of the non-starch polysaccharide fractions were determined at the Dunn Clinical Nutrition Centre, Cambridge, under the direction of Dr H A Englyst. In addition, a number of new foods were analysed at the Leatherhead Food Research Association under the direction of Dr P Farnell and the fatty acid values for pork are from analyses conducted at the Meat and Livestock Commission.

We would like to thank British Bakeries Ltd, Ferrero Ltd, The Kellogg Company of Great Britain Ltd, SmithKline Beecham, Van den Berghs and Jurgens Ltd, Weetabix Ltd and Weston Research Laboratories Ltd for providing information specifically for this fifth edition. We are also indebted to numerous manufacturers, retailers, other organisations and individuals who helped with the fourth edition and supplements.

We would also like to thank members of the British Dietetic Association for their suggestions on useful inclusions to the Tables.

The final preparation of this book was overseen by a committee which, besides the authors, comprised Miss P J Brereton (Northwick Park Hospital, Harrow), Dr M C Edwards (Campden Food and Drink Research Association, Chipping Campden) and Dr A M Fehily (MRC Epidemiology Unit, Cardiff). The committee was joined at various stages by Dr C A Berry (formerly Flour Milling and Baking Research Association, Chorleywood), Dr J L Buttriss (National Dairy Council, London) and Mr R M Faulks and Mr P M Finglas (AFRC Institute of Food Research, Norwich).

We would like to express our appreciation for all the help given to us by so many people in the Ministry of Agriculture, Fisheries and Food, the Royal Society of Chemistry and elsewhere who were involved in the work leading to the production of the book.

INTRODUCTION

1.1 General introduction

This fifth edition of the UK food composition tables extends and updates a series which began with the vision of R A McCance and Elsie Widdowson in the 1930s. Each thorough revision has greatly improved the coverage and value of the preceding edition, and the present book is no exception: the number of nutrients in the main tables has been extended, 1188 foods have now been included compared with 969 in 1978, and some 65 per cent of the nutrient values are new.

The need for this new updated table of food composition arises for a number of reasons. In the 13 years since the fourth edition was published, many new fresh and manufactured foods have become familiar items in our shops, and values for these have been included wherever possible. In addition, the nutritional value of many of the more traditional foods has changed. This can happen when there are new varieties or new sources of supply for the raw materials; with new farming practices which can affect the nutritional value of both plant and animal products; with new manufacturing practices including changes in the type and amounts of ingredients (including reductions in the amount of fat, sugar and salt added or new fortification practices); and with new methods of preparation and cooking in the home.

To ensure that the UK food composition tables could continue to have as wide a coverage and be as up to date as possible, the Ministry of Agriculture, Fisheries and Food (MAFF) decided in the early 1980s to set up a rolling programme of food analysis, largely at the Laboratory of the Government Chemist and the Norwich and Reading laboratories of the Institute of Food Research. The detailed results are published in the scientific literature, and have also been brought together with new manufacturers' information in a series of supplements to the 1978 edition of these tables. The supplements published up to 1991 are listed in Table 1, and are detailed reference works which include an even wider range of nutrients than in the present book. They also include approximately three times as many foods in each of the subgroups of foods covered so far.

Table 1   *Supplements to 'The Composition of Foods' since 1978*

Amino acids and fatty acids per 100 grams of food	HMSO, 1980
Immigrant Foods	HMSO, 1985
Cereals and Cereal Products	RSC, 1988
Milk Products and Eggs	RSC, 1989
Vegetables, Herbs and Spices	RSC, 1991
Fruit and Nuts	RSC, in preparation

The present comprehensive edition includes a wide selection of values from these supplements, some of which have been updated still further. It also includes many new results in most other groups of foods. More analytical work is, however, needed and is being undertaken to determine the nutritional value of, for example, the leaner cuts of meat now available and of the wider range of alcoholic and other drinks. The results of this and other work will continue to be made available in future supplements from the Royal Society of Chemistry (RSC), but this edition has been produced at this time in response to the widely expressed need for a convenient book which includes in one volume the most recent nutrient values for the whole range of common foods.

1.2 Sources of data and methods of evaluation

Many of the values included in these Tables have been derived from the series of analytical studies commissioned by MAFF. Where new analytical data were not available the values have been taken both from the 1978 edition and from more recent sources including the scientific literature, manufacturers' data and by calculation.

Where the values in the Tables were derived by direct analysis of the foods, great care was taken when designing sampling protocols to ensure that the foods analysed were representative of those used by the UK population. For most foods a number of samples were purchased at different shops, supermarkets or other retail outlets. The samples were not analysed separately but were pooled before analysis. When the composite sample was made up from a number of different brands of food, the numbers of the individual brands purchased were related to their relative shares of the retail market. If the food required preparation prior to analysis, techniques such as washing, soaking, cooking, etc. were as similar as possible to normal domestic practices.

Details of the analytical techniques used for this and previous editions are given in Section 3.2.

Where data from literature sources were included in the Tables preference was given to reports where the food was similar to that in the UK, where the publication gave full details of the sample and its method of preparation and analysis, and where the results were presented in a detailed and acceptable form. The criteria for assessing literature values are summarised in Table 2.

Where manufactured foods with proprietary names are included in the database they are restricted to leading brands with an established composition. It should be noted that manufacturers can change their products from time to time and this will influence nutrient content. This is particularly relevant for foods where nutrients are added for fortification purposes, as antioxidants or as colouring agents. The inclusion of a particular brand does not imply that it has a special nutritional value.

The final selection of values published here is dependent on the judgement of the compilers and their interpretation of the available data. There can be no guarantee that a particular item will have precisely the same composition as that in these Tables because of the natural variability of foods.

2

Table 2 *Criteria applied before acceptance of literature values*[a]

Name of food	Common name, with local and foreign synonyms
	Systematic name with variety where known
Origin	*Plants:*
	Country of origin
	Locality, with details of growth conditions if available
	Animals:
	Country of origin
	Locality and method of husbandry and slaughter (if available)
Sampling	Place and time of collection
	Number of samples and how these were obtained
	Nature of sample (e.g. raw, prepared, deep frozen, prepacked etc.)
	Ingredient list details
Treatment of samples before analysis	Conditions and length of storage
	Preparative treatment e.g. material discarded as waste and whether washed or drained
	Cooking details (where applicable) e.g. length of cooking, temperature and the cooking medium
Analysis	Details of material analysed
	Methods used, with appropriate references and details of any modifications
Method of expression of results	Statistical treatment of analytical values
	Whether expressed on an 'as purchased', 'edible matter' or 'dry matter', etc. basis

[a] Modified from Southgate (1974)

1.3 Arrangement of the Tables

This book is composed of three parts, the Introduction, the Tables and a number of Appendices.

The Tables contain four pages of information for each food. The first page gives the food number, name and description along with data for edible proportion and the major nutrients. As in previous editions the foods have been numbered in a continuous sequence from 1 onwards. The food name has been chosen as that most recognisable and descriptive of the food referenced. Information given under the description and number of samples describes the number and nature of the samples taken for analysis. Sources of values derived either from the literature or by calculation are also indicated under this heading.

The second page gives nitrogen, fatty acid totals, cholesterol, starch, total sugars and dietary fibre.

The third page gives data for inorganic elements and the fourth page data for the vitamin composition of the foods.

All nutrients are quoted per 100g of food with the exception of the alcoholic beverages group where they are per 100ml.

Foods have been arranged in groups with common characteristics. The arrangement of the food groups in the Tables are as follows:- cereals and cereal products, milk and milk products, eggs, fats and oils, meat and meat products, fish and fish products, vegetables, herbs and spices, fruits, nuts, sugars, preserves and snacks, beverages, alcoholic beverages and sauces, soups, and miscellaneous foods. Generally the order within the groups is similar to that in the corresponding supplement where the food group has been revised (and the 1978 edition where not). A few foods have been placed in different groups from those in which they previously appeared where this is more appropriate for a general work covering all food groups. Each food group is preceded by text covering points of specific relevance to the foods in that group.

Information contained in the Appendices covers organic acids, a summary of analytical techniques, weight changes on the preparation of foods, cooked foods and dishes, the recipes, and a table of alternative and taxonomic names for foods. These sections provide useful supporting information for the data in the Tables.

A combined food index and coding list is provided at the end of the appendices. This also includes cross-references from alternative food names to the food names used in the Tables.

1.4 The definition and expression of nutrients

The expression of nutrient values

Almost all the nutrient values in the tables apply to the edible part of the food and are expressed per 100g. The only exceptions are where a food is usually served with the inedible matter as an integral part of the food (for example a chop with its bone), when an **extra** line of nutrient values is given which apply to the composition of 100g of the food weighed with its inedible matter.

Generally the values have been expressed to a constant number of decimal places for each nutrient. However, exceptions have been made where appropriate, either within groups of foods or for individual values. For example, the iron content of liquid milks has been expressed to 2 decimal places, because the amounts that can be drunk render this value significant. The values of the more variable vitamins such as vitamin D and biotin have been expressed to less than their usual number of places where large values render the extra places non-significant.

Edible proportion

Many foods are purchased or served with inedible material and a factor is given which shows the proportion of the edible matter in the food. For raw food this refers to the edible material remaining after the inedible waste has been trimmed away, e.g. the outer leaves of a cabbage. For canned foods such as vegetables the factor refers to the edible contents after the liquid has been drained off. For foods that are normally served with waste (for example a chop with its bone) the edible proportion shown for a food is given in association with the nutrient values for 100g of that food weighed **with** its waste.

Protein

For most foods, protein has been calculated by multiplying the total nitrogen value by the factors shown in Table 3.

Table 3 *Factors for converting total grams of nitrogen in foods to protein[a]*

Cereals			Nuts	
Wheat			Peanuts, Brazil nuts	5.41
Wholemeal flour	5.83		Almonds	5.18
Flours, except wholemeal	5.70		All other nuts	5.30
Pasta	5.70			
Bran	6.31		Milk and milk products	6.38
Maize	6.25		Gelatin	5.55
Rice	5.95		All other foods	6.25
Barley, oats, rye	5.83			
Soya	5.70			

[a] FAO/WHO (1973)

The proportion of non-protein nitrogen is high in many foods, notably fish, fruits and vegetables. In most of these, however, this is amino acid in nature and therefore little error is involved in the use of a factor applied to the total nitrogen, although protein in the strictest sense is overestimated. For those foods which contain a measurable amount of non-protein nitrogen in the form of urea, purines and pyrimidines (e.g. mushrooms) the non-protein nitrogen has been subtracted before multiplication by the appropriate factor.

Fat

The fat in most foods is a mixture of triglycerides, phospholipids, sterols and related compounds. The values in the Tables refer to this total fat and not just to the triglycerides.

Carbohydrates

Total carbohydrate and its components, starch and total sugar, but not fibre, are wherever possible expressed as their monosaccharide equivalent. The values for total carbohydrate in the Tables have generally been obtained from the sum of analysed values for these components of 'available carbohydrate', contrasting with figures for carbohydrate 'by difference' which are sometimes used in other food tables or on the labels of manufactured foods. Such figures are obtained by subtracting the measured weights of the other proximates from the total weight and many include the contribution from any dietary fibre present as well as errors from the other analyses. A few values have been included from other tables and are printed in italics to distinguish them from direct analyses.

Available carbohydrate is the sum of the free sugars (glucose, fructose, galactose, sucrose, maltose, lactose and oligosaccharides) and complex carbohydrates (dextrins, starch and glycogen). These are the carbohydrates which are digested and absorbed, and are glucogenic in man.

Carbohydrate values expressed as monosaccharide equivalents can exceed 100g per 100g of food because on hydrolysis 100g of a disaccharide such as sucrose gives 105g monosaccharide (glucose and fructose). 100g of a polysaccharide such as starch gives 110g of the corresponding monosaccharide (glucose). Thus white sugar appears to contain 105g carbohydrate (expressed as monosaccharide) per 100g sugar. For conversion between carbohydrate weights and monosaccharide equivalents, the values shown in Table 4 should be used.

Table 4 *Conversion of carbohydrate weights to monosaccharide equivalents*

Carbohydrate	Equivalents after hydrolysis g/100g	Conversion to monosaccharide equivalents
Monosaccharides e.g. glucose, fructose and galactose	100	no conversion necessary
Disaccharides e.g. sucrose, lactose and maltose	105	× 1.05 or ÷ 0.95
Oligosaccharides e.g.		
raffinose(trisaccharide)	107	× 1.07 or ÷ 0.93
stachyose(tetrasaccharide)	108	× 1.08 or ÷ 0.93
verbascose (pentasaccharide)	109	× 1.09 or ÷ 0.92
Polysaccharides e.g. starch	110	× 1.10 or ÷ 0.90

Any known or measured contribution from oligosaccharides and/or maltodextrins has been included in the total carbohydrate value but not in the columns for starch or total sugars. In most foods oligosaccharides are present in relatively low quantities. In vegetables however, and some processed foods where glucose syrups and maltodextrins are added, oligosaccharides will make a significant contribution to carbohydrate content. Because of this the sum of starch and total sugars will be less than the total carbohydrate for these foods and where this occurs the values have been marked in the Tables with footnotes.

Dietary fibre

Different methods give different estimates of the total fibre content of food. Two values are shown in the Tables, that obtained by the Southgate method (Southgate, 1969) and total non-starch polysaccharides (NSP) (Englyst and Cummings, 1988). In the original Southgate procedure used in the 1978 edition of 'The

Composition of Foods', the dietary fibre constituents were given as mono-saccharide equivalents. These values have now been multiplied by 0.9 to put them on the same basis as the NSP data. The relationships between the various forms and fractions of dietary fibre are shown in Table 5, and further details of the fractions present in NSP are given in the Supplements.

Table 5 *Relationships between the dietary fibre fractions*

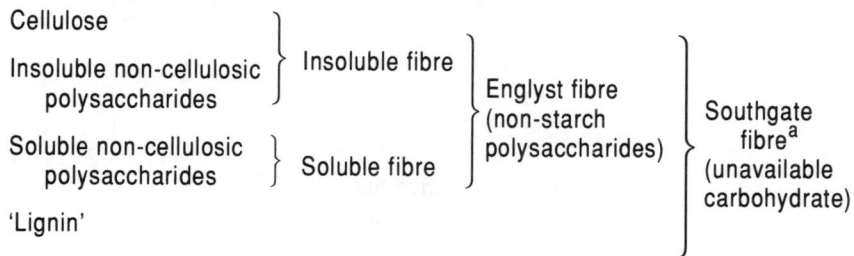

[a] The Southgate values are generally higher than NSP values because they include substances measuring as lignin and also because the enzymatic preparation used leaves some enzymatically resistant starch in the dietary fibre residue. A 'resistant starch' value can be obtained from the NSP procedures but because this uses different conditions and enzymes this may or may not be the same as the enzymatically resistant starch in the Southgate method

Alcohol

The values for alcohol are given as g/100ml of alcoholic beverages. Pure ethyl alcohol has a specific gravity of 0.79 and dividing the values by 0.79 converts them to alcohol by volume (i.e. ml/100ml). The specific gravities of the alcoholic beverages are given in the introduction to that section of the Tables so that calculations can be made if the beverages are measured by weight. The alcohol contents of a range of strengths 'by volume' are given in Table 6.

Table 6 *Alcohol contents of various strengths 'by volume'*

% Alcohol by volume	Alcohol (g/100ml)
5	4.0
10	7.9
15	11.9
20	15.8
25	19.8
30	23.7
35	27.7
40	31.6

Energy value - kcal and kJ

The energy values of all foods are given in both kilocalories (kcal) and kilojoules (kJ). These energy values have been calculated from the amounts of protein, fat, carbohydrate and alcohol in the foods using the energy conversion factors shown in Table 7.

Table 7 *Energy conversion factors used in these Tables[a,b]*

	kcal/g	kJ/g
Protein	4	17
Fat	9	37
Available carbohydrate		
expressed as monosaccharide	3.75	16
Alcohol	7	29

[a] Royal Society (1972)
[b] See Section 1.9 for the conversion factors which should be used in food labelling

These factors permit the calculation of the metabolisable energy of a typical United Kingdom mixed diet with a level of accuracy which compares well with values obtained in human subjects using calorimetry (Southgate and Durnin, 1970).

The energy value of foods in kilojoules can also be calculated from the kilocalorie value using the conversion factor 4.184 kJ/kcal. Whilst it is more accurate to apply the kilojoule factors in Table 7 to protein, fat, carbohydrate and alcohol, a direct kcal/kJ conversion produces differences of little dietetic significance (1–2 per cent).

Fatty acids

For this edition, only total saturated, monounsaturated, and polyunsaturated fatty acids are given. A complete revision of values for the individual fatty acids is planned for the future.

The fat in most foods contains non fatty acid material such as phospholipids and sterols. To allow the calculation of the total fatty acids in a given weight of food, the conversion factors shown in Table 8 were applied.

A worked example is shown below (TFA = total fatty acids, taken from Paul *et al.*, 1980):

	Total fat in beef (sirloin, lean only, roast)	=	9.1g/100g
	Conversion factor	=	0.916
	Total fatty acids in beef = 9.1 × 0.916	=	**8.3g/100g**
Saturates	at 44.9g/100g TFA × 8.3 ÷100	=	3.7g/100g food
Monounsaturates	at 50.8g/100g TFA × 8.3 ÷100	=	4.2g/100g food
Polyunsaturates	at 4.3g/100g TFA × 8.3 ÷100	=	0.4g/100g food

Table 8 *Conversion factors to give total fatty acids in fat*

Wheat, barley and rye		Beef lean	0.916
whole grain	0.720	Beef fat	0.953
flour	0.670	Lamb, take as beef	
bran	0.820	Pork lean	0.910
		Pork fat	0.953
Oats, whole	0.940	Poultry	0.945
Rice, milled	0.850	Brain	0.561
Milk and milk products	0.945	Heart	0.789
Eggs	0.830	Kidney	0.747
		Liver	0.741
Fats and oils		Fish, fatty	0.900
all except coconut oil	0.956	white	0.700
Coconut oil	0.942	Vegetables and fruit	0.800
		Avocado pears	0.956
		Nuts	0.956

Cholesterol

Cholesterol values are included for all foods in this publication and are expressed as mg/100g food. To convert to mmol cholesterol, divide the values by 386.6.

It should be noted that the cholesterol values in the Tables are derived in part from analysis of basic foods but many calculated values have also been included.

Inorganic constituents

Details of the inorganic constituents covered in the Tables are given in Table 9.

Table 9 *Inorganic constituents*

Atomic symbol	Name	Units	Atomic weight[a]
Na	Sodium	mg/100g	23
K	Potassium	mg/100g	39
Ca	Calcium	mg/100g	40
Mg	Magnesium	mg/100g	24
P	Phosphorus[b]	mg/100g	31
Fe	Iron	mg/100g	56
Cu	Copper	mg/100g	64
Zn	Zinc	mg/100g	65
Cl	Chloride	mg/100g	35
Mn	Manganese	mg/100g	55
Se	Selenium	μg/100g	79
I	Iodine	μg/100g	127

[a] To convert the weight of a mineral to mmol or μmol divide by the atomic weight
[b] To convert mg P to mg PO_4 multiply by 3.06

Selenium

The selenium content of soil has a large effect on the foods harvested from it. The levels of selenium in UK soils are low and analysed values reflect this. Data from literature sources have been taken from those countries with similar soil profiles to the UK. Where the values selected are of non-UK origin (or a food is from an overseas source) the values appear in brackets.

Vitamins

Details of vitamins covered in the Tables are given in Table 10.

Table 10   *Vitamins*

Vitamin	Units	International Units (IU)[a]
Vitamin A		
Retinol	µg/100g	0.3µg
Carotene equivalents[b]	µg/100g	0.6µg
Vitamin D	µg/100g	0.025µg
Cholecalciferol, ergocalciferol		
Vitamin E	mg/100g	0.67mg
α-Tocopherol equivalents[b]		
Thiamin (vitamin B_1)	mg/100g	
Riboflavin (vitamin B_2)	mg/100g	
Niacin		
Total preformed niacin	mg/100g	
Tryptophan (mg) divided by 60	mg/100g	
Vitamin B_6	mg/100g	
All forms (pyridoxine, pyridoxal, pyridoxamine and phosphates of these)		
Vitamin B_{12}	µg/100g	
Folate	µg/100g	
Total folate		
Pantothenate	mg/100g	
Biotin	µg/100g	
Vitamin C	mg/100g	
Total ascorbic and dehydroascorbic acid		

[a] Amount equivalent to one International Unit
[b] Values for these vitamins now include contributions from additional fractions not included in the values in the 1978 edition

Vitamin A: retinol and carotene

The two main components of the vitamin are given separately in the Tables.

Retinol is found in many animal products, the main forms being all-*trans* retinol and 13-*cis* retinol. The latter has about 75% of the activity of the former (Sivell *et al.*, 1984). Eggs and fish roe also contain retinaldehyde which has 90% of the activity of all-*trans* retinol. Retinol is expressed in the Tables as the weight of all-*trans* retinol equivalent, i.e. the sum of all-*trans* retinol plus contributions from the other two forms after correction to account for their relative activities.

Approximately 600 carotenoids are found in plant products and milks but few have vitamin A activity (Olson, 1989). Of these, the most important is β-carotene. The other main forms with vitamin A activity are α-carotene and α- and β-crypto-xanthins, which have approximately half the activity of β-carotene. Carotene is expressed in the Tables in the form of β-carotene equivalents, that is the sum of the β-carotene and half the amounts of α-carotene and α- and β-cryptoxanthins present. Where the carotenoid profile was incomplete, it has been assumed that all is β-carotene. This may result in an overestimate but as α-carotene and cryptoxanthin are usually present in low levels in foods without complete caroten-oid profiles, it is likely that any error is small.

Retinol equivalents

In the UK Dietary Reference Values and other Recommended Daily Intakes the requirement for vitamin A is expressed as retinol equivalents. This measure of the overall potency of vitamin A relates to the lower biological efficiency of carotenoids compared with retinol. Although the absorption and utilisation of carotenes vary, for example with the amount of fat in the diet and with β-carotene concentration (Brubacher and Weiser, 1985), the generally accepted relationship is that 6µg β-carotene or 12µg of all other active caroteniods are equivalent to 1µg retinol (Department of Health, 1991), so that:-

Vitamin A potency as µg retinol equivalents $= $µg retinol $+ \dfrac{\text{µg β-carotene equivalents}}{6}$

The relationship between the different units used to express vitamin A is shown in Table 11.

Table 11   *Relationship and conversion between the units used to express retinol and carotene.*

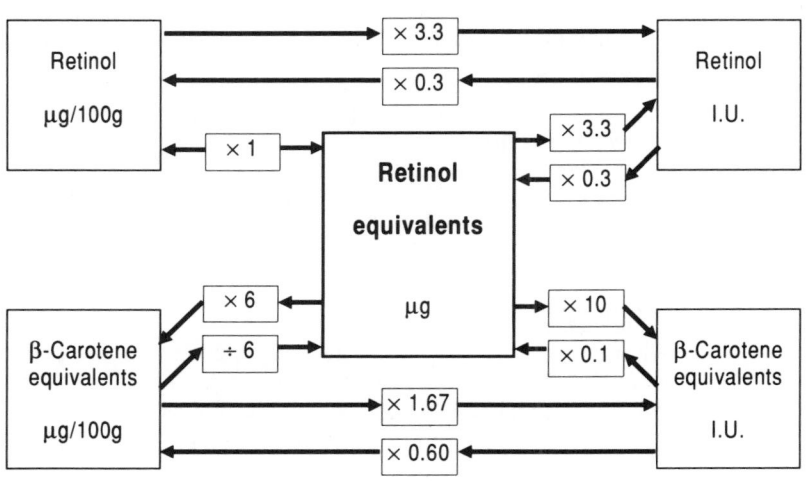

Table 12 *Conversion factors for vitamin E activity*[a]

alpha-tocopherol	×	1.00
beta-tocopherol	×	0.40
gamma-tocopherol	×	0.10
delta-tocopherol	×	0.01
alpha-tocotrienol	×	0.30
beta-tocotrienol	×	0.05
gamma-tocotrienol	×	0.01

[a] McLaughlin and Weihrauch (1979)

Vitamin D

Few foods contain vitamin D. All those which do so naturally are animal products and contain vitamin D_3 (cholecalciferol) derived, as in humans, from the action of sunlight on the animal's skin or from its own food. Vitamin D_2 (ergocalciferol) made commercially has the same potency in man. Both vitamin D_2 and vitamin D_3 are used to fortify a number of foods.

Vitamin E

The vitamin E in food is present as various tocopherols and tocotrienols, each having a different level of vitamin E activity. In most animal products the α-form is the only significant form present but in plant products, especially seeds and their oils, γ-tocopherol and other forms are present in significant amounts. The values for vitamin E are expressed as α-tocopherol equivalents, using the factors shown in Table 12.

In the 1978 edition the Table values for vitamin E were for α-tocopherol only, although data for other tocopherols or for tocotrienols were given as footnotes. In this edition, the values given for vitamin E include the contribution from all the forms adjusted for their activity in terms of α-tocopherol equivalents. Therefore, where a significant proportion of vitamin E is present as vitamers other than α-tocopherol, the figures for total vitamin E are higher than those in the 1978 edition, for example in the fish products and fats and oils food groups.

Thiamin

The majority of values for thiamin are expressed as thiamin hydrochloride.

Niacin

The values are the sum of nicotinic acid and nicotinamide which are collectively known as niacin.

Tryptophan is converted in the body to nicotinic acid with varying efficiency. On average, 60mg tryptophan is equivalent to 1mg niacin, so the tryptophan content of the protein in each food has been shown after division by 60. This may be added to the amount of niacin to give the niacin equivalent for the food.

Vitamin B_6

Vitamin B_6 occurs in foods as pyridoxine, pyridoxal, pyridoxamine and their phosphates. However, the active form in the tissues is pyridoxal phosphate. In the main, pyridoxine is expressed in the Tables as pyridoxine hydrochloride.

Folate

For folates, the value refers to total folates measured by microbiological assay after deconjugation of the polyglutamyl forms. The values in this edition are generally higher than in the 1978 edition because conditions of the micro-biological assay have been modified to ensure that all forms of folate give equivalent responses in the assay.

Pantothenate

The majority of values for pantothenate are expressed as calcium D-pantothen-ate.

Vitamin C

Values include both ascorbic acid and dehydroascorbic acid, as both forms are biologically active. In fresh foods the reduced form is the major one present but the amount of the dehydro-form increases during cooking and processing.

1.5   The variability of nutrients in foods

Although the values in these Tables have been derived from careful analyses of representative samples of each food, it is important to appreciate that the composition of any individual sample may differ considerably from this. There are two main reasons for the variability, apart from the apparent differences caused by analytical variations.

Natural variation

All natural products vary in composition. Two samples from the same animal or plant may well be different, but the composition of meat, milk and eggs are also affected by season and by the feeding regime and age of the animal. Different varieties of the same plant may differ in composition, and their nutritional value will also vary with the country of origin, growing conditions and subsequent storage. In general, those nutrients that are closely associated with structure and metabolic function show rather less variation than those which accumulate in particular locations of the plant or animal or those which are unstable. For instance, nitrogen and phosphorus tend to show less variation than vitamin A, iron or vitamin C.

A major influence on the nutrient concentration in foods is the water content and this is particularly important in plant foods where water is the main constituent. As the length and conditions of food storage affect the water content of foods, these will have an effect on their nutrient content per 100g. Many individual nutrients will also be affected by storage conditions with the greatest effect being on the more labile vitamins such as vitamin C, vitamin E and folate. Thus if the storage conditions of a food item differ from those for the samples analysed for the Tables, the nutrient values may differ from those given.

The level of fat in food can vary greatly and result in large variations in the nutrient content of each 100g of the food. It will also influence energy and the level of fat soluble vitamins.

Extrinsic differences

Further differences in composition can be introduced by food manufacturers, caterers, and in the home. For example, manufacturers may change both their recipes and their fortification practices, and dishes prepared in the home or by caterers may vary widely in the amounts and types of ingredient used and thus differ in nutritional value from those included here.

Examples of some external influences on nutrient contents are shown below:

Sodium
The level found in many foods will depend upon the amount of salt and other sodium-containing compounds used in cooking or added by manufacturers, and can therefore be very variable. The majority of vegetables analysed for the food tables were cooked in distilled water without salt, although there are a few to which salt was added and these are indicated in the Tables. For planning low-sodium diets the Table values are adequate.

Potassium
The potassium content of boiled vegetables is dependent on the amount of water, length of cooking time and the state of preparation of the vegetable. The user should refer to the description and main data sources for the foods in the Tables to ensure sample foods are comparable.

Calcium
Most vegetables in the Tables were cooked in distilled water. Foods cooked in and prepared with tap water, which contains variable amounts of calcium, may not have the same levels as in these Tables.
The concentration of calcium in baking powder is high and variations in the quantity used will affect the calcium content of some cereal products.

Iron
Food can become contaminated with iron from knives, pans, soil particles and processing machinery. This has most effect on the iron content of ground foods such as spices.

Chloride
Chloride variation will be similar to that of sodium.

Iodine
Iodine levels in milk are affected by the levels in animal feedstuffs, and to a lesser degree by the iodine levels in the solutions used for teat dips, sanitizers and the lactation promoter iodinated casein (Phillips et al., 1988).

β-Carotene
This is sometimes used as a food colouring additive (E160a). In certain manufactured foods such as orange squash, samples may contain added β-carotene.

Vitamin C
This is added to a number of foods for fortification or antioxidant purposes (E300, L-ascorbic acid) and so may be present in unexpectedly high levels in some foods, including some meat products and soft drinks.

1.6   Bioavailability of nutrients

The term bioavailability means the proportion of a nutrient capable of being absorbed and available for use or storage (Bender, 1989).

Many nutrients are not fully absorbed and utilised, but because the interactions are so variable it is not possible to provide an accurate measure of bioavailability in these Tables. Bioavailability depends not only on the chemical form of the nutrients that occur in food, but also on the level of intake, the presence of binding agents and other nutrients in foods that are ingested at the same time, and on factors associated with the individual. For many minerals the homeostatic mechanisms which regulate absorption prevent build-up of potentially toxic concentrations. In addition, the potentially inhibiting effects on absorption of factors such as dietary fibre need to be viewed in the context of the diet as a whole (Southgate, 1987).

Allowance has been made for the reduced activities of different forms of three of the vitamins, i.e. 13-cis retinol and retinaldehyde, carotenes other than β-carotene, and tocopherols and tocotrienols other than α-tocopherol, as described in section 1.4. Other nutrients in the Tables which are absorbed and utilised with varying degrees of efficiency include iron, calcium, magnesium, zinc, copper, manganese, selenium, niacin and folate. For all these, no allowance is made in these Tables for the lower availability and the values quoted represent the actual value in foods.

Iron

Dietary iron occurs in two well recognised forms, haem and non-haem. These exhibit different levels of absorption via separate pathways (FAO, 1988). Haem iron, which is contained in the haemoglobin and myoglobin of animal foods, is relatively available and its absorption is relatively unaffected by other food items or by the iron status of an individual.

Non-haem iron easily forms complexes which are less readily solubilised and absorbed than haem iron. The absorption of non-haem iron is highly variable depending on the nature of the meal. Ascorbic acid enhances absorption, as do animal protein and haem iron. But absorption is impaired by a large number of substances, such as tannins, phytates, some forms of protein and dietary fibre, and by interactions with other inorganic constituents. Tea and eggs are notable inhibitors of non-haem iron absorption (FAO, 1988; Fairweather-Tait, 1991).

Calcium

The amount of calcium absorbed is dependent on individual vitamin D status, the customary level of calcium intake and needs of the individual and the presence of binding substances in the food (Allen, 1982). Phytic acid is the principal inhibitory factor, but owing to the intestinal breakdown of phytate, it is unlikely that its presence is of practical significance in western diets.

There is some evidence that calcium absorption is enhanced from milk, and is adversely affected by oxalates in foods such as spinach, but the effect of these and other factors on the bioavailability of calcium in normal mixed diets is likely to be small (Miller, 1989).

Magnesium, zinc, copper, manganese and selenium

Absorption of these elements is usually low and can be influenced by phytate (particularly for zinc), fibre, interactions with each other and further mineral

15

elements. Zinc is better absorbed from animal tissues than from plants (Solomons, 1982; Schwartz et al., 1986; Sanstead, 1982; Young et al., 1982).

Niacin

Much of the niacin occurring naturally in cereals is in a bound form and may be unavailable (Carter and Carpenter, 1982).

Folate

Folate bioavailability is very variable in different foods and depends, amongst other factors, on the form in which it occurs, the presence of conjugase inhibitors, and the degree of heating and cooking. Some studies suggest that dietary fibre may exert an inhibitory effect.

1.7 Calculation of nutrient intakes using the Tables

Calculation

There are several steps involved in the calculation of nutrient intake from the Tables. The first is to choose the item in the Tables which corresponds most closely with the food consumed. The index includes many alternative names and it should be noted that a food may be found in a different food group from the one in which it is expected.

If the food consumed is not in the Tables then it is necessary to choose a suitable alternative by consideration of the food type, general characteristics and likely nutrient profile. The results, however, are likely to be less accurate.

Once the food has been chosen, calculation of nutrient intake is achieved by multiplying the nutrient figure quoted in the Tables by the weight of the food consumed (nutrients are expressed either per 100g of the edible portion of the food, or per 100ml for alcoholic beverages), e.g. if 80g food has been consumed, the nutrient should be multiplied by 0.8, and if 120g consumed multiplied by 1.2. The results from these calculations are then summed to give the total intake.

Computerised calculation

Calculation of nutrient intake 'by hand' is a time consuming process which has largely been superseded by the use of computers. Information concerning the datafiles and packages available both for personal computers and mainframes can be obtained from the Royal Society of Chemistry.

Recipes

If the sample of food consumed is a cooked dish prepared with a different recipe from any of those in this book, the nutrients for the new recipe can be calculated using the methods given in Section 3.4.

Portion sizes

If the weight of food consumed has not been recorded or if an estimate is required, publications such as Bingham and Day (1987), Crawley (1988), and Davies and Dickerson (1991) may be used to provide information on portion sizes. In field-work, representations such as pictures, models or household measures may also be used to obtain estimates of portion size.

1.8 Potential pitfalls when using the Tables

Those who are unfamiliar the uses of these tables should note the following points which can reflect on the accuracy of the information obtained from them.

- When comparing the nutrient values in these Tables with those of other countries or literature reports, the expression of units and conversion factors used in calculation may vary.

- As nutrients are increasingly added to foods for fortification, antioxidant and colorant purposes, users should check the labels of manufactured products.

- Missing nutrient values in food composition tables should not be treated as zero values during calculation otherwise an underestimation of nutrient intake will result. However, the major sources of any nutrient are likely to have been analysed and included in these Tables.

- Errors will arise if food is classified incorrectly, for instance it may be assumed that milk has been consumed in the full fat form when it was in fact skimmed.

- Misclassification of foods may arise as a result of a food having several names. It is therefore important to be familiar with local and alternative names when using food tables, e.g. roast potatoes are known as baked potatoes in some parts of the country.

- In manual coding systems incorrect food code numbers may be used. Computerised systems which avoid the use of numbers and input information only by the food name tend to reduce this problem. However, it is still possible for names to be identified incorrectly during the use of the Tables and calculation software.

- It is possible that errors can be made both in the measurement and recording of food weights which will affect the calculation of nutrient intakes.

- Sources of estimated weight are more prone to error than the recorded weight of food because the portion size chosen by an investigator may not give a true indication of the actual amount eaten or an individual may misinterpret the amount shown in a representation of a portion size.

- There are several methods for collecting food intake data which range from weighed intakes to food frequency questionnaires giving information which is either quantitative or qualitative. It is worthwhile consulting appropriate publications, e.g. Bingham (1987), Cameron and van Staveren (1988), to find which method is the most suitable for the level of information required.

- As any one person exhibits a great deal of variation in diet, varied lengths of recording time are needed to assess representative intakes of nutrients. For example a 7-day weighed record collection (not necessarily consecutive days) may be necessary to assess energy and protein intakes assuming that an accuracy of $\pm10\%$ standard error is acceptable. It may be possible to observe people with very stable eating habits for a shorter time but those with greater variation may require longer. For most other nutrients the recording period would need to be longer than for energy and protein, particularly for those concentrated in only a few foods. For example, vitamin C may require 36 days of recording to be within $\pm10\%$ of the true intake. This topic is covered in greater detail in Bingham (1987) and Cameron and van Staveren (1988).

1.9   Food labelling

Nutrition information is increasingly being given on food labels. Values from these food composition tables may be used for this purpose, but only if certain conditions are met.

The rules which govern nutrition labelling are the (UK) Food Labelling Regulations 1984, and the (EEC) Directive on Nutrition Labelling for Foodstuffs of 1990. These rules are there to ensure consistency and accuracy, and to prevent misleading claims. Nutrition labelling is not compulsory unless a nutrition claim is made, but when such information is given the details in one of the following groups must be shown per 100 grams or per 100ml of the food as sold:

Either

energy value in kJ and kcal, **and**
protein, carbohydrate and fat, in grams, **and**
the amount of any other nutrients for which a claim is made

Or (optionally until October 1995)

energy value in kJ and kcal, **and**
protein, carbohydrate, sugars, fat, saturates, fibre and sodium, all in grams, **and**
the amount of any other nutrient for which a claim is made

Preference should be given to values derived from analyses of representative samples of the food. However, if the product or its ingredients are similar to those described in this book or the supplements to the 1978 Edition, these values may be used instead. Nevertheless, it is important to note the following differences:–

1. Protein should be given as total nitrogen $\times$ 6.25 for every food, whereas more specific factors have been used in this book.

2. Carbohydrate is to be declared as the weight of the carbohydrates themselves and not their monosaccharide equivalents. The following factors may be used to convert monosaccharide equivalents from these Tables to actual weights:

Total carbohydrate	Divide by 1.05 unless it is known to be mainly starch
Starch	Divide by 1.10
Sucrose and lactose	Divide by 1.05
Glucose, etc.	As given

3. Different factors are to be used to calculate energy values. These are shown in Table 13.

Table 13   *Energy conversion factors to be used in food labelling*

	kcal/g	kJ/g
Protein	4	17
Carbohydrate expressed as weight	4	17
Fat	9	37
Alcohol	7	29
All organic acids	3	13
Sorbitol and other polyols	2.4	10

The Tables

Symbols and abbreviations used in the Tables

Symbols

0	None of the nutrient is present
Tr	Trace
N	The nutrient is present in significant quantities but there is no reliable information on the amount
( )	Estimated value
Italic text	Carbohydrate estimated 'by difference', and energy values based upon these quantities

Abbreviations

IFR	Institute of Food Research, Norwich
LGC	Laboratory of The Government Chemist, Teddington
calcd.	calculated

Cereals and cereal products

The majority of the data and foods in this section of the Tables have been taken from the *'Cereals and Cereal Products'* supplement.

Values from the literature for wheat flours and their products were restricted to those from the UK because flours are required to be fortified by law (The Bread and Flour Regulations, 1984). UK flour should contain at least 1.65mg iron, 0.24mg thiamin and 1.60mg niacin per 100g and so these nutrients are added to all white flours and most brown flours in this country. Calcium carbonate must also be added to all flours except wholemeal and certain self-raising flours at a rate equivalent to 94 – 156mg calcium per 100g flour.

Fatty acid values were derived either from new analyses, data in the 1978 edition or by recipe calculation. The fatty acid profile of margarine used in the recipes was an average of hard, soft and polyunsaturated varieties.

Modifications to the nutrient values published in the *'Cereals and Cereal Products'* supplement include changes for breakfast cereals to reflect more recent fortification practices and corrections as noted in the erratum slip to *'Cereals and Cereal Products'*. Where data in other food groups has been revised since that publication and the changes have a significant effect on the nutrient values for cereal-based recipes included in this section, revised nutrient values have been included.

Losses of labile vitamins assigned on recipe calculation were estimated using the figures in Section 3.4. Changes in weight on toasting bread and boiling rice and pastas are shown in Section 3.3. Taxonomic names for foods included in this part of the Tables can be found in Section 3.6.

Cereals and cereal products

Composition of food per 100g

No.	Food	Description and main data sources	Edible proportion	Water g	Protein g	Fat g	Carbohydrate g	Energy value kcal	Energy value kJ
	Flours, grains and starches								
1	Bran, wheat	Analytical and literature sources	1.00	8.3	14.1	5.5	26.8	206	872
2	Chapati flour, brown	1 sample, single supplier	1.00	12.2	11.5	1.2	73.7	333	1419
3	white	2 samples, different suppliers, same weights	1.00	12.0	9.8	0.5	77.6	335	1426
4	Cornflour	3 samples from different shops	1.00	12.5	0.6	0.7	92.0	354	1508
5	Custard powder	Taken as cornflour except Na, Cl and Cu	1.00	12.5	0.6	0.7	92.0	354	1508
6	Oatmeal, quick cook, *raw*	10 samples, 8 brands	1.00	8.2	11.2	9.2	66.0	375	1587
7	Rye flour, *whole*	Analytical and literature sources	1.00	15.0	8.2	2.0	75.9	335	1428
8	Sago, *raw*	2 samples from different shops	1.00	12.6	0.2	0.2	94.0	355	1515
9	Soya flour, *full fat*	Analytical and literature sources	1.00	7.0	36.8	23.5	23.5	447	1871
10	*low fat*	Analytical and literature sources	1.00	7.0	45.3	7.2	28.2	352	1488
11	Tapioca, *raw*	4 varieties, medium pearl, seed pearl, coarse and flake	1.00	12.2	0.4	0.1	95.0	359	1531
12	Wheat flour, brown	VFSS, 1977-81, and literature sources	1.00	14.0	12.6	1.8	68.5	323	1377
13	white, *breadmaking*	Data from Voluntary Flour Sampling Scheme	1.00	14.0	11.5	1.4	75.3	341	1451
14	white, *plain*	(VFSS), 1977-81 plus literature sources.	1.00	14.0	9.4	1.3	77.7	341	1450
15	white, *self-raising*	Biscuit and cake flours are similar in composition	1.00	14.0	8.9	1.2	75.6	330	1407
16	wholemeal	to plain flour	1.00	14.0	12.7	2.2	63.9	310	1318
17	Wheatgerm	Literature sources	1.00	11.7	26.7	9.2	(44.7)	357	1509

Cereals and cereal products

No.	Food	Total nitrogen g	Fatty acids			Cholest-erol mg	Starch g	Total sugars g	Dietary fibre	
			Satd g	Mono unsatd g	Poly unsatd g				Southgate method g	Englyst method g

Flours, grains and starches

No.	Food	Total nitrogen g	Satd g	Mono unsatd g	Poly unsatd g	Cholest-erol mg	Starch g	Total sugars g	Southgate method g	Englyst method g
1	**Bran**, wheat	2.24	0.9	0.7	2.9	0	23.0	3.8	39.6	36.4
2	**Chapati flour**, brown	2.02	0.2	0.1	0.5	0	70.5	3.2[a]	10.3	N
3	white	1.72	0.1	Tr	0.2	0	75.5	2.1[a]	4.1	N
4	**Cornflour**	0.09	0.1	0.1	0.3	0	92.0	Tr	N	0.1
5	**Custard powder**	0.09	0.1	0.1	0.3	0	92.0	Tr	N	(0.1)
6	**Oatmeal**, quick cook, raw	1.92	1.6	3.3	3.7	0	64.9	1.1	6.8	7.1
7	**Rye flour**, whole	1.40	0.3	0.2	0.9	0	75.9	Tr	N	11.7
8	**Sago**, raw	0.04	0.1	0.1	Tr	0	94.0	Tr	N	0.5
9	**Soya flour**, full fat	6.45	2.9	4.5	11.4	0	12.3	11.2	10.7	11.2
10	low fat	7.94	0.9	1.4	3.5	0	14.8	13.4	13.3	(13.5)
11	**Tapioca**, raw	0.07	Tr	Tr	Tr	0	95.0	Tr	N	0.4
12	**Wheat flour**, brown	2.20	0.2	0.2	0.8	0	66.8	1.7[a]	7.0	6.4
13	white, breadmaking	2.02	0.2	0.1	0.6	0	73.9	1.4[a]	3.7	(3.1)
14	white, plain	1.64	0.2	0.1	0.6	0	76.2	1.5[a]	3.6	3.1
15	white, self-raising	1.56	0.2	0.1	0.5	0	74.3	1.3[a]	4.1	(3.1)
16	wholemeal	2.18	0.3	0.3	1.0	0	61.8	2.1[a]	8.6	9.0
17	**Wheatgerm**	4.54	1.3	1.1	4.2	0	(28.7)	(16.0)	N	15.6

[a] Includes the glucofructan levosin

25

Cereals and cereal products

1 to 17

Inorganic constituents per 100g

Flours, grains and starches

No.	Food	Na	K	Ca	Mg	P	Fe	Cu	Zn	Cl	Mn	Se	I
						mg						µg	
1	**Bran, wheat**	28	1160	110	520	1200	12.9	1.34	16.2	150	9.0	(2)	N
2	**Chapati flour,** brown	39	280	86	69	250	3.4	0.33	2.1	67	2.0	N	N
3	white	15	200	84	29	140	2.5	0.25	1.3	68	1.0	N	N
4	**Cornflour**	52	61	15	7	39	1.4	0.13	0.3	71	N	N	N
5	**Custard powder**	320	61	15	7	39	1.4	0.05	0.3	480	N	N	N
6	**Oatmeal,** quick cook, raw	9	350	52	110	380	3.8	0.49	3.3	25	3.9	3	N
7	**Rye flour,** whole	(1)	410	32	92	360	2.7	0.42	3.0	N	0.7	N	N
8	**Sago,** raw	3	5	10	3	29	1.2	0.03	N	13	N	N	N
9	**Soya flour,** full fat	9	1660	210	240	600	6.9	2.92	3.9	110	2.3	9	N
10	low fat	14	2030	240	290	640	9.1	3.12	3.2	N	2.9	(11)	N
11	**Tapioca,** raw	4	20	8	2	30	0.3	0.07	N	13	N	N	N
12	**Wheat flour,** brown	4	250	130[a]	80	230	3.2[a]	0.32	1.9	45	1.9	N	N
13	white, breadmaking	3	130	140[b]	31	120	2.1[b]	0.18	0.9	62	0.7	42	10
14	white, plain	3	150	140[b]	20	110	2.0[b]	0.15	0.6	81	0.6	4	N
15	white, self-raising	360[c]	150	350[c]	20	450[c]	2.0[b]	0.17	0.6	88	0.6	4	10
16	wholemeal	3	340	38	120	320	3.9	0.45	2.9	38	3.1	53	N
17	**Wheatgerm**	5	950	55	270	1050	8.5	0.90	17.0	80	12.3	(3)	N

[a] These are levels for fortified flour. Unfortified brown flour would contain about 20mg Ca and 2.5mg Fe per 100g
[b] These are levels for fortified flour. Unfortified white flours would contain about 15mg Ca and 1.5mg Fe per 100g
[c] The amount present will depend on the nature and level of the raising agent used

Cereals and cereal products

Vitamins per 100g

No.	Food	Retinol μg	Carotene μg	Vitamin D μg	Vitamin E mg	Thiamin mg	Riboflavin mg	Niacin mg	Trypt 60 mg	Vitamin B6 mg	Vitamin B12 μg	Folate μg	Pantothenate mg	Biotin μg	Vitamin C mg
Flours, grains and starches															
1	**Bran**, wheat	0	0	0	2.60	0.89	0.36	29.6	3.0	1.38	0	260	2.4	45	0
2	**Chapati flour**, brown	0	0	0	(0.60)	0.26	0.05	3.8	2.4	0.29	0	29	(0.4)	(3)	0
3	white	0	0	0	(0.30)	0.36	0.06	1.9	2.0	0.17	0	20	(0.3)	(1)	0
4	**Cornflour**	0	0	0	Tr	Tr	Tr	Tr	0.1	Tr	0	Tr	Tr	Tr	0
5	**Custard powder**	0	0	0	Tr	Tr	Tr	Tr	0.1	Tr	0	Tr	Tr	Tr	0
6	**Oatmeal**, quick cook, *raw*	0	0	0	1.50	0.90	0.09	0.8	2.6	0.33	0	60	1.2	21	0
7	**Rye flour**, *whole*	0	0	0	1.60	0.40	0.22	1.0	1.6	0.35	0	78	1.0	6	0
8	**Sago**, *raw*	0	0	0	Tr	Tr	Tr	Tr	Tr	Tr	0	Tr	Tr	Tr	0
9	**Soya flour**, *full fat*	0	N	0	1.50	0.75	0.28	2.0	8.6	0.46	0	345	1.6	N	0
10	*low fat*	0	N	0	N	0.90	0.29	2.4	10.6	0.52	0	410	1.8	N	0
11	**Tapioca**, *raw*	0	0	0	Tr	Tr	Tr	Tr	0.1	Tr	0	Tr	Tr	Tr	0
12	**Wheat flour**, brown	0	0	0	0.60	0.39[a]	0.07	4.0[a]	2.6	(0.30)	0	51	(0.4)	(3)	0
13	white, *breadmaking*	0	0	0	(0.30)	0.32[b]	0.03	2.0[b]	2.3	0.15	0	31	0.3	1	0
14	white, *plain*	0	0	0	0.30	0.31[b]	0.03	1.7[b]	1.9	0.15	0	22	0.3	1	0
15	white, *self-raising*	0	0	0	(0.30)	0.30[b]	0.03	1.5[b]	1.8	0.15	0	19	0.3	1	0
16	wholemeal	0	0	0	1.40	0.47[b]	0.09	5.7[b]	2.5	0.50	0	57	0.8	7	0
17	**Wheatgerm**	0	0	0	22.00	2.01	0.72	4.5	5.3	3.30	0	331	1.9	25	0

[a] These are levels for fortified flour. Unfortified brown flour would contain 0.30mg thiamin and 1.7mg niacin per 100g
[b] These are levels for fortified flour. Unfortified white flours would contain 0.10mg thiamin and 0.7mg niacin per 100g

Cereals and cereal products *continued*

18 to 32

Composition of food per 100g

No.	Food	Description and main data sources	Edible proportion	Water g	Protein g	Fat g	Carbo-hydrate g	Energy value kcal	kJ
Rice									
18	**Brown rice**, *raw*	5 assorted samples	1.00	13.9	6.7	2.8	81.3	357	1518
19	*boiled*	Water content weighed, other nutrients calculated from raw	1.00	66.0	2.6	1.1	32.1	141	597
20	**Savoury rice**, *raw*	10 samples, 5 varieties, meat and vegetable	1.00	7.0	8.4	10.3	77.4	415	1755
21	*cooked*	Calculation from raw, boiled in water	1.00	68.7	2.9	3.5[a]	26.3	142	599
22	**White rice**, easy cook, *raw*	10 samples, 9 different brands, parboiled	1.00	11.4	7.3	3.6	85.8	383	1630
23	easy cook, *boiled*	Calculation from raw	1.00	68.0	2.6	1.3	30.9	138	587
24	fried in lard/dripping	Recipe	1.00	70.3	2.2	3.2	25.0	131	554
Pasta									
25	**Macaroni**, *raw*	10 samples, 7 brands; literature sources	1.00	9.7	12.0	1.8	75.8	348	1483
26	*boiled*	10 samples, 7 brands boiled in water	1.00	78.1	3.0	0.5	18.5	86	365
27	**Noodles**, egg, *raw*	10 samples, 8 brands	1.00	9.1	12.1	8.2	71.7	391	1656
28	egg, *boiled*	10 samples, 8 brands boiled in water	1.00	84.3	2.2	0.5	13.0	62	264
29	**Spaghetti**, white, *raw*	10 samples, 7 brands	1.00	9.8	12.0	1.8	74.1	342	1456
30	white, *boiled*	10 samples, 7 brands boiled in water	1.00	73.8	3.6	0.7	22.2	104	442
31	wholemeal, *raw*	10 samples, 5 brands	1.00	10.5	13.4	2.5	66.2	324	1379
32	wholemeal, *boiled*	Water content weighed, other nutrients calculated from raw	1.00	69.1	4.7	0.9	23.2	113	485

[a] Calculated assuming water only was added; savoury rice cooked with fat contains approximately 8.8g fat per 100g

Cereals and cereal products *continued*

Composition of food per 100g

No.	Food	Total nitrogen g	Fatty acids Satd g	Mono unsatd g	Poly unsatd g	Cholest-erol mg	Starch g	Total sugars g	Dietary fibre Southgate method g	Englyst method g
Rice										
18	**Brown rice**, *raw*	1.10	0.7	0.7	1.0	0	80.0	1.3	3.8	1.9
19	*boiled*	0.43	0.3	0.3	0.4	0	31.6	0.5	1.5	0.8
20	**Savoury rice**, *raw*	1.41	3.2	3.7	1.8	1	73.8	3.6	4.0	N
21	*cooked*	0.48	1.1	1.3	0.6	Tr	25.1	1.2	1.3	1.4
22	**White rice**, easy cook, *raw*	1.23	0.9	0.9	1.3	0	85.8	Tr	2.7	0.4
23	easy cook, *boiled*	0.44	0.3	0.3	0.5	0	30.9	Tr	1.0	0.1
24	fried in lard/dripping	0.37	1.4	1.2	0.5	3	23.1	1.9	1.2	0.6
Pasta										
25	**Macaroni**, *raw*	2.11	0.3	0.1	0.8	0	73.6	2.2	5.0	3.1[a]
26	*boiled*	0.52	0.1	Tr	0.2	0	18.2	0.3	1.5	0.9[a]
27	**Noodles**, egg, *raw*	2.12	2.3	3.5	0.9	30	69.8	1.9	5.0	(2.9)
28	egg, *boiled*	0.40	0.1	0.2	0.1	6	12.8	0.2	1.0	(0.6)
29	**Spaghetti**, white, *raw*	2.11	0.2	0.2	0.8	0	70.8	3.3	5.1	2.9
30	white, *boiled*	0.63	0.1	0.1	0.3	0	21.7	0.5	1.8	1.2
31	wholemeal, *raw*	2.30	0.4	0.3	1.1	0	62.5	3.7	11.5	8.4
32	wholemeal, *boiled*	0.81	0.1	0.1	0.4	0	21.9	1.3	4.0	3.5

[a] Wholemeal macaroni contains 8.3g (raw) and 2.8g (boiled) Englyst fibre per 100g

Cereals and cereal products *continued*

No.	Food	Na	K	Ca	Mg	P	Fe	Cu	Zn	Cl	Mn	Se	I
						mg						µg	
Rice													
18	**Brown rice**, *raw*	3	250	10	110	310	1.4	0.85	1.8	230	2.3	(2)	N
19	*boiled*	1	99	4	43	120	0.5	0.33	0.7	91	0.9	Tr	N
20	**Savoury rice**, *raw*	1440	340	73	45	200	1.5	0.14	1.3	2520	1.2	N	N
21	*cooked*	490	110	25	15	67	0.5	0.05	0.4	860	0.4	N	N
22	**White rice**, easy cook, *raw*	4	150	51	32	150	0.5	0.37	1.8	10	1.2	10	(14)
23	easy cook, *boiled*	1	54	18	11	54	0.2	0.13	0.7	4	0.2	4	5
24	fried in lard/dripping	56	85	7	5	38	0.3	0.06	0.5	98	0.3	N	N
Pasta													
25	**Macaroni**, *raw*	11	230	25	53	180	1.6	0.30	1.5	20	0.9	16	Tr
26	*boiled*	1	25	6	14	42	0.5	0.09	0.5	5	0.3	4	Tr
27	**Noodles**, egg, *raw*	180	260	28	43	200	1.5	0.24	1.3	180	0.8	N	N
28	egg, *boiled*	15	23	5	8	31	0.3	0.06	0.3	10	0.2	N	N
29	**Spaghetti**, white, *raw*	3	250	25	56	190	2.1	0.32	1.5	25	0.9	(1)	Tr
30	white, *boiled*	Tr	24	7	15	44	0.5	0.10	0.5	Tr	0.3	Tr	Tr
31	wholemeal, *raw*	130	390	31	120	330	3.9	0.51	3.0	210	2.6	N	N
32	wholemeal, *boiled*	45	140	11	42	110	1.4	0.18	1.1	73	0.9	N	N

Cereals and cereal products *continued*

No.	Food	Retinol µg	Carotene µg	Vitamin D µg	Vitamin E mg	Thiamin mg	Ribo-flavin mg	Niacin mg	Trypt 60 mg	Vitamin B6 mg	Vitamin B12 µg	Folate µg	Panto-thenate mg	Biotin µg	Vitamin C mg
Rice															
18	**Brown rice**, *raw*	0	0	0	0.80	0.59	0.07	5.3	1.5	N	0	49	N	N	0
19	*boiled*	0	0	0	0.30	0.14	0.02	1.3	0.6	N	0	10	N	N	0
20	**Savoury rice**, *raw*	0	N	Tr	N	0.46	0.06	5.2	1.9	0.37	Tr	25	N	N	0
21	*cooked*	0	N	Tr	N	0.10	0.01	1.1	0.6	0.07	Tr	4	N	N	0
22	**White rice**, easy cook, *raw*	0	0	0	(0.10)	0.41	0.02	4.2	1.6	0.31	0	20	(0.6)	(3)	0
23	easy cook, *boiled*	0	0	0	Tr	0.01	Tr	0.9	0.6	0.07	0	4	(0.1)	(1)	0
24	fried in lard/dripping	Tr	0	Tr	N	0.03	0.01	0.3	0.5	0.06	Tr	3	0.1	1	1
Pasta															
25	**Macaroni**, *raw*	0	0	0	Tr	0.18	0.05	2.9	2.5	0.10	(0)	23	(0.3)	(1)	0
26	*boiled*	0	0	0	Tr	0.03	Tr	0.5	0.6	0.01	(0)	3	Tr	Tr	0
27	**Noodles**, egg, *raw*	37	0	0.3	N	0.26	0.10	2.2	2.5	0.10	Tr	29	N	N	0
28	egg, *boiled*	2	0	Tr	N	0.01	0.01	0.2	0.5	0.01	Tr	1	N	N	0
29	**Spaghetti**, white, *raw*	0	0	0	Tr	0.22	0.03	3.1	2.5	0.17	(0)	34	(0.3)	(1)	0
30	white, *boiled*	0	0	0	Tr	0.01	0.01	0.5	0.7	0.02	(0)	4	Tr	Tr	0
31	wholemeal, *raw*	0	0	0	Tr	0.99	0.11	6.2	2.7	0.39	(0)	40	(0.8)	(1)	0
32	wholemeal, *boiled*	0	0	0	Tr	0.21	0.02	1.3	1.0	0.08	(0)	7	(0.2)	Tr	0

Cereals and cereal products *continued*

33 to 47

Composition of food per 100g

No.	Food	Description and main data sources	Edible proportion	Water g	Protein g	Fat g	Carbo-hydrate g	Energy value kcal	kJ
	Breads								
33	**Brown bread,** *average*	Average of 2 types of brown bread, sliced and unsliced	1.00	39.5	8.5	2.0	44.3	218	927
34	*toasted*	Calculated using 22% weight loss	1.00	24.4	10.4	2.1	56.5	272	1158
35	**Chapatis,** *made with fat*[a]	6 samples	1.00	28.5	8.1	12.8	48.3	328	1383
36	*made without fat*	Analysed and calculated values	1.00	45.8	7.3	1.0	43.7	202	860
37	**Currant bread**	10 samples, 10 different shops	1.00	29.4	7.5	7.6	50.7	289	1220
38	*toasted*	Calculated using 12% weight loss	1.00	25.9	8.4	8.5	56.8	323	1366
39	**Granary bread**	10 samples, 10 different shops	1.00	35.4	9.3	2.7	46.3	235	999
40	**Hovis,** *average*	Average of 3 types of Hovis (wheatgerm) bread	1.00	40.3	9.5	2.0	41.5	212	899
41	*toasted*	Calculated using 22% weight loss	1.00	23.5	12.1	2.6	53.2	271	1151
42	**Malt bread**	10 samples, 5 brands	1.00	25.8	8.3	2.4	56.8	268	1139
43	**Naan bread**	Recipe	1.00	28.8	8.9	12.5	50.1	336	1415
44	**Papadums,** *fried in vegetable oil*	Calculated from raw using weighed fat uptake	1.00	10.3	17.5	16.9	39.1	369	1548
45	**Pitta bread,** white	10 samples, 4 brands	1.00	32.7	9.2	1.2	57.9	265	1127
46	**Rye bread**	15 samples, different shops; literature sources	1.00	37.4	8.3	1.7	45.8	219	932
47	**Vitbe,** *average*	Average of 3 types of Vitbe (wheatgerm) bread	1.00	37.5	9.7	3.1	43.4	229	974

[a] Puris (deep fried chapatis) contain 19.1g water, 7.0g protein, 25.0g fat and 43.3g carbohydrate per 100g

Cereals and cereal products *continued*

Breads

No.	Food	Total nitrogen g	Fatty acids Satd g	Mono unsatd g	Poly unsatd g	Cholest- erol mg	Starch g	Total sugars g	Dietary fibre Southgate method g	Englyst method g
33	**Brown bread**, *average*	1.48	0.4	0.3	0.6	0	41.3	3.0	5.9	(3.5)
34	*toasted*	1.82	0.4	0.4	0.6	0	52.1	4.5	7.1	4.5
35	**Chapatis**, *made with fat*	1.42	N	N	N	N	46.5	1.8	7.0[a]	N
36	*made without fat*	1.28	0.1	0.1	0.4	0	42.1	1.6	6.4	N
37	**Currant bread**	1.32	(1.6)	(1.5)	(2.0)	0	36.3	14.4	3.8	N
38	*toasted*	1.48	(1.8)	(1.6)	(2.2)	0	40.7	16.1	4.2	N
39	**Granary bread**	1.62	0.5	0.6	0.7	0	44.1	2.2	6.5	4.3
40	**Hovis**, *average*	1.67	0.3	0.3	0.7	0	39.7	1.8	5.1	3.3
41	*toasted*	2.04	0.4	0.4	0.8	0	50.9	2.3	6.5	(4.2)
42	**Malt bread**	1.46	0.3	0.3	1.0	1	30.7	26.1	6.5	N
43	**Naan bread**	1.53	N	N	N	6	44.6	5.5	2.2	1.9
44	**Papadums**, *fried in vegetable oil*	2.80	1.7	5.7	7.8	0	39.1	Tr	9.1	N
45	**Pitta bread**, white	1.61	0.2	0.1	0.5	0	55.5	2.4	3.9[b]	2.2[b]
46	**Rye bread**	1.46	0.3	0.3	0.6	0	44.0	1.8	5.8	4.4[c]
47	**Vitbe**, *average*	1.70	0.6	0.6	0.9	0	40.4	3.0	5.8	(3.3)

[a] Puris (deep fried chapatis) contain 4.4g Southgate fibre per 100g
[b] Wholemeal pitta bread contains 9.0g Southgate fibre and 5.2g Englyst fibre per 100g
[c] Pumpernickel contains approximately 7.5g Englyst fibre per 100g

No.	Food	Na	K	Ca	Mg	P	Fe	Cu	Zn	Cl	Mn	Se	I
						mg						µg	
Breads													
33	**Brown bread**, *average*	540	170	100	53	150	2.2	0.16	1.1	890	1.2	N	N
34	*toasted*	690	210	140	62	180	2.7	0.15	1.3	1170	1.3	N	N
35	**Chapatis**, *made with fat*	130	160	66	41	130	2.3	0.20	1.1	250	(1.4)	N	N
36	*made without fat*	120	150	60	37	120	2.1	0.20	1.0	230	(1.24)	N	N
37	**Currant bread**	290	220	86	26	93	1.6	0.32	0.7	480	0.4	N	29
38	*toasted*	330	250	96	29	100	1.8	0.36	0.8	540	0.5	N	33
39	**Granary bread**	580	190	77	59	180	2.7	0.18	1.5	930	1.4	N	N
40	**Hovis**, *average*	600	200	120	56	190	3.7	0.24	2.1	900	1.9	N	22
41	*toasted*	770	250	150	72	240	4.7	0.31	2.7	1150	2.4	N	28
42	**Malt bread**	280	280	110	45	160	2.8	0.26	1.1	260	0.8	N	27
43	**Naan bread**	380	180	160	28	130	1.3	0.12	0.8	630	0.4	25	19
44	**Papadums**, *fried in vegetable oil*	2420	750	69	170	250	11.0	0.50	2.5	2640	1.3	N	N
45	**Pitta bread**, white	520[a]	110	91[a]	24	92	1.7[a]	0.21	0.6[a]	830	0.4	N	N
46	**Rye bread**	580	190	80	48	160	2.5	0.18	1.3	1410	1.0	N	N
47	**Vitbe**, *average*	550	180	150	52	170	2.1	0.18	1.7	950	1.7	N	(22)

[a] Wholemeal pitta bread contains 460mg Na, 48mg Ca, 2.7mg Fe and 1.8mg Zn per 100g

Cereals and cereal products *continued*

No.	Food	Retinol µg	Carotene µg	Vitamin D µg	Vitamin E mg	Thiamin mg	Ribo-flavin mg	Niacin mg	Trypt 60 mg	Vitamin B6 mg	Vitamin B12 µg	Folate µg	Panto-thenate mg	Biotin µg	Vitamin C mg
	Breads														
33	**Brown bread**, *average*	0	0	0	Tr	0.27	0.09	2.5	1.7	0.13	0	40	0.3	3	0
34	*toasted*	0	0	0	Tr	0.26	0.12	3.1	2.1	0.14	0	44	(0.4)	(4)	0
35	**Chapatis**, *made with fat*	N	0	N	N	0.26	0.04	1.7	1.7	(0.20)	0	15	(0.2)	(2)	0
36	*made without fat*	0	0	0	Tr	0.23	0.04	1.5	1.5	(0.18)	0	14	(0.2)	(2)	0
37	**Currant bread**	0	Tr	0	Tr	0.19	0.09	1.5	1.5	0.09	Tr	19	N	N	0
38	*toasted*	0	Tr	0	Tr	0.16	0.10	1.7	1.7	0.10	Tr	21	N	N	0
39	**Granary bread**	0	0	0	N	0.30	0.11	3.0	1.9	0.17	0	90	N	N	0
40	**Hovis**, *average*	0	0	0	N	0.80	0.09	4.2	1.9	0.11	0	39	(0.3)	(2)	0
41	*toasted*	0	0	0	N	0.87	0.12	5.4	2.6	0.14	0	50	(0.4)	(3)	0
42	**Malt bread**	0	Tr	(0)	N	0.45	0.13	2.8	1.7	0.11	(0)	30	N	N	0
43	**Naan bread**	86	60	0.2	1.38	0.19	0.10	1.2	1.8	0.13	Tr	14	0.3	2	Tr
44	**Papadums**, *fried in vegetable oil*	N	N	0	N	0.13	0.09	1.0	2.3	N	0	28	N	N	0
45	**Pitta bread**, white	0	0	0	N	0.24	0.05	1.4	1.9	N	0	21	N	N	0
46	**Rye bread**	0	0	0	1.20	0.29	0.05	2.3	1.7	0.09	0	24	0.5	N	0
47	**Vitbe**, *average*	0	0	0	N	0.23	0.10	1.7	2.0	0.18	0	53	(0.3)	(2)	0

Cereals and cereal products *continued*

Composition of food per 100g

No.	Food	Description and main data sources	Edible proportion	Water g	Protein g	Fat g	Carbo-hydrate g	Energy value kcal	kJ
Breads									
48	**White bread**, *average*	Weighted average of 5 main types of white bread	1.00	37.3	8.4	1.9	49.3	235	1002
49	*sliced*	42 samples, 6 batches	1.00	40.4	7.6	1.3	46.8	217	926
50	*fried in blended oil*	Calculated from fried in lard	1.00	7.4	7.9	32.2[a]	48.5	503	2102
51	*fried in lard*	Calculated on white sliced bread using analysed fat and water changes	1.00	7.4	7.9	32.2[a]	48.5	503	2102
52	*toasted*	Calculated using water loss of 18%	1.00	27.3	9.3	1.6	57.1	265	1129
53	French stick	10 samples, 10 different shops	1.00	29.2	9.6	2.7	55.4	270	1149
54	'with added fibre'	Manufacturer's data for Mighty White (Allied Bakeries) and Champion (British Bakeries)	1.00	40.0	7.6	1.5	49.6	230	978
55	'with added fibre', *toasted*	Calculated using water loss of 16%	1.00	26.2	9.0	1.8	59.0	273	1164
56	**Wholemeal bread**, *average*	Average of 3 types of wholemeal bread	1.00	38.3	9.2	2.5	41.6	215	914
57	*toasted*	Calculated using water loss of 14.6%	1.00	27.8	10.8	2.9	48.7	252	1070
Rolls									
58	**Brown rolls**, *crusty*	12 samples of 6 rolls, different shops	1.00	30.5	10.3	2.8	50.4	255	1085
59	*soft*	14 samples of 6 rolls, different shops	1.00	31.6	10.0	3.8	51.8	268	1139
60	**Croissants**	Recipe	1.00	31.1	8.3	20.3	38.3	360	1505
61	**Hamburger buns**	5 packets of 6 buns including frozen	1.00	32.9	9.1	5.0	48.8	264	1121
62	**White rolls**, *crusty*	14 samples of 6 rolls, different shops	1.00	26.4	10.9	2.3	57.6	280	1192
63	*soft*	14 samples of 6 rolls, different shops	1.00	32.7	9.2	4.2	51.6	268	1137
64	**Wholemeal rolls**	2 samples of 6 rolls, different shops	1.00	31.2	9.0	2.9	48.3	241	1025

[a] The fat content depends on the conditions of frying; thin slices pick up proportionately more fat than thick ones

Cereals and cereal products *continued*

Composition of food per 100g

No.	Food	Total nitrogen g	Fatty acids			Cholest-erol mg	Starch g	Total sugars g	Dietary fibre	
			Satd g	Mono unsatd g	Poly unsatd g				Southgate method g	Englyst method g
Breads										
48	**White bread**, *average*	1.47	0.4	0.4	0.5	0	46.7	2.6	3.8	1.5
49	*sliced*	1.33	0.3	0.2	0.4	0	43.8	3.0	3.7	1.5
50	*fried in blended oil*	1.38	2.8	16.0	11.8	0	45.3	3.1	3.8	(1.6)
51	*fried in lard*	1.38	12.5	13.4	2.9	30	45.3	3.1	3.8	(1.6)
52	*toasted*	1.62	0.4	0.2	0.5	0	53.4	3.7	4.5	1.8
53	French stick	1.68	0.6	0.5	0.7	0	53.5	1.9	5.1	(1.5)
54	'with added fibre'	1.33	0.4	0.6	0.3	0	46.3	3.3	N	3.1
55	'with added fibre', *toasted*	1.58	0.5	0.7	0.3	0	55.1	3.9	N	3.7
56	**Wholemeal bread**, *average*	1.58	0.5	0.5	0.7	0	39.8	1.8	7.4	(5.8)
57	*toasted*	1.85	0.6	0.6	0.9	0	46.6	2.1	8.7	(5.9)
Rolls										
58	**Brown rolls**, *crusty*	1.81	0.6	0.6	0.7	0	48.5	1.9	7.1	(3.5)
59	*soft*	1.75	0.9	0.8	0.8	0	49.3	2.5	6.4	(3.5)
60	**Croissants**	1.44	6.5	8.1	4.9	75	37.2	1.0	2.5	1.6
61	**Hamburger buns**	1.60	1.1	1.3	1.1	0	46.6	2.2	4.0	(1.5)
62	**White rolls**, *crusty*	1.92	0.5	0.4	0.6	0	55.4	2.2	4.3	(1.5)
63	*soft*	1.61	1.0	1.0	0.7	0	49.4	2.2	3.9	(1.5)
64	**Wholemeal rolls**	1.46	0.7	0.8	0.6	0	46.8	1.5	8.8	5.9

Cereals and cereal products *continued*

Inorganic constituents per 100g

No.	Food	Na	K	Ca	Mg	P	Fe	Cu	Zn	Cl	Mn	Se	I
						mg						µg	
Breads													
48	**White bread**, *average*	520	110	110	24	91	1.6	0.19	0.6	820	0.5	28	6
49	*sliced*	530	99	100	20	79	1.4	0.13	0.5	830	0.4	(28)	(6)
50	*fried in blended oil*	550	100	100	21	82	1.5	0.14	0.5	860	0.4	(29)	(7)
51	*fried in lard*	550	100	100	21	82	1.5	0.14	0.5	860	0.4	(29)	(7)
52	*toasted*	650	120	120	24	96	1.7	0.16	0.6	1010	0.5	(34)	(7)
53	French stick	570	130	130	28	110	2.1	0.16	0.7	870	0.5	(28)	(6)
54	'with added fibre'	450	160	150	30	100	2.3	0.10	0.9	790	0.5	(28)	(6)
55	'with added fibre', *toasted*	540	190	180	36	120	2.7	0.12	1.1	940	0.6	(32)	(7)
56	**Wholemeal bread**, *average*	550	230	54	76	200	2.7	0.26	1.8	880	1.9	35	Tr
57	*toasted*	640	270	63	89	230	3.2	0.30	2.1	1030	2.2	(41)	Tr
Rolls													
58	**Brown rolls**, *crusty*	570	200	100	65	190	3.2	0.34	1.5	1040	(1.5)	N	N
59	*soft*	560	190	110	59	170	3.4	0.23	1.5	1130	1.4	N	N
60	**Croissants**	390	140	80	26	130	2.0	0.26	(0.9)	610	(0.3)	N	N
61	**Hamburger buns**	550	110	130	31	150	2.3	0.13	0.7	890	0.5	(28)	(19)
62	**White rolls**, *crusty*	640	130	140	34	120	2.1	0.19	0.9	1180	0.6	(28)	(19)
63	*soft*	560	120	120	26	100	2.2	0.13	0.7	930	0.5	(28)	19
64	**Wholemeal rolls**	460	230	55	69	170	3.5	0.25	1.6	850	1.6	(35)	Tr

No.	Food	Retinol µg	Carotene µg	Vitamin D µg	Vitamin E mg	Thiamin mg	Ribo-flavin mg	Niacin mg	Trypt 60 mg	Vitamin B6 mg	Vitamin B12 µg	Folate µg	Panto-thenate mg	Biotin µg	Vitamin C mg
Breads															
48	**White bread**, *average*	0	0	0	Tr	0.21	0.06	1.7	1.7	0.07	0	29	0.3	1	0
49	*sliced*	0	0	0	Tr	0.20	0.05	1.5	1.6	0.07	0	17	(0.3)	(1)	0
50	*fried in blended oil*	0	Tr	0	N	0.15	0.04	1.4	1.7	0.05	0	9	(0.2)	(1)	0
51	*fried in lard*	Tr	0	N	Tr	0.15	0.04	1.4	1.7	0.05	Tr	9	(0.2)	(1)	0
52	*toasted*	0	0	0	Tr	0.21	0.06	2.8	1.9	0.09	0	21	(0.4)	(1)	0
53	French stick	0	0	0	Tr	0.19	0.07	1.3	2.0	0.08	0	24	(0.3)	(1)	0
54	'with added fibre'	0	0	0	Tr	0.20[a]	(0.05)	1.6[a]	1.5	(0.07)	0[a]	(17)[a]	(0.30)	(1)	0
55	'with added fibre', *toasted*	0	0	0	Tr	0.20[a]	(0.06)	1.9[a]	1.8	(0.08)	0[a]	(20)[a]	(0.35)	(1)	0
56	**Wholemeal bread**, *average*	0	0	0	0.20	0.34	0.09	4.1	1.8	0.12	0	39	0.6	6	0
57	*toasted*	0	0	0	0.20	0.34	0.10	4.7	2.2	0.14	0	46	0.7	7	0
Rolls															
58	**Brown rolls**, *crusty*	0	0	0	Tr	0.43	0.07	3.5	2.1	0.09	0	54	(0.3)	(3)	0
59	*soft*	0	0	0	Tr	0.41	0.08	3.4	2.0	0.09	0	29	(0.3)	(3)	0
60	**Croissants**	21	0	0.2	Tr	(0.18)	(0.16)	(2.0)	1.8	(0.11)	Tr	(73)	(0.5)	(9)	0
61	**Hamburger buns**	N	0	N	Tr	0.23	0.10	1.5	1.9	0.06	Tr	48	(0.3)	(1)	0
62	**White rolls**, *crusty*	0	0	0	Tr	0.26	0.05	2.1	2.2	0.04	0	32	(0.3)	(1)	0
63	*soft*	0	0	0	Tr	0.28	0.04	1.9	1.9	0.04	0	24	(0.3)	(1)	0
64	**Wholemeal rolls**	0	0	0	(0.20)	0.30	0.09	4.1	1.8	0.10	0	62	(0.6)	(6)	0

[a] May be present at higher levels as a result of fortification

Cereals and cereal products *continued*

Composition of food per 100g

No.	Food	Description and main data sources	Edible proportion	Water g	Protein g	Fat g	Carbohydrate g	Energy value kcal	kJ
	Breakfast cereals								
65	**All-Bran**	Analysis and manufacturer's data (Kelloggs)	1.00	3.0	14.0	3.4	46.6	261	1109
66	**Bran Flakes**	Manufacturer's data (Kelloggs)	1.00	3.0	10.2	1.9	69.3	318	1353
67	**Coco Pops**	Manufacturer's data (Kelloggs)	1.00	3.0	5.3	1.0	94.3	384	1636
68	**Common Sense Oat Bran Flakes**	Manufacturer's data (Kelloggs)	1.00	3.0	11.0	4.0	74.0	357	1519
69	**Corn Flakes**	Analysis and manufacturer's data (Kelloggs)	1.00	3.0	7.9	0.7	85.9	360	1535
70	**Crunchy Nut Corn Flakes**	Manufacturer's data (Kelloggs)	1.00	3.0	7.4	4.0	88.6	398	1691
71	**Frosties**	Manufacturer's data (Kelloggs)	1.00	3.0	5.3	0.5	93.7	377	1608
72	**Fruit 'n Fibre**	Manufacturer's data (Kelloggs)	1.00	5.7	9.0	4.7	72.1	349	1481
73	**Muesli**, *Swiss style*[a]	Analysis and manufacturers' data (Kelloggs, Weetabix)	1.00	7.2	9.8	5.9	72.2	363	1540
74	*with no added sugar*	Analysis and manufacturers' data (Kelloggs, Weetabix)	1.00	7.6	10.5	7.8	67.1	366	1552
75	**Oat and Wheat Bran**	Manufacturer's data (Weetabix)	1.00	2.6	10.6	3.5	67.7	325	1381
76	**Porridge**, *made with water*	Recipe. Ref. Wiles *et al.* (1980)	1.00	87.4	1.5	1.1	9.0	49	209
77	*made with whole milk*	Recipe	1.00	74.8	4.8	5.1	13.7	116	488
78	**Puffed Wheat**	Analytical and literature sources	1.00	2.5	14.2	1.3	67.3	321	1366

[a] Muesli composition is very variable

Cereals and cereal products continued

Composition of food per 100g

No.	Food	Total nitrogen g	Fatty acids Satd g	Mono unsatd g	Poly unsatd g	Cholest- erol mg	Starch g	Total sugars g	Dietary fibre Southgate method g	Englyst method g
Breakfast cereals										
65	**All-Bran**	2.22	0.6	0.4	1.5	0	27.6	19.0	30.0	24.5
66	**Bran Flakes**	1.79	0.4	0.2	1.0	0	50.6	18.7	17.3	13.0
67	**Coco Pops**	0.90	0.4	0.2	0.3	0	56.1	38.2	1.1	0.6
68	**Common Sense Oat Bran Flakes**	1.76	0.7	1.2	1.7	0	57.2	16.8	N	10.0
69	**Corn Flakes**	1.26	0.1	0.1	0.3	0	77.7	8.2	3.4	0.9
70	**Crunchy Nut Corn Flakes**	1.18	0.8	1.8	1.2	0	52.3	36.3	1.6	0.8
71	**Frosties**	0.85	0.1	0.1	0.2	0	51.8	41.9	1.2	0.6
72	**Fruit 'n Fibre**	1.58	2.5	1.3	1.0	0	47.7	24.4	10.1	7.0
73	**Muesli**, Swiss style	1.57	0.8	2.8	1.6	Tr	46.0	26.2	8.1	6.4
74	*with no added sugar*	1.68	1.5	3.5	2.4	Tr	51.4	15.7	10.5	7.6
75	**Oat and Wheat Bran**	1.70	0.6	0.9	1.4	0	51.0	16.7	N	17.9
76	**Porridge**, *made with water*	0.26	0.2	0.4	0.4	0	9.0	Tr	0.8	0.8
77	*made with whole milk*	0.77	2.7	1.5	0.5	14	9.0	4.7	0.8	0.8
78	**Puffed Wheat**	2.44	0.2	0.2	0.6	0	67.0	0.3	8.8	5.6

Cereals and cereal products continued

Inorganic constituents per 100g

No.	Food	Na	K	Ca	Mg	P	Fe	Cu	Zn	Cl	Mn	Se	I
						mg						µg	
												Se	I

Breakfast cereals

No.	Food	Na	K	Ca	Mg	P	Fe	Cu	Zn	Cl	Mn	Se	I
65	**All-Bran**	900	1000	69	210	700	12.0	0.44	6.7	1390	N	N	N
66	**Bran Flakes**	1000	540	50	130	370	20.0	0.35	3.3	1540	N	N	N
67	**Coco Pops**	800	190	20	40	120	6.7	0.20	0.8	1230	N	N	N
68	**Common Sense Oat Bran Flakes**	900	400	50	100	400	6.7	N	2.5	1390	N	N	N
69	**Corn Flakes**	1110	100	15	14	50	6.7	0.03	0.2	1820	0.1	2	10
70	**Crunchy Nut Corn Flakes**	770	150	18	20	50	6.7	0.08	0.3	1210	N	N	N
71	**Frosties**	800	63	11	10	50	6.7	Tr	0.1	1230	N	N	N
72	**Fruit 'n Fibre**	700	450	40	60	200	6.7	0.24	1.7	1080	N	N	N
73	**Muesli,** *Swiss style*	380	440	110	85	280	5.8	0.10	2.3	790	N	N	N
74	*with no added sugar*	47	530	47	90	330	3.5	0.36	2.1	10	2.6	N	N
75	**Oat and Wheat Bran**	600	(730)	(70)	(180)	(620)	45.0	(1.00)	(4.0)	920	N	N	N
76	**Porridge,** *made with water*	560	46	7	18	47	0.5	0.03	0.4	870	0.5	Tr	N
77	*made with whole milk*	620	190	120	29	140	0.6	0.03	0.8	970	0.5	Tr	N
78	**Puffed Wheat**	4	390	26	140	350	4.6	0.56	2.8	50	N	N	N

Cereals and cereal products *continued*

Breakfast cereals

No.	Food	Retinol µg	Carotene µg	Vitamin D µg	Vitamin E mg	Thiamin mg	Ribo-flavin mg	Niacin mg	Trypt 60 mg	Vitamin B6 mg	Vitamin B12 µg	Folate µg	Panto-thenate mg	Biotin µg	Vitamin C mg
65	**All-Bran**	0	0	1.6	2.20	0.80	1.00	11.3	3.0	1.30	1.3	190	1.7	25	0
66	**Bran Flakes**	0	0	2.1	N	1.00	1.30	15.0	2.4	1.80	1.7	250	0.9	11	25
67	**Coco Pops**	0	0	2.1	N	1.00	1.30	15.0	1.2	1.80	1.7	250	N	N	0
68	**Common Sense Oat Bran Flakes**	0	0	2.1	N	1.00	1.30	15.0	2.3	1.80	1.7	250	N	N	0
69	**Corn Flakes**	0	0	2.1	0.40	1.00	1.30	15.0	0.9	1.80	1.7	250	0.3	2	0
70	**Crunchy Nut Corn Flakes**	0	0	2.1	N	1.00	1.30	15.0	0.8	1.80	1.7	250	N	N	N
71	**Frosties**	0	0	2.1	N	1.00	1.30	15.0	0.6	1.80	1.7	250	(0.3)	(1)	0
72	**Fruit 'n Fibre**	0	Tr	2.1	N	1.00	1.30	15.0	1.7	1.80	2	250	N	N	N
73	**Muesli,** Swiss style	Tr	Tr	0	3.20	0.50	0.70	6.5	2.3	1.60	0	(140)	1.2	15	Tr
74	with no added sugar	Tr	Tr	0	(2.90)	0.25	0.40	4.6	2.2	0.30	0	N	N	N	Tr
75	**Oat and Wheat Bran**	0	0	0	N	1.80	1.40	15.0	2.0	N	0	N	N	N	0
76	**Porridge,** made with water	0	0	0	0.21	0.06	0.01	0.1	0.3	0.01	0	4	0.1	2	0
77	made with whole milk	53	21	0	0.29	0.10	0.17	0.2	1.1	0.06	0	7	0.4	3	1
78	**Puffed Wheat**	0	0	0	2.00	Tr	0.06	5.2	2.9	0.14	0	19	0.5	7	0

Cereals and cereal products *continued*

Composition of food per 100g

Breakfast cereals

No.	Food	Description and main data sources	Edible proportion	Water g	Protein g	Fat g	Carbo-hydrate g	Energy value kcal	Energy value kJ
79	Raisin Splitz	Manufacturer's data (Kelloggs)	1.00	9.0	9.0	2.0	75.4	337	1433
80	Ready Brek	6 packets of the same brand (Weetabix)	1.00	8.6	11.4	7.8	68.6	373	1580
81	Rice Krispies	Analysis and manufacturer's data (Kelloggs)	1.00	3.0	6.1	0.9	89.7	369	1572
82	Ricicles	Manufacturer's data (Kelloggs)	1.00	3.0	4.3	0.5	95.7	381	1623
83	Shredded Wheat	6 packets of the same brand (Nabisco)	1.00	7.6	10.6	3.0	68.3	325	1384
84	Shreddies	10 samples (Nabisco)	1.00	4.0	10.0	1.5	74.1	331	1411
85	Smacks	Manufacturer's data (Kelloggs)	1.00	3.0	8.0	2.0	89.6	386	1644
86	Special K	Analysis and manufacturer's data (Kelloggs)	1.00	3.0	15.3	1.0	81.7	377	1604
87	Start	Manufacturer's data (Kelloggs)	1.00	3.0	7.9	1.7	81.7	355	1504
88	Sugar Puffs	6 packets of the same brand (Quaker)	1.00	1.8	5.9	0.8	84.5	348	1482
89	Sultana Bran	Manufacturer's data (Kelloggs)	1.00	7.0	8.5	1.6	67.8	303	1289
90	Weetabix	Manufacturer's data (Weetabix)	1.00	5.6	11.0	2.7	75.7	352	1498
91	Weetaflake	Manufacturer's data (Weetabix)	1.00	3.6	9.2	2.8	79.3	359	1529
92	Weetos	Manufacturer's data (Weetabix)	1.00	4.4	6.1	2.7	86.1	372	1581

Cereals and cereal products *continued*

Composition of food per 100g

No.	Food	Total nitrogen g	Fatty acids Satd g	Mono unsatd g	Poly unsatd g	Cholest-erol mg	Starch g	Total sugars g	Dietary fibre Southgate method g	Englyst method g
Breakfast cereals										
79	**Raisin Splitz**	1.54	0.4	0.2	0.8	0	57.2	18.2	N	8.0
80	**Ready Brek**	1.95	1.2	2.9	2.5	0	66.9	1.7[a]	N	7.2
81	**Rice Krispies**	1.03	0.3	0.2	0.3	0	79.1	10.6	1.1	0.7
82	**Ricicles**	0.72	0.2	0.1	0.2	0	53.8	41.9	0.9	0.4
83	**Shredded Wheat**	1.81	0.4	0.4	1.4	0	67.5	0.8	10.1	9.8
84	**Shreddies**	1.72	0.2	0.2	0.7	0	63.9	10.2	10.9	9.5
85	**Smacks**	1.28	0.4	0.2	0.9	0	39.6	50.0	N	3.0
86	**Special K**	2.46	0.3	0.2	0.4	0	64.5	17.2	2.7	2.0
87	**Start**	1.26	0.3	0.4	0.7	0	52.6	29.1	9.3	5.7
88	**Sugar Puffs**	1.01	0.1	0.1	0.4	0	28.0	56.5	4.8	3.2
89	**Sultana Bran**	1.49	0.4	0.3	1.0	0	39.2	28.6	15.5	10.0
90	**Weetabix**	1.90	0.4	0.4	1.2	0	70.5	5.2	11.6	9.7
91	**Weetaflake**	1.58	0.4	0.4	1.3	0	59.0	20.3	(9.7)	8.8
92	**Weetos**	1.04	0.6	Tr	1.3	0	52.9	33.2	N	5.3

a Flavoured instant oat varieties contain approximately 8.5g total sugars per 100g

Cereals and cereal products *continued*

79 to 92

Inorganic constituents per 100g

No.	Food	Na	K	Ca	Mg	P	Fe	Cu	Zn	Cl	Mn	Se	I
						mg						µg	
Breakfast cereals													
79	Raisin Splitz	10	500	50	70	250	5.1	N	1.8	N	N	N	N
80	Ready Brek	12	390	65	120	420	13.2	0.41	2.7	18	N	N	N
81	Rice Krispies	1260	150	10	30	130	6.7	0.10	0.9	1980	1.0	N	N
82	Ricicles	800	96	10	20	89	6.7	0.13	0.6	1230	N	N	N
83	Shredded Wheat	8	330	38	130	340	4.2	0.40	2.3	53	N	N	N
84	Shreddies	550	210	40	88	320	2.8	0.44	2.5	220	2.3	N	N
85	Smacks	20	200	20	40	150	6.7	N	N	N	N	N	N
86	Special K	1000	230	70	52	150	13.3	0.13	1.9	1540	N	N	N
87	Start	500	300	40	60	190	15.0	0.13	18.7	770	N	N	N
88	Sugar Puffs	9	160	14	55	140	2.1	0.23	1.5	41	N	N	N
89	Sultana Bran	700	660	51	120	350	15.0	0.13	2.8	1080	N	N	N
90	Weetabix	270	370	35	120	290	7.4	0.54	2.0	420	N	N	N
91	Weetaflake	660	(310)	(29)	(100)	(240)	11.5	(0.45)	(1.7)	1020	N	N	N
92	Weetos	300	490	65	120	290	10.9	Tr	2.3	460	N	N	N

Cereals and cereal products *continued*

Vitamins per 100g

No.	Food	Retinol µg	Carotene µg	Vitamin D µg	Vitamin E mg	Thiamin mg	Riboflavin mg	Niacin mg	Trypt 60 mg	Vitamin B6 mg	Vitamin B12 µg	Folate µg	Pantothenate mg	Biotin µg	Vitamin C mg
	Breakfast cereals														
79	Raisin Splitz	0	Tr	1.6	N	0.80	1.00	11.3	1.8	1.30	1.3	190	N	N	0
80	Ready Brek	0	0	0	1.20	1.80	0.50	9.4	2.3	1.50	0	53	1.3	23	27
81	Rice Krispies	0	0	2.1	0.60	1.00	1.30	15.0	1.4	1.80	1.7	250	0.7	2	0
82	Ricicles	0	0	2.1	N	1.00	1.30	15.0	1.0	1.80	1.7	250	(0.4)	(1)	0
83	Shredded Wheat	0	0	0	1.20	0.27	0.05	4.5	2.1	0.24	0	29	0.8	9	0
84	Shreddies	0	0	0	N	1.20	2.20	21.1	2.0	0.64	1	28	0.8	7	0
85	Smacks	0	0	2.1	N	1.00	1.30	15.0	1.7	1.80	1.7	250	N	N	0
86	Special K	0	0	2.8	0.55	1.30	1.80	20.0	2.8	2.20	2.2	330	0.5	3	0
87	Start	0	0	3.1	18.70	1.50	2.00	22.5	1.1	2.70	2.5	375	N	N	37
88	Sugar Puffs	0	0	0	0.34	Tr	0.03	2.5	1.2	0.05	0	12	N	N	0
89	Sultana Bran	0	Tr	2.1	N	1.00	1.30	15.0	2.0	1.80	1.7	250	N	N	N
90	Weetabix	0	0	0	1.03	0.90	1.50	11.2	2.2	0.22	0	50	0.7	8	0
91	Weetaflake	0	0	0	1.03	1.80	1.40	36.0	1.8	(0.18)	0	(42)	(0.6)	(7)	0
92	Weetos	7	0	2.8	Tr	1.90	1.60	16.0	1.2	Tr	3	2	N	N	30

Cereals and cereal products *continued*

Composition of food per 100g

Biscuits

No.	Food	Description and main data sources	Edible proportion	Water g	Protein g	Fat g	Carbo-hydrate g	Energy value kcal	kJ
93	**Chocolate biscuits**, full coated	7 different kinds	1.00	2.2	5.7	27.6	67.4	524	2197
94	**Cream crackers**	6 packets	1.00	4.3	9.5	16.3	68.3	440	1857
95	**Crispbread**, rye	Analytical and literature sources	1.00	6.4	9.4	2.1	70.6	321	1367
96	**Digestive biscuits**, chocolate	10 packets, 5 plain chocolate, 5 milk chocolate	1.00	2.5	6.8	24.1	66.5	493	2071
97	plain	10 samples, 3 brands	1.00	2.5	6.3	20.9	68.6	471	1978
98	**Flapjacks**	Recipe	1.00	6.4	4.5	26.6	60.4	484	2028
99	**Gingernut biscuits**	10 packets, 6 brands	1.00	3.4	5.6	15.2	79.1	456	1923
100	**Homemade biscuits**, *creaming method*	Recipe	1.00	9.0	6.2	21.9	64.3	463	1943
101	**Jaffa cakes**	Recipe	1.00	18.0	3.5	10.5	67.8	363	1532
102	**Oatcakes**, *retail*	6 packets, 4 brands	1.00	5.5	10.0	18.3	63.0	441	1855
103	**Sandwich biscuits**	10 packets, custard creams and similar types	1.00	2.6	5.0	25.9	69.2	513	2151
104	**Semi-sweet biscuits**	10 packets, Osborne, Rich Tea, Marie	1.00	2.5	6.7	16.6	74.8	457	1925
105	**Short-sweet biscuits**	10 packets, shortcake and Lincoln	1.00	2.6	6.2	23.4	62.2	469	1966
106	**Shortbread**	Recipe	1.00	5.8	5.9	26.1	63.9	498	2087
107	**Wafer biscuits**, filled	9 packets, assorted	1.00	2.3	4.7	29.9	66.0	535	2242
108	**Wholemeal crackers**	Farmhouse-type, recipe	1.00	4.4	10.1	11.3	72.1	413	1744

Cereals and cereal products *continued*

93 to 108

Composition of food per 100g

No.	Food	Total nitrogen g	Fatty acids Satd g	Mono unsatd g	Poly unsatd g	Cholest- erol mg	Starch g	Total sugars g	Dietary fibre Southgate method g	Englyst method g
Biscuits										
93	**Chocolate biscuits**, full coated	1.00	16.7	8.0	1.1	(22)	24.0	43.4	2.9	2.1
94	**Cream crackers**	1.66	N	N	N	N	68.3	Tr	6.1	2.2
95	**Crispbread**, rye	1.61	0.3	0.3	0.9	0	67.4	3.2	11.6[a]	11.7[a,b]
96	**Digestive biscuits**, chocolate	1.17	12.2	8.9	1.8	51	38.0	28.5	3.1	2.2
97	plain	1.10	8.6	9.6	1.7	41	55.0	13.6	4.6	2.2
98	**Flapjacks**	0.77	7.6	10.4	7.4	43	25.0	35.5	2.6	2.7
99	**Gingernut biscuits**	0.98	7.2	5.8	1.4	N	43.3	35.8	1.8	1.4
100	**Homemade biscuits**, *creaming method*	1.07	6.6	8.5	5.7	84	37.6	26.7	1.8	1.5
101	**Jaffa cakes**	0.58	N	N	N	21	10.6	57.2	N	N
102	**Oatcakes**, *retail*	1.71	3.9	8.2	5.2	(8)	59.9	3.1	3.6	N
103	**Sandwich biscuits**	0.87	14.5	8.0	1.8	(51)	39.0	30.2	1.1	N
104	**Semi-sweet biscuits**	1.18	8.0	5.9	1.7	(31)	52.5	22.3	2.1	1.7
105	**Short-sweet biscuits**	1.08	11.7	7.9	2.4	(37)	38.1	24.1	1.5	1.5
106	**Shortbread**	1.04	17.3	6.4	1.2	74	46.9	17.1	2.2	1.9
107	**Wafer biscuits**, filled	0.82	18.6	8.2	0.9	N	21.3	44.7	1.4	N
108	**Wholemeal crackers**	1.76	N	N	N	N	70.5	1.6	4.8	4.4

[a] Cracotte type crispbread contains 9.2g Southgate fibre and 3.5g Englyst fibre per 100g

[b] High fibre varieties contain approximately 17.9g Englyst fibre per 100g

Cereals and cereal products continued

Inorganic constituents per 100g

No.	Food	Na	K	Ca	Mg	P	Fe	Cu	Zn	Cl	Mn	Se	I
						mg						μg	
Biscuits													
93	**Chocolate biscuits**, full coated	160	230	110	42	130	1.7	0.25	0.8	250	N	N	N
94	**Cream crackers**	610	120	110	25	110	1.7	(0.20)	(0.7)	830	(0.0)	(4)	(13)
95	**Crispbread**, rye	220[a]	500	45[a]	100	310	3.5[a]	0.38	3.0[a]	370	3.5	(3)	15
96	**Digestive biscuits**, chocolate	450	210	84	41	130	2.1	0.24	1.0	410	N	N	N
97	plain	600	170	92	23	88	3.2	0.28	0.5	540	0.5	N	N
98	**Flapjacks**	280	200	37	48	160	2.1	0.23	1.5	370	2.0	1	N
99	**Gingernut biscuits**	330	220	130	25	87	4.0	0.16	0.5	320	(0.9)	N	N
100	**Homemade biscuits**, creaming method	220	92	78	12	82	1.3	0.10	0.6	360	0.3	4	18
101	**Jaffa cakes**	130	170	55	34	130	1.5	0.30	0.3	170	N	N	48
102	**Oatcakes**, retail	1230	340	54	100	420	4.5	0.37	2.3	1290	(3.2)	N	N
103	**Sandwich biscuits**	220	120	100	13	82	1.6	0.07	0.5	290	N	N	N
104	**Semi-sweet biscuits**	410	140	120	17	84	2.1	0.08	0.6	520	N	N	N
105	**Short-sweet biscuits**	360	110	87	15	85	1.8	0.11	0.6	490	N	N	N
106	**Shortbread**	230	97	91	13	75	1.3	0.10	0.4	460	0.4	2	18
107	**Wafer biscuits**, filled	70	160	73	22	83	1.6	0.16	0.6	150	N	N	N
108	**Wholemeal crackers**	700	200	110	49	170	2.5	0.25	1.2	1040	1.2	N	N

[a] Cracotte type crispbread contains 640mg Na, 80mg Ca, 2.1mg Fe and 0.6mg Zn per 100g

Cereals and cereal products *continued*

Vitamins per 100g

No.	Food	Retinol µg	Carotene µg	Vitamin D µg	Vitamin E mg	Thiamin mg	Ribo-flavin mg	Niacin mg	Trypt 60 mg	Vitamin B6 mg	Vitamin B12 µg	Folate µg	Panto-thenate mg	Biotin µg	Vitamin C mg
Biscuits															
93	**Chocolate biscuits**, full coated	Tr	Tr	Tr	1.43	0.03	0.13	0.5	1.2	0.04	Tr	N	N	N	0
94	**Cream crackers**	0	0	0	(1.30)	(0.23)	(0.05)	(1.7)	1.9	(0.12)	0	(22)	(0.3)	(2)	0
95	**Crispbread**, rye	0	0	0	0.50	0.28	0.14	1.1	1.8	0.29	0	35	(1.1)	(7)	0
96	**Digestive biscuits**, chocolate	Tr	Tr	Tr	1.10	0.08	0.11	1.3	1.4	0.08	Tr	N	N	N	0
97	plain	0	0	0	N	0.14	0.11	1.1	1.3	0.09	0	13	N	N	0
98	**Flapjacks**	220	215	2.3	2.85	0.26	0.03	0.3	1.0	0.09	0	11	0.3	8	0
99	**Gingernut biscuits**	0	N	0	1.50	0.10	0.03	0.9	1.1	0.07	0	(4)	(0.1)	(1)	0
100	**Homemade biscuits**, *creaming method*	215	185	2.2	2.32	0.12	0.06	0.8	1.4	0.07	Tr	9	0.3	4	0
101	**Jaffa cakes**	14	0	0.1	0.81	0.05	0.05	0.3	0.7	0.03	0	5	0.2	3	2
102	**Oatcakes**, *retail*	0	0	0	2.14	0.32	0.09	0.7	2.3	0.10	0	(26)	(1.0)	(17)	0
103	**Sandwich biscuits**	0	0	0	3.40	0.14	0.13	1.1	1.0	0.04	0	N	N	N	0
104	**Semi-sweet biscuits**	0	0	0	1.40	0.13	0.08	1.5	1.4	0.06	0	(13)	N	N	0
105	**Short-sweet biscuits**	0	0	0	1.30	0.16	0.04	0.9	1.3	0.05	0	(13)	N	N	0
106	**Shortbread**	250	130	0.2	0.80	0.14	0.02	1.0	1.2	0.07	0	7	0.1	1	0
107	**Wafer biscuits**, filled	0	0	0	1.90	0.09	0.08	0.5	1.0	0.03	0	N	N	N	0
108	**Wholemeal crackers**	0	0	0	1.53	0.26	0.06	2.7	2.0	0.18	0	26	0.4	3	0

No.	Food	Description and main data sources	Edible proportion	Water g	Protein g	Fat g	Carbo-hydrate g	Energy value kcal	kJ
Cakes									
109	**Battenburg cake**	Recipe. Ref. Wiles *et al.* (1980)	1.00	25.3	5.9	17.5	50.0	370	1551
110	**Cake mix**, *made up*	Recipe; made as packet directions	1.00	31.5	5.3	3.3	52.4	248	1052
111	**Crispie cakes**	Chocolate-coated; recipe	1.00	1.6	5.6	18.6	73.1	464	1951
112	**Fancy iced cakes**, individual	10 different types	1.00	12.7	3.8	14.9	68.8	407	1717
113	**Fruit cake**, plain, *retail*	10 cakes, 4 brands	1.00	19.5	5.1	12.9	57.9	354	1490
114	rich	Recipe	1.00	17.6	3.8	11.0	59.6	341	1438
115	rich, iced	Coated with marzipan and Royal icing; recipe	1.00	15.7	4.1	11.4	62.7	356	1504
116	wholemeal	Recipe	1.00	21.5	6.0	15.7	52.8	363	1525
117	**Gateau**	Recipe. Ref. Wiles *et al.* (1980)	1.00	35.1	5.7	16.8	43.4	337	1413
118	**Madeira cake**	10 cakes, 4 brands	1.00	20.2	5.4	16.9	58.4	393	1652
119	**Sponge cake**	Basic recipe, creaming method	1.00	15.2	6.4	26.3	52.4	459	1920
120	*fatless*	Basic recipe, whisking method	1.00	31.5	10.1	6.1	53.0	294	1245
121	*jam filled*	10 cakes, 3 brands; sandwich and Swiss roll	1.00	24.5	4.2	4.9	64.2	302	1280
122	*with butter icing*	Recipe. Ref. Wiles *et al.* (1980)	1.00	13.0	4.5	30.6	52.4	490	2046
123	**Swiss rolls**, chocolate, individual	10 samples, 5 brands, 4 bakeries	1.00	17.5	4.3	11.3	58.1	337	1421

Cereals and cereal products continued

Composition of food per 100g

No.	Food	Total nitrogen g	Fatty acids Satd g	Mono unsatd g	Poly unsatd g	Cholest- erol mg	Starch g	Total sugars g	Southgate method g	Englyst method g
									Dietary fibre	
Cakes										
109	**Battenburg cake**	1.02	4.7	7.7	4.4	94	16.1	34.0	1.5	N
110	**Cake mix**, *made up*	0.89	1.4	1.2	0.3	67	24.2	28.3	3.3	N
111	**Crispie cakes**	0.90	10.7	5.9	0.9	5	32.4	40.7	0.8	0.3
112	**Fancy iced cakes**, individual	0.66	9.3	3.9	0.8	N	14.8	54.0	2.2	N
113	**Fruit cake**, plain, *retail*	0.89	5.8	5.2	1.1	N	14.8	43.1	2.5	N
114	rich	0.63	3.4	4.4	2.7	63	11.2	48.4	3.2	1.7
115	rich, iced	0.71	2.6	5.5	2.8	47	7.5	55.1	3.1	1.7
116	wholemeal	1.01	4.8	6.1	4.1	74	23.5	29.3	3.0	2.4
117	**Gateau**	0.93	9.5	5.3	0.8	148	11.0	32.3	0.5	0.4
118	**Madeira cake**	0.94	8.8	5.7	1.5	N	21.9	36.5	1.3	0.9
119	**Sponge cake**	1.06	8.0	10.5	6.5	152	22.0	30.4	1.0	0.9
120	*fatless*	1.65	1.7	2.5	0.8	223	22.1	30.9	1.0	0.9
121	*jam filled*	0.74	1.6	1.7	0.7	N	16.5	47.7	1.1	1.8
122	*with butter icing*	0.74	9.4	12.3	7.6	128	15.4	37.1	0.7	0.6
123	**Swiss rolls**, chocolate, individual	0.75	N	N	N	86	16.3	41.8	2.4	N

53

Cereals and cereal products *continued*

Inorganic constituents per 100g

109 to 123

No.	Food	Na	K	Ca	Mg	P	Fe	Cu	Zn	Cl	Mn	Se	I
						mg						µg	
Cakes													
109	**Battenburg cake**	440	140	87	24	190	1.1	0.09	0.7	500	0.2	4	21
110	**Cake mix**, *made up*	370	82	59	9	260	0.9	0.14	0.5	110	0.2	N	17
111	**Crispie cakes**	450	230	28	75	120	4.0	0.47	0.4	770	0.5	N	N
112	**Fancy iced cakes**, individual	250	170	44	30	120	1.4	0.25	0.7	230	N	N	N
113	**Fruit cake**, plain, *retail*	250	390	60	25	110	1.7	0.25	0.5	320	N	N	N[a]
114	**rich**	200	380	82	21	71	1.9	0.25	0.5	240	0.4	2	N[a]
115	**rich, iced**	140	330	75	33	84	1.6	0.19	0.6	170	0.4	2	N[a]
116	**wholemeal**	310	240	85	33	220	1.9	0.22	1.0	250	0.9	11	N
117	**Gateau**	56	88	60	8	94	0.9	0.05	0.6	79	0.1	N	17
118	**Madeira cake**	380	120	42	12	120	1.1	0.10	0.5	500	N	N	N
119	**Sponge cake**	350	82	66	9	150	1.2	0.10	0.7	410	0.2	5	25
120	*fatless*	82	120	75	13	150	1.7	0.10	1.0	120	0.2	10	34
121	*jam filled*	420	140	44	14	220	1.6	0.20	0.5	260	N	(10)	14
122	*with butter icing*	360	59	47	7	110	0.9	0.08	0.6	470	0.1	4	21
123	**Swiss rolls**, chocolate, individual	350	210	77	19	200	1.1	0.25	0.5	510	0.2	N	13

[a] Iodine from erythrosine is present but largely unavailable

54

Cereals and cereal products *continued*

Cakes

No.	Food	Retinol μg	Carotene μg	Vitamin D μg	Vitamin E mg	Thiamin mg	Ribo-flavin mg	Niacin mg	Trypt 60 mg	Vitamin B_6 mg	Vitamin B_{12} μg	Folate μg	Panto-thenate mg	Biotin μg	Vitamin C mg
109	**Battenburg cake**	46	Tr	0.4	2.68	0.08	0.16	0.5	1.4	0.05	1	13	0.4	8	0
110	**Cake mix**, *made up*	31	0	N	N	0.14	0.07	0.5	1.3	0.03	Tr	8	N	N	0
111	**Crispie cakes**	Tr	25	0.8	0.75	0.41	0.53	5.8	0.9	0.68	0.6	99	0.6	3	0
112	**Fancy iced cakes**, individual	0	N	0	N	0.01	0.04	0.2	0.8	N	0	N	N	N	0
113	**Fruit cake**, plain, *retail*	N	N	N	N	0.08	0.07	0.6	1.0	(0.11)	0	(8)	(0.2)	(5)	0
114	rich	110	91	1.1	1.26	0.08	0.07	0.6	0.8	0.11	Tr	8	0.2	5	0
115	rich, iced	78	61	0.8	2.34	0.07	0.13	0.5	0.8	0.08	Tr	13	0.2	4	0
116	wholemeal	155	125	1.5	1.85	0.12	0.09	1.3	1.3	0.12	Tr	11	0.3	5	0
117	**Gateau**	250	88	0.5	0.77	0.07	0.18	0.3	1.5	0.05	1	11	0.5	8	0
118	**Madeira cake**	N	N	N	N	0.06	0.11	0.5	1.1	N	Tr	N	N	N	0
119	**Sponge cake**	275	215	2.8	2.83	0.09	0.12	0.5	1.6	0.06	1	10	0.5	7	0
120	*fatless*	110	0	1.0	1.02	0.11	0.24	0.5	2.7	0.08	1	18	0.9	15	0
121	*jam filled*	N	N	N	Tr	0.04	0.07	0.4	0.9	N	(1)	N	N	N	0
122	*with butter icing*	305	265	3.1	3.18	0.06	0.08	0.3	1.1	0.04	1	7	0.3	5	0
123	**Swiss rolls**, chocolate, individual	N	N	N	N	0.12	0.19	0.3	0.9	0.03	Tr	10	N	N	0

55

Cereals and cereal products *continued*

Composition of food per 100g

No.	Food	Description and main data sources	Edible proportion	Water g	Protein g	Fat g	Carbo-hydrate g	Energy value kcal	kJ
	Pastry								
124	**Flaky pastry**, *raw*	Recipe	1.00	30.1	4.2	30.7	34.8	424	1765
125	*cooked*	Recipe	1.00	7.7	5.6	40.6	45.9	560	2332
126	**Shortcrust pastry**, *raw*	Recipe	1.00	20.0	5.7	27.9	46.8	449	1874
127	*cooked*	Recipe	1.00	7.2	6.6	32.3	54.2	521	2174
128	**Wholemeal pastry**, *raw*	Recipe. Ref. Wiles *et al.* (1980)	1.00	20.0	7.7	28.4	38.5	431	1797
129	*cooked*	Recipe. Ref. Wiles *et al.* (1980)	1.00	7.4	8.9	32.9	44.6	499	2080
	Buns and pastries								
130	**Chelsea buns**	Recipe. Ref. Wiles *et al.* (1980)	1.00	20.1	7.8	13.8	56.1	366	1542
131	**Cream horns**	Recipe. Ref. Wiles *et al.* (1980)	1.00	34.4	3.8	35.8	25.8	435	1803
132	**Crumpets**, *toasted*	Calculated using 11% weight loss	1.00	46.5	6.7	1.0	43.4	199	846
133	**Currant buns**	10 samples, 5 brands, 5 bakeries	1.00	27.7	7.6	7.5	52.7	296	1250
134	**Custard tarts**, individual	10 samples, 2 brands, 8 bakeries	1.00	44.7	6.3	14.5	32.4	277	1161
135	**Danish pastries**	10 samples, different shops	1.00	21.6	5.8	17.6	51.3	374	1571
136	**Doughnuts**, jam	10 samples, different shops	1.00	26.9	5.7	14.5	48.8	336	1414
137	ring	10 samples, different shops	1.00	23.8	6.1	21.7	47.2	397	1662
138	**Eccles cake**	Recipe. Ref. Wiles *et al.* (1980)	1.00	4.2	3.9	26.4	59.3	475	1991
139	**Eclairs**, *frozen*	10 samples of the same brand (Birds Eye)	1.00	38.7	5.6	30.6	26.1	396	1647

Cereals and cereal products *continued*

Composition of food per 100g

No.	Food	Total nitrogen g	Fatty acids Satd g	Mono unsatd g	Poly unsatd g	Cholest-erol mg	Starch g	Total sugars g	Dietary fibre Southgate method g	Englyst method g
	Pastry									
124	**Flaky pastry**, *raw*	0.74	11.1	12.6	5.6	41	34.1	0.7	1.6	1.4
125	*cooked*	0.97	14.7	16.7	7.4	54	45.0	0.9	2.1	1.8
126	**Shortcrust pastry**, *raw*	0.99	10.1	11.5	5.1	37	45.8	0.9	2.2	1.9
127	*cooked*	1.15	11.7	13.3	5.9	43	53.2	1.1	2.5	2.2
128	**Wholemeal pastry**, *raw*	1.32	10.2	11.7	5.3	37	37.2	1.3	5.2	5.4
129	*cooked*	1.52	11.8	13.5	6.1	43	43.1	1.5	6.0	6.3
	Buns and pastries									
130	**Chelsea buns**	1.34	4.2	5.3	3.7	55	34.7	21.4	2.2	1.7
131	**Cream horns**	0.66	16.7	13.3	4.1	69	19.6	6.2	1.0	0.9
132	**Crumpets**, *toasted*	1.18	0.1	0.1	0.5	0	41.5	1.9	2.9	(2.0)
133	**Currant buns**	1.34	N	N	N	17	37.6	15.1	1.8	N
134	**Custard tarts**, *individual*	1.00	5.6	6.2	1.7	95	19.6	12.8	1.2	1.2
135	**Danish pastries**	1.01	5.6	7.7	3.2	41	22.8	28.5	2.7	1.6
136	**Doughnuts**, *jam*	1.00	4.3	5.4	3.6	15	30.0	18.8	2.5	N
137	*ring*	1.07	6.3	8.3	5.5	24	31.9	15.3	3.1	N
138	**Eccles cake**	0.68	10.1	9.3	5.8	53	18.2	41.1	2.0	1.6
139	**Eclairs**, *frozen*	0.98	16.1	10.2	1.9	150	19.5	6.6	1.5	0.8

Cereals and cereal products *continued*

Inorganic constituents per 100g

No.	Food	Na	K	Ca	Mg	P	Fe	Cu	Zn	Cl	Mn	Se	I
						mg						µg	
Pastry													
124	**Flaky pastry**, *raw*	350	71	64	11	52	1.0	0.08	0.3	570	0.3	2	10
125	*cooked*	460	94	84	15	69	1.3	0.11	0.5	750	0.4	2	13
126	**Shortcrust pastry**, *raw*	410	91	85	15	68	1.3	0.10	0.4	680	0.4	2	11
127	*cooked*	480	110	99	17	79	1.5	0.12	0.5	790	0.4	3	13
128	**Wholemeal pastry**, *raw*	360	210	24	74	190	2.4	0.28	1.8	570	1.9	32	N
129	*cooked*	410	240	28	86	230	2.8	0.32	2.1	650	2.2	37	N
Buns and pastries													
130	**Chelsea buns**	330	220	110	25	120	1.5	0.26	0.8	530	0.4	N	N
131	**Cream horns**	200	71	62	11	57	0.7	0.07	0.4	310	0.2	N	6
132	**Crumpets**, *toasted*	810	93	120	18	180	1.1	0.20	0.6	980	0.3	(27)	(1)
133	**Currant buns**	230	210	110	27	100	1.9	0.18	0.6	210	0.4	N	N
134	**Custard tarts**, individual	130	110	95	14	98	0.8	0.07	0.5	390	0.2	N	N
135	**Danish pastries**	190	170	92	24	98	1.3	0.06	0.5	340	0.3	N	N
136	**Doughnuts**, jam	180	110	72	19	71	1.2	0.09	0.5	290	0.3	N	15
137	ring	230	87	76	21	81	1.2	0.14	0.6	360	0.3	N	(17)
138	**Eccles cake**	240	360	79	22	66	1.2	0.43	0.5	380	N	N	N
139	**Eclairs**, *frozen*	73	160	87	20	120	1.1	0.22	0.8	75	0.1	N	N

Cereals and cereal products *continued*

No.	Food	Retinol µg	Carotene µg	Vitamin D µg	Vitamin E mg	Thiamin mg	Ribo-flavin mg	Niacin mg	Trypt 60 mg	Vitamin B6 mg	Vitamin B12 µg	Folate µg	Panto-thenate mg	Biotin µg	Vitamin C mg
Pastry															
124	**Flaky pastry**, *raw*	125	125	1.3	1.47	0.14	0.01	0.8	0.9	0.07	0	10	0.1	1	1
125	*cooked*	170	165	1.8	1.95	0.14	0.02	1.0	1.1	0.07	0	7	0.1	1	Tr
126	**Shortcrust pastry**, *raw*	115	115	1.2	1.38	0.19	0.02	1.0	1.1	0.09	0	13	0.2	1	0
127	*cooked*	135	130	1.4	1.60	0.16	0.02	1.1	1.3	0.08	0	8	0.2	1	0
128	**Wholemeal pastry**, *raw*	115	115	1.2	2.05	0.28	0.05	3.4	1.5	0.30	0	34	0.5	4	0
129	*cooked*	135	130	1.4	2.37	0.25	0.05	3.8	1.7	0.26	0	20	0.4	5	0
Buns and pastries															
130	**Chelsea buns**	17	1	0.1	1.49	0.16	0.13	1.4	1.6	0.11	Tr	32	0.4	6	Tr
131	**Cream horns**	200	195	0.1	2.17	0.07	0.06	0.5	0.8	0.04	Tr	7	0.1	1	1
132	**Crumpets**, *toasted*	0	0	0	(0.19)	0.17	0.03	1.0	1.4	0.06	0	9	(0.3)	(4)	0
133	**Currant buns**	0	N	0	N	0.37	0.16	1.5	1.6	0.11	Tr	40	N	N	0
134	**Custard tarts**, individual	32	Tr	N	N	0.14	0.16	0.5	1.4	0.03	Tr	13	N	N	0
135	**Danish pastries**	N	N	N	Tr	0.13	0.07	0.9	1.2	0.07	Tr	20	(0.5)	(7)	0
136	**Doughnuts**, jam	N	N	N	Tr	0.22	0.07	1.3	1.2	0.03	Tr	21	N	N	N
137	ring	N	N	N	Tr	0.22	0.07	1.2	1.2	0.02	Tr	19	N	N	0
138	**Eccles cake**	55	32	0.1	1.99	0.11	0.03	0.9	0.7	0.11	0	5	0.1	2	0
139	**Eclairs**, *frozen*	240	Tr	(0)	(1.25)	0.10	0.19	0.3	1.1	0.03	1	11	(0.3)	(5)	Tr

Cereals and cereal products *continued*

Composition of food per 100g

No.	Food	Description and main data sources	Edible proportion	Water g	Protein g	Fat g	Carbo-hydrate g	Energy value kcal	kJ
	Buns and pastries								
140	**Greek pastries**	4 assorted samples, baclava, tangos, tsamika, shredded type	1.00	17.5	4.7	17.0	40.0	322	1349
141	**Hot cross buns**	Recipe	1.00	25.2	7.4	6.8	58.5	310	1313
142	**Jam tarts**	Recipe	1.00	19.6	3.3	14.9	62.0	380	1598
143	*retail*	10 samples, 6 brands, 4 bakeries	1.00	14.4	3.3	13.0	63.4	368	1551
144	**Mince pies**, individual	Recipe	1.00	12.0	4.3	20.4	59.0	423	1772
145	**Scones**, fruit	10 samples, 2 brands, 8 bakeries	1.00	25.3	7.3	9.8	52.9	316	1333
146	plain	Recipe	1.00	22.9	7.2	14.6	53.8	362	1523
147	wholemeal	Recipe. Ref. Wiles *et al.* (1980)	1.00	26.9	8.7	14.4	43.1	326	1368
148	**Scotch pancakes**	Drop scones; recipe	1.00	39.1	5.8	11.7	43.6	292	1228
149	**Teacakes**, *toasted*	Calculated using weight loss of 10%	1.00	18.6	8.9	8.3	58.3	329	1392
	Puddings								
150	**Blackcurrant pie**, *pastry top and bottom*	Recipe. Ref. Wiles *et al.* (1980)	1.00	42.3	3.1	13.3	34.5	262	1099
151	**Bread pudding**	Recipe	1.00	29.3	5.9	9.6	49.7	297	1252
152	**Christmas pudding**	Recipe	1.00	30.4	4.6	9.7	49.5	291	1227
153	*retail*	10 samples, 4 brands	1.00	(23.6)	3.0	11.8	56.3	329	1388
154	**Crumble**, fruit	Recipe. Apple, gooseberry, plum, rhubarb	1.00	54.8	2.0	6.9	34.0	198	835
155	fruit, wholemeal	Recipe. Apple, gooseberry, plum, rhubarb	1.00	54.8	2.6	7.1	31.7	193	813

Cereals and cereal products *continued*

140 to 155

Composition of food per 100g

No.	Food	Total nitrogen g	Fatty acids Satd g	Mono unsatd g	Poly unsatd g	Cholest- erol mg	Starch g	Total sugars g	Dietary fibre Southgate method g	Englyst method g
	Buns and pastries									
140	**Greek pastries**	0.82	N	N	N	N	21.6	18.4	1.9	N
141	**Hot cross buns**	1.27	2.1	2.3	1.6	31	35.1	23.4	2.2	1.7
142	**Jam tarts**	0.57	5.1	6.4	2.8	22	24.5	37.5	1.7	1.6
143	*retail*	0.53	4.8	5.3	2.0	42	27.4	36.0	2.5	N
144	**Mince pies**, individual	0.75	7.4	8.4	3.7	25	30.9	28.1	2.8	2.1
145	**Scones**, fruit	1.28	(3.3)	(3.7)	(2.4)	27	36.0	16.9	3.6	N
146	plain	1.23	4.9	5.5	3.6	29	47.9	5.9	2.2	1.9
147	wholemeal	1.48	4.8	5.4	3.6	27	37.1	5.9	5.0	5.2
148	**Scotch pancakes**	0.99	4.1	4.4	2.7	61	34.8	8.9	1.6	1.4
149	**Teacakes**, *toasted*	1.56	N	N	N	20	41.9	16.4	4.7	N
	Puddings									
150	**Blackcurrant pie**, *pastry top and bottom*	0.55	4.8	5.5	2.4	18	22.0	12.5	4.8[a]	2.6[a]
151	**Bread pudding**	0.98	5.9	2.6	0.5	53	16.6	33.1	3.0	1.2
152	**Christmas pudding**	0.79	4.5	4.0	0.7	45	15.2	34.3	2.7	1.3
153	*retail*	0.53	6.1	4.1	0.6	36	10.1	46.2	3.4	1.7
154	**Crumble**, fruit	0.35	2.1	2.6	1.8	12	12.7	21.3	2.2	1.7
155	fruit, wholemeal	0.44	2.1	2.7	2.0	12	10.3	21.4	3.0	2.7

[a] Blackcurrant pie made with wholemeal pastry contains 6.2g Southgate fibre and 4.3g Englyst fibre per 100g

Cereals and cereal products *continued*

Inorganic constituents per 100g

No.	Food	mg										µg	
		Na	K	Ca	Mg	P	Fe	Cu	Zn	Cl	Mn	Se	I
Buns and pastries													
140	**Greek pastries**	310	90	44	22	70	0.9	0.14	0.4	390	0.3	N	N
141	**Hot cross buns**	120	200	110	24	110	1.6	0.24	0.7	190	0.4	N	N
142	**Jam tarts**	230	110	55	12	46	1.4	0.14	0.3	370	N	1	10
143	retail	130	120	72	14	50	1.7	0.18	0.6	160	0.3	N	2
144	**Mince pies**, individual	310	180	75	15	58	1.5	0.15	0.3	550	0.3	N	N
145	**Scones**, fruit	710	220	150	24	360	1.5	0.22	0.8	450	0.3	N	N
146	plain	770	150	180	18	460	1.3	0.19	0.7	450	0.4	2	19
147	wholemeal	730	250	110	75	560	2.3	0.35	2.0	400	1.8	28	N
148	**Scotch pancakes**	430	300	120	16	95	1.0	0.07	0.5	560	0.3	2	18
149	**Teacakes**, *toasted*	300	240	98	32	110	2.9	0.27	0.8	490	0.5	N	N
Puddings													
150	**Blackcurrant pie**, *pastry top and bottom*	200	220	70	15	53	1.2	0.12	N	330	0.4	1	N
151	**Bread pudding**	310	310	120	24	110	1.6	0.20	0.6	450	0.4	11	N
152	**Christmas pudding**	200	350	79	27	76	1.5	0.22	0.5	300	0.4	7	N
153	retail	170	340	35	18	92	1.2	0.14	0.7	180	0.5	N	N
154	**Crumble**, fruit	68	190	49	9	33	0.6	0.10	0.2	130	0.2	1	4
155	fruit, wholemeal	68	220	32	25	68	0.9	0.15	0.6	120	0.6	9	N

Cereals and cereal products *continued*

No.	Food	Retinol µg	Carotene µg	Vitamin D µg	Vitamin E mg	Thiamin mg	Ribo-flavin mg	Niacin mg	Trypt 60 mg	Vitamin B6 mg	Vitamin B12 µg	Folate µg	Panto-thenate mg	Biotin µg	Vitamin C mg
Buns and pastries															
140	**Greek pastries**	N	N	N	N	0.09	0.04	1.0	1.0	N	N	N	N	N	N
141	**Hot cross buns**	65	43	0.6	0.71	0.15	0.11	1.3	1.5	0.10	Tr	28	0.3	5	0
142	**Jam tarts**	62	68	0.6	0.74	0.07	0.01	0.5	0.6	0.04	0	4	0.1	Tr	2
143	*retail*	N	N	N	N	0.06	0.02	0.5	0.7	0.03	Tr	5	(0.1)	Tr	Tr
144	**Mince pies**, individual	77	82	0.8	0.93	0.11	0.02	0.8	0.8	0.08	0	6	0.1	Tr	0
145	**Scones**, fruit	N	N	N	N	0.24	0.10	1.2	1.5	0.05	Tr	6	N	N	Tr
146	plain	140	125	1.2	1.44	0.16	0.07	1.0	1.5	0.09	Tr	8	0.2	1	Tr
147	wholemeal	130	115	1.2	1.99	0.22	0.10	3.1	1.8	0.23	Tr	18	0.4	5	Tr
148	**Scotch pancakes**	110	86	1.0	1.10	0.13	0.09	0.8	1.2	0.08	Tr	6	0.3	1	Tr
149	**Teacakes**, *toasted*	N	N	0	N	0.20	0.17	2.0	1.8	0.06	Tr	40	N	N	Tr
Puddings															
150	**Blackcurrant pie**, *pastry top and bottom*	55	105	0.6	1.14	0.08	0.03	0.6	0.6	0.06	0	N	0.2	1	72
151	**Bread pudding**	100	45	0.2	0.38	0.10	0.12	0.8	1.3	0.09	Tr	8	0.3	4	Tr
152	**Christmas pudding**	23	9	0.2	0.64	0.08	0.08	0.7	1.0	0.08	Tr	8	0.2	4	0
153	*retail*	N	N	N	N	Tr	0.03	0.4	0.6	0.07	Tr	9	N	N	Tr
154	**Crumble**, fruit	64	145	0.7	0.98	0.05	0.02	0.5	0.4	0.03	0	3	0.1	Tr	3
155	fruit, wholemeal	64	145	0.7	1.16	0.07	0.03	1.1	0.5	0.08	0	6	0.2	1	3

No.	Food	Description and main data sources	Edible proportion	Water g	Protein g	Fat g	Carbo-hydrate g	Energy value kcal	kJ
Puddings									
156	**Fruit pie,** *one crust*	Recipe. Apple, gooseberry, plum, rhubarb	1.00	59.0	2.0	7.9	28.7	186	784
157	*pastry top and bottom*	Recipe. Ref. Wiles *et al.* (1980)	1.00	47.9	3.0	13.3	34.0	260	1089
158	individual	10 pies, as purchased, 3 brands; apple, blackcurrant, blackberry, apricot	1.00	22.9	4.3	15.5	56.7	369	1554
159	wholemeal, *one crust*	Recipe. Ref. Wiles *et al.* (1980). Apple, gooseberry, plum, rhubarb	1.00	58.6	2.6	8.1	26.6	183	770
160	wholemeal, *pastry top and bottom*	Recipe. Ref. Wiles *et al.* (1980). Apple, gooseberry, plum, rhubarb	1.00	47.9	4.0	13.6	30.0	251	1052
161	**Lemon meringue pie**	Recipe	1.00	35.2	4.5	14.4	45.9	319	1342
162	**Pancakes, sweet,** *made with whole milk*	Recipe	1.00	43.4	5.9	16.2	35.0	301	1260
163	**Pie,** *with pie filling*	Recipe	1.00	47.5	3.2	14.5	34.6	273	1145
164	**Sponge pudding**	Recipe	1.00	32.8	5.8	16.3	45.3	340	1426
165	**Treacle tart**	Recipe	1.00	21.4	3.7	14.1	60.4	368	1550
Savouries									
166	**Cauliflower cheese**	Recipe	1.00	78.6	5.9	6.9	5.1	105	438
167	**Dumplings**	Recipe	1.00	60.5	2.8	11.7	24.5	208	871
168	**Macaroni cheese**	Recipe	1.00	67.1	7.3	10.8	13.6	178	743
169	**Pancakes, savoury,** *made with whole milk*	Recipe	1.00	51.9	6.3	17.5	24.0	273	1138

Cereals and cereal products *continued*

Composition of food per 100g

No.	Food	Total nitrogen g	Fatty acids			Cholest-erol mg	Starch g	Total sugars g	Dietary fibre	
			Satd g	Mono unsatd g	Poly unsatd g				Southgate method g	Englyst method g
Puddings										
156	**Fruit pie,** *one crust*	0.35	2.9	3.2	1.4	11	13.0	15.6	2.1	1.7
157	*pastry top and bottom*	0.53	4.8	5.5	2.4	18	22.0	12.0	2.2	1.8
158	*individual*	0.75	N	N	N	0	25.8	30.9	2.3	N
159	*wholemeal, one crust*	0.45	2.9	3.3	1.5	11	10.7	15.9	3.0	2.7
160	*wholemeal, pastry top and bottom*	0.68	4.9	5.6	2.5	18	17.9	12.2	3.6	3.5
161	**Lemon meringue pie**	0.75	5.0	6.1	2.8	89	21.1	24.8	0.8	0.7
162	**Pancakes,** *sweet, made with whole milk*	0.97	7.1	6.8	1.5	68	18.8	16.2	0.9	0.8
163	**Pie,** *with pie filling*	0.55	5.3	6.0	2.6	19	26.5	8.1	2.0	1.5
164	**Sponge pudding**	0.97	5.1	6.4	4.1	92	26.4	18.9	1.2	1.1
165	**Treacle tart**	0.64	5.1	5.8	2.6	19	26.8	33.6	1.4	1.1
Savouries										
166	**Cauliflower cheese**	0.95	3.4	2.1	1.0	16	2.1	3.0	1.4	1.3
167	**Dumplings**	0.48	6.4	4.2	0.4	8	24.0	0.4	1.0	0.9
168	**Macaroni cheese**	1.18	5.6	3.4	1.2	28	10.7	2.9	0.8	0.5
169	**Pancakes,** *savoury, made with whole milk*	1.04	7.7	7.3	1.7	67	20.2	3.8	1.0	0.8

Cereals and cereal products continued

Inorganic constituents per 100g

No.	Food	Na	K	Ca	Mg	P	Fe	Cu	Zn	Cl	Mn	Se	I
		mg										µg	
Puddings													
156	**Fruit pie**, *one crust*	120	180	48	9	33	0.5	0.10	0.2	210	0.2	1	5
157	*pastry top and bottom*	200	160	59	11	43	0.8	0.10	0.2	340	0.2	1	7
158	individual	210	120	51	12	64	1.2	0.10	0.5	260	2.0	N	N
159	wholemeal, *one crust*	100	210	31	26	69	0.9	0.15	0.6	180	0.6	9	12
160	wholemeal, *pastry top and bottom*	170	210	29	39	100	1.3	0.19	0.9	280	0.9	15	19
161	**Lemon meringue pie**	200	82	45	9	66	0.9	0.08	0.5	320	0.1	4	15
162	**Pancakes, sweet**, *made with whole milk*	53	150	110	14	110	0.8	0.05	0.6	100	0.1	3	24
163	**Pie**, *with pie filling*	240	91	60	10	44	1.4	0.06	0.2	380	0.3	1	N
164	**Sponge pudding**	310	89	84	10	180	1.1	0.10	0.6	270	0.2	4	19
165	**Treacle tart**	360	150	62	13	50	1.4	0.10	0.3	430	0.2	4	N
Savouries													
166	**Cauliflower cheese**	200	300	120	17	120	0.6	0.03	0.7	320	0.2	1	8
167	**Dumplings**	400	44	52	8	120	0.6	0.07	0.2	460	0.2	1	4
168	**Macaroni cheese**	310	110	170	17	150	0.4	0.05	0.8	490	0.1	4	16
169	**Pancakes, savoury**, *made with whole milk*	150	160	130	16	120	0.8	0.05	0.6	250	0.2	3	26

Cereals and cereal products *continued*

No.	Food	Retinol µg	Carotene µg	Vitamin D µg	Vitamin E mg	Thiamin mg	Ribo-flavin mg	Niacin mg	Trypt 60 mg	Vitamin B6 mg	Vitamin B12 µg	Folate µg	Panto-thenate mg	Biotin µg	Vitamin C mg
Puddings															
156	**Fruit pie**, *one crust*	33	110	0.3	0.65	0.05	0.02	0.5	0.4	0.03	0	3	0.1	Tr	5
157	*pastry top and bottom*	55	115	0.6	0.85	0.08	0.02	0.6	0.6	0.04	0	4	0.1	Tr	3
158	*individual*	0	Tr	0	N	0.05	0.02	0.4	0.9	N	0	N	N	N	Tr
159	*wholemeal, one crust*	33	110	0.3	0.84	0.08	0.03	1.1	0.5	0.08	0	6	0.2	1	5
160	*wholemeal, pastry top and bottom*	55	115	0.6	1.17	0.11	0.03	1.7	0.8	0.12	0	9	0.2	2	4
161	**Lemon meringue pie**	99	65	0.5	1.03	0.07	0.08	0.4	1.1	0.05	Tr	7	0.3	5	3
162	**Pancakes, sweet,** *made with whole milk*	56	12	0.2	0.33	0.10	0.17	0.5	1.4	0.09	1	8	0.5	5	1
163	**Pie,** *with pie filling*	60	68	0.6	0.72	0.08	0.01	0.6	0.6	0.04	0	4	0.1	Tr	3
164	**Sponge pudding**	170	130	1.6	1.75	0.09	0.09	0.6	1.4	0.06	Tr	8	0.3	5	Tr
165	**Treacle tart**	58	57	0.6	0.69	0.08	0.01	0.6	0.7	0.04	0	4	0.1	Tr	0
Savouries															
166	**Cauliflower cheese**	63	79	0.22	0.41	0.10	0.10	0.5	1.4	0.16	0.2	7	0.40	2	7
167	**Dumplings**	9	0	Tr	0.09	0.05	0.01	0.3	0.6	0.03	Tr	3	0.1	Tr	0
168	**Macaroni cheese**	110	74	0.36	0.45	0.04	0.16	0.3	1.7	0.05	0.4	5	0.26	2	Tr
169	**Pancakes, savoury,** *made with whole milk*	60	14	0.2	0.34	0.11	0.19	0.5	1.5	0.10	1	8	0.5	5	1

Cereals and cereal products continued

Composition of food per 100g

Savouries

No.	Food	Description and main data sources	Edible proportion	Water g	Protein g	Fat g	Carbo-hydrate g	Energy value kcal	Energy value kJ
170	**Pizza**	Cheese and tomato, recipe	1.00	51.7	9.0	11.8	24.8	235	984
171	*frozen*	10 samples, 2 brands, cheese and tomato	1.00	49.3	7.5	10.7	32.9	250	1050
172	**Ravioli,** *canned in tomato sauce*	10 samples, 4 brands	1.00	79.9	3.0	2.2	10.3	70	297
173	**Risotto,** *plain*	Recipe	1.00	55.1	3.0	9.3	34.4	224	943
174	**Samosas,** meat	Recipe	1.00	20.4	5.1	56.1	17.9	593	2451
175	vegetable	Recipe	1.00	31.5	3.1	41.8	22.3[a]	472	1954
176	**Spaghetti,** *canned in tomato sauce*	10 samples, 3 brands	1.00	81.9	1.9	0.4	14.1	64	273
177	**Stuffing,** sage and onion	Recipe	1.00	56.5	5.2	14.8	20.4	231	962
178	**Stuffing mix**	10 samples, 4 brands; assorted flavours	1.00	5.9	9.9	5.2	67.2	338	1436
179	*made up with water*	Calculated from No. 178; made up and cooked according to packet directions	1.00	76.4	2.8	1.5	19.3	97	412
180	**Yorkshire pudding**	Recipe	1.00	57.4	6.6	9.9	24.7	208	874

[a] Including oligosaccharides

Cereals and cereal products *continued*

No.	Food	Total nitrogen g	Fatty acids			Cholesterol mg	Starch g	Total sugars g	Dietary fibre	
			Satd g	Mono unsatd g	Poly unsatd g				Southgate method g	Englyst method g
Savouries										
170	**Pizza**	1.46	5.5	3.7	2.0	16	22.6	2.2	1.8	1.5
171	*frozen*	1.31	4.3	3.6	1.9	26	26.0	6.9	1.9	(1.5)
172	**Ravioli,** *canned in tomato sauce*	0.53	0.8	0.8	0.3	60	8.1	2.2	1.0	0.9
173	**Risotto,** *plain*	0.51	2.8	3.5	2.6	14	33.2	1.2	1.3	0.4
174	**Samosas,** meat	0.85	7.8	19.8	25.7	20	16.8	1.0	1.9	1.2
175	vegetable	0.52	5.2	14.8	20.0	4	20.1	1.9[a]	2.4	1.8
176	**Spaghetti,** *canned in tomato sauce*	0.33	0.1	0.1	0.2	0	8.6	5.5	2.8[b]	0.7[b]
177	**Stuffing,** sage and onion	0.87	4.5	5.9	3.7	76	14.6	5.8	2.4	1.7
178	**Stuffing mix**	1.58	2.4	1.6	0.1	5	62.8	4.4	N	4.7
179	*made up with water*	0.45	0.8	0.5	Tr	1	18.0	1.3	N	1.3
180	**Yorkshire pudding**	1.09	5.3	3.6	0.6	68	21.0	3.7	1.0	0.9

[a] Not including oligosaccharides

[b] Wholemeal types contain 6.2g Southgate fibre and 2.0g Englyst fibre per 100g

Cereals and cereal products *continued*

Inorganic constituents per 100g

No.	Food	Na	K	Ca	Mg	P	Fe	Cu	Zn	Cl	Mn	Se	I
						mg						μg	
												Se	I
Savouries													
170	**Pizza**	570	150	210	18	160	1.0	0.12	0.9	940	0.2	4	14
171	*frozen*	540	170	180	16	130	1.0	0.11	1.0	910	0.3	(4)	(14)
172	**Ravioli,** *canned in tomato sauce*	490	150	16	12	43	0.8	0.08	0.5	760	0.2	N	N
173	**Risotto,** *plain*	410	82	24	16	64	0.3	0.16	0.8	640	0.5	4	4
174	**Samosas,** meat	33	120	34	11	62	0.8	0.07	0.7	67	0.2	1	4
175	vegetable	200	200	32	15	47	0.8	0.08	0.3	340	0.8	N	4
176	**Spaghetti,** *canned in tomato sauce*	420	110	12	10	29	0.3	0.06	0.3	500	0.1	N	N
177	**Stuffing,** sage and onion	420	150	58	13	77	1.0	0.11	0.6	650	0.3	11	15
178	**Stuffing mix**	1460	240	960	41	130	5.1	0.17	0.8	2820	1.0	N	N
179	*made up with water*	420	69	280	12	37	1.5	0.05	0.2	810	0.3	N	N
180	**Yorkshire pudding**	590	160	130	19	120	0.9	0.05	0.6	940	0.2	3	27

Cereals and cereal products *continued*

No.	Food	Retinol µg	Carotene µg	Vitamin D µg	Vitamin E mg	Thiamin mg	Ribo-flavin mg	Niacin mg	Trypt 60 mg	Vitamin B6 mg	Vitamin B12 µg	Folate µg	Panto-thenate mg	Biotin µg	Vitamin C mg
Savouries															
170	**Pizza**	65	250	0.1	1.27	0.10	0.13	0.9	2.0	0.09	Tr	23	0.3	3	3
171	*frozen*	N	N	N	N	0.16	0.14	0.9	2.0	0.13	Tr	20	N	N	N
172	**Ravioli**, *canned in tomato sauce*	N	N	0	N	0.05	0.04	0.9	0.6	0.10	Tr	3	N	N	Tr
173	**Risotto**, *plain*	74	73	0.8	0.81	0.15	0.01	1.5	0.7	0.12	0	5	0.2	1	Tr
174	**Samosas**, meat	21	45	Tr	11.85	0.09	0.05	1.0	1.1	0.07	Tr	6	0.1	Tr	1
175	vegetable	21	94	Tr	9.75	0.12	0.02	0.6	0.6	0.09	0	11	0.1	Tr	4
176	**Spaghetti**, *canned in tomato sauce*	N	N	0	N	0.07	0.01	0.6	0.4	0.07	Tr	5	Tr	Tr	Tr
177	**Stuffing**, sage and onion	145	120	1.5	1.48	0.12	0.07	0.8	1.2	0.10	Tr	13	0.3	4	2
178	**Stuffing mix**	Tr	Tr	Tr	N	1.42	0.90	1.8	1.8	N	Tr	N	N	N	0
179	*made up with water*	Tr	Tr	Tr	Tr	0.31	0.22	0.5	0.5	N	Tr	N	N	N	0
180	**Yorkshire pudding**	64	14	0.3	0.37	0.10	0.16	0.5	1.6	0.07	1	9	0.4	5	1

Section 2.2

Milk and Milk Products

The majority of the data and foods in this section of the Tables have been taken from the *'Milk Products and Eggs'* supplement. Although the values for most of the nutrients presented in this section were obtained by analysis, a few were derived by calculation. Where analytical values were not available for the fat soluble vitamins, fatty acids and cholesterol in some dairy products these were calculated from the levels found in milk fat.

Where summer and winter values are given separately (for whole and Channel Island milks), summer is from May to October and winter from November to April. Recipe calculations use the average of the values for summer and winter milks. Some loss of vitamins is inevitable when milk is stored. On the doorstep, milk exposed for several hours to bright sunlight can lose up to 70 per cent of its riboflavin. Vitamin C can also decline under these conditions from the 1 – 1.5mg per 100g in the original milk to almost zero. There will also be gradual losses of folate and vitamin B_{12} from UHT and sterilised milks even under ideal storage conditions because of reactions with small amounts of oxygen in the pack. For those soft and blue cheeses where the rind may be eaten, mineral and vitamin values are for the whole cheese.

As many dairy products are sold or measured by volume, typical specific gravities (densities) of some of these products are given below. For the majority of purposes, the values given on a weight basis may be regarded as the same as those expressed by volume.

Specific gravities of selected dairy products

Skimmed milk	1.036	Single cream		1.00
Semi-skimmed milk	1.034	Whipping cream		0.99
Whole milk	1.031	Double cream		0.99
Condensed milk (sweetened)	1.160	Yogurts		1.08
Evaporated milk (unsweetened)	1.066			(range 1.03 - 1.2)
		Ice cream		variable 0.5 - 0.6 approx

Where new data have become available since the *'Milk Products and Eggs'* supplement significantly affecting some of the nutrients in the recipes, revised nutrient values have been included where appropriate. Losses of labile vitamins assigned on recipe calculation were estimated using the figures in Section 3.4.

Milk and milk products

Composition of food per 100g

No.	Food	Description and main data sources	Edible proportion	Water g	Protein g	Fat g	Carbo-hydrate g	Energy value kcal	Energy value kJ
181	**Skimmed milk**, *average*	Weighted average of pasteurised, sterilised and UHT	1.00	91.1	3.3	0.1	5.0	33	140
182	pasteurised	10 samples	1.00	91.1	3.3	0.1	5.0	33	140
183	*fortified plus SMP*	10 samples, own label and Vitapint	1.00	89.3	3.8	0.1	6.0	39	164
184	UHT, *fortified*	9 samples	1.00	90.9	3.5	0.2	5.0	35	147
185	**Semi-skimmed milk**, *average*	Weighted average of pasteurised and UHT	1.00	89.8	3.3	1.6	5.0	46	195
186	pasteurised	10 samples	1.00	89.8	3.3	1.6	5.0	46	195
187	*fortified plus SMP*	10 samples, own label and Vitapint	1.00	88.4	3.7	1.6	5.8	51	215
188	UHT	10 samples	1.00	89.7	3.3	1.7	4.8	46	194
189	**Whole milk**, *average*	Weighted average of pasteurised, sterilised and UHT	1.00	87.8	3.2	3.9	4.8	66	275
190	pasteurised[a]	186 samples, bottles and cartons. Fat from Milk Marketing Board	1.00	87.8	3.2	3.9	4.8	66	275
191	*summer*	Selected nutrients only	1.00	87.8	3.2	3.9	4.8	66	275
192	*winter*	Selected nutrients only	1.00	87.8	3.2	3.9	4.8	66	275
193	sterilised	10 samples, 2 brands, polybottles	1.00	87.6	3.5	3.9	4.5	66	277
194	**Channel Island milk**, whole,	Samples from dairy and retail outlets. Fat from Milk Marketing Board	1.00	86.4	3.6	5.1	4.8	78	327
195	*summer*	Selected nutrients only	1.00	86.4	3.6	5.1	4.8	78	327
196	*winter*	Selected nutrients only	1.00	86.4	3.6	5.1	4.8	78	327
197	semi-skimmed, UHT	10 samples	1.00	89.4	3.6	1.6	4.8	47	197

[a] All the values for pasteurised milk are equally applicable to unpasteurised milk

Milk and milk products

Composition of food per 100g

No.	Food	Total nitrogen g	Fatty acids Satd g	Mono unsatd g	Poly unsatd g	Cholest-erol mg	Starch g	Total sugars g	Dietary fibre Southgate method g	Englyst method g
181	**Skimmed milk**, *average*	0.52	0.1	Tr	Tr	2	0	5.0	0	0
182	pasteurised	0.52	0.1	Tr	Tr	2	0	5.0	0	0
183	*fortified plus SMP*	0.60	0.1	Tr	Tr	2	0	6.0	0	0
184	UHT, *fortified*	0.54	0.1	0.1	Tr	2	0	5.0	0	0
185	**Semi-skimmed milk**, *average*	0.52	1.0	0.5	Tr	7	0	5.0	0	0
186	pasteurised	0.52	1.0	0.5	Tr	7	0	5.0	0	0
187	*fortified plus SMP*	0.59	1.0	0.5	Tr	7	0	5.8	0	0
188	UHT	0.52	1.1	0.5	Tr	7	0	4.8	0	0
189	**Whole milk**, *average*	0.50	2.4	1.1	0.1	14	0	4.8	0	0
190	pasteurised	0.50	2.4	1.1	0.1	14	0	4.8	0	0
191	*summer*	0.50	2.4	1.2	0.1	14	0	4.8	0	0
192	*winter*	0.50	2.5	1.1	0.1	14	0	4.8	0	0
193	sterilised	0.55	2.4	1.1	0.1	14	0	4.5	0	0
194	**Channel Island milk**, whole, pasteurised	0.57	3.3	1.3	0.1	16	0	4.8	0	0
195	*summer*	0.57	3.2	1.4	0.1	16	0	4.8	0	0
196	*winter*	0.57	3.3	1.3	0.1	16	0	4.8	0	0
197	semi-skimmed, UHT	0.57	1.0	0.4	Tr	7	0	4.8	0	0

Milk and milk products

Inorganic constituents per 100g

No.	Food	Na	K	Ca	Mg	P	Fe	Cu	Zn	Cl	Mn	Se	I
						mg						µg	
181	**Skimmed milk**, *average*	54	150	120	12	94	0.06	Tr	0.4	100	Tr	(1)	(15)
182	pasteurised	55	150	120	12	95	0.05	Tr	0.4	100	Tr	(1)	(15)
183	*fortified plus SMP*	61	170	140	13	110	0.04	Tr	0.4	110	Tr	(1)	(15)
184	*UHT, fortified*	54	150	110	10	89	0.08	Tr	0.3	100	Tr	(1)	(15)
185	**Semi-skimmed milk**, *average*	55	150	120	11	95	0.05	Tr	0.4	100	Tr	(1)	(15)
186	pasteurised	55	150	120	11	95	0.05	Tr	0.4	100	Tr	(1)	(15)
187	*fortified plus SMP*	59	150	130	12	100	0.03	Tr	0.4	110	Tr	(1)	(15)
188	UHT	50	150	110	11	90	0.17	Tr	0.4	100	Tr	(1)	(15)
189	**Whole milk**, *average*	55	140	115	11	92	0.06	Tr	0.4	100	Tr	1	15
190	pasteurised	55	140	115	11	92	0.05	Tr	0.4	100	Tr	1	15
191	*summer*	55	140	115	11	92	0.05	Tr	0.4	100	Tr	1	7
192	*winter*	55	140	115	11	92	0.05	Tr	0.4	100	Tr	1	37
193	sterilised	57	140	120	13	91	0.18	Tr	0.3	100	Tr	(1)	(15)
194	**Channel Island milk**, whole, pasteurised	54	140	130	12	100	0.05	Tr	0.4	100	Tr	(1)	N
195	*summer*	54	140	130	12	100	0.06	Tr	0.4	100	Tr	(1)	N
196	*winter*	54	140	130	12	100	0.06	Tr	0.4	100	Tr	(1)	N
197	semi-skimmed, UHT	55	140	120	11	100	0.14	Tr	0.4	97	Tr	N	N

Milk and milk products

No.	Food	Retinol µg	Carotene µg	Vitamin D µg	Vitamin E mg	Thiamin mg	Riboflavin mg	Niacin mg	Trypt/60 mg	Vitamin B6 mg	Vitamin B12 µg	Folate µg	Pantothenate mg	Biotin µg	Vitamin C mg
181	**Skimmed milk**, *average*	1	Tr	Tr	Tr	0.04	0.17	0.1	0.8	0.06	0.4	5	0.32	1.9	1
182	pasteurised	1	Tr	Tr	Tr	0.04	0.18	0.1	0.8	0.06	0.4	6	0.32	2.0	1
183	*fortified plus SMP*	43	5	0.26	0.01	0.04	0.19	0.1	0.9	0.06	0.4	5	0.40	2.4	1
184	UHT, *fortified*	61	18	0.1	0.02	0.04	0.18	0.1	0.8	0.05	Tr	4	0.33	1.5	35[a]
185	**Semi-skimmed milk**, *average*	21	9	0.01	0.03	0.04	0.18	0.1	0.8	0.06	0.4	6	0.32	2.0	1
186	pasteurised	21	9	0.01	0.03	0.04	0.18	0.1	0.8	0.06	0.4	6	0.32	2.0	1
187	*fortified plus SMP*	90	5	0.13	0.04	0.04	0.19	0.1	0.9	0.06	0.4	5	0.37	2.3	1
188	UHT	20	11	0.01	0.03	0.04	0.18	0.1	0.8	0.05	0.2	2	0.33	1.8	Tr
189	**Whole milk**, *average*	52	21	0.03	0.09	0.03	0.17	0.1	0.7	0.06	0.4	6	0.35	1.9	1
190	pasteurised	52	21	0.03	0.09	0.04	0.17	0.1	0.7	0.06	0.4	6	0.35	1.9	1
191	*summer*	62	31	0.03	0.10	0.04	0.17	0.1	0.7	0.06	0.4	4	0.35	1.9	1
192	*winter*	41	11	0.03	0.07	0.04	0.17	0.1	0.7	0.06	0.4	7	0.35	1.9	1
193	sterilised	52	21	0.03	0.09	0.03	0.14	0.1	0.8	0.04	0.1	Tr	0.28	1.8	Tr
194	**Channel Island milk**, whole, pasteurised	46	71	0.03	0.11	0.04	0.19	0.1	0.9	0.06	0.4	6	0.36	1.9	1
195	*summer*	65	115	0.04	0.13	0.04	0.19	0.1	0.9	0.06	0.4	5	0.36	1.9	1
196	*winter*	27	27	0.03	0.09	0.04	0.19	0.1	0.9	0.06	0.4	7	0.36	1.9	1
197	semi-skimmed, UHT	14	22	0.01	0.04	0.04	0.19	0.1	0.9	0.05	0.2	1	0.34	1.5	Tr

[a] Unfortified milk would contain only traces of vitamin C

Milk and milk products *continued*

198 to 210

Composition of food per 100g

No.	Food	Description and main data sources	Edible proportion	Water g	Protein g	Fat g	Carbo-hydrate g	Energy value kcal	kJ
198	**Condensed milk,** skimmed, *sweetened*	10 cans (Fussells)	1.00	29.7	10.0	0.2	60.0	267	1137
199	whole, *sweetened*	10 cans, 2 brands	1.00	25.9	8.5	10.1	55.5	333	1406
200	**Dried skimmed milk**	20 samples, 7 brands, fortified	1.00	3.0	36.1	0.6	52.9	348	1482
201	*with vegetable fat*	12 samples, 5 brands, fortified	1.00	2.0	23.3	25.9	42.6	487	2038
202	**Evaporated milk,** whole	12 samples, Ideal, Carnation and own brands	1.00	69.1	8.4	9.4	8.5	151	629
203	**Flavoured milk**	32 samples in polybottles; mixed flavours, sterilised; skimmed and whole milk	1.00	85.4	3.6	1.5	10.6[a]	68	287
204	**Goats milk,** pasteurised	20 samples from one herd and literature sources	1.00	88.9	3.1	3.5	4.4	60	253
205	**Human milk,** colostrum	Literature sources	1.00	88.2	2.0	2.6	6.6	56	236
206	transitional	Mixed sample, 15 mothers at 10th day post partum and literature sources	1.00	(87.4)	1.5	3.7	6.9	67	281
207	mature	Department of Health and literature sources	1.00	87.1	1.3[b]	4.1	7.2	69	289
208	**Sheeps milk,** *raw*	30 samples from 2 herds and literature sources	1.00	83.0	5.4	6.0	5.1	95	396
209	**Soya milk,** plain	6 samples, 4 brands	1.00	89.7	2.9	1.9	0.8	32	132
210	flavoured	4 brands, assorted flavours	1.00	89.4	2.8	1.7	3.6	40	168

[a] Including oligosaccharides from the glucose syrup/maltodextrins in the product

[b] N x 6.38. Excluding the non-protein nitrogen, true protein = 0.85g per 100g

Milk and milk products *continued*

Composition of food per 100g

No.	Food	Total nitrogen g	Fatty acids Satd g	Mono unsatd g	Poly unsatd g	Cholest- erol mg	Starch g	Total sugars g	Dietary fibre Southgate method g	Englyst method g
198	**Condensed milk**, skimmed, *sweetened*	1.57	0.1	0.1	Tr	1	0	60.0	0	0
199	whole, *sweetened*	1.33	6.3	2.9	0.3	36	0	55.5	0	0
200	**Dried skimmed milk**	5.70	0.4	0.2	Tr	12	0	52.9	0	0
201	*with vegetable fat*	3.70	16.8	7.3	0.7	17	0	42.6	0	0
202	**Evaporated milk**, whole	1.32	5.9	2.7	0.3	34	0	8.5	0	0
203	**Flavoured milk**	0.56	0.9	0.4	Tr	7	Tr	9.4[a]	0	0
204	**Goats milk**, pasteurised	0.49	2.3	0.8	0.1	10	0	4.4	0	0
205	**Human milk**, colostrum	0.31	1.1	1.1	0.3	31	0	6.6	0	0
206	transitional	0.23	1.5	1.5	0.5	24	0	6.9	0	0
207	mature	0.20	1.8	1.6	0.5	16	0	7.2	0	0
208	**Sheeps milk**, *raw*	0.85	3.8	1.5	0.3	11	0	5.1	0	0
209	**Soya milk**, plain	0.52	0.3	0.4	1.1	0	0	0.8	Tr	Tr
210	flavoured	0.49	(0.2)	(0.4)	(1.0)	0	0	3.6	Tr	Tr

[a] Not including oligosaccharides from the glucose syrup/maltodextrins in the product

Milk and milk products *continued*

Inorganic constituents per 100g

No.	Food	Na	K	Ca	Mg	P	Fe	Cu	Zn	Cl	Mn	Se	I
						mg						μg	
198	**Condensed milk**, skimmed, *sweetened*	150	450	330	33	270	0.33	Tr	1.2	300	Tr	(3)	(89)
199	whole, *sweetened*	140	360	290	29	240	0.23	Tr	1.0	230	Tr	(3)	74
200	**Dried skimmed milk**	550	1590	1280	130	970	0.27	Tr	4.0	1070	Tr	(11)	(150)
201	*with vegetable fat*	440	1030	840	74	680	0.19	Tr	0.6	760	Tr	(7)	N
202	**Evaporated milk**, whole	180	360	290	29	260	0.26	0.02	0.9	250	Tr	(3)	11
203	**Flavoured milk**	61	150	110	13	89	0.23	Tr	0.5	110	Tr	N	N
204	**Goats milk**, pasteurised	42	170	100	13	90	0.12	0.03	0.5	150	Tr	N	N
205	**Human milk**, colostrum	47	70	28	3	14	0.07	0.05	0.6	N	Tr	N	N
206	transitional	30	57	25	3	16	0.07	0.04	(0.3)	86	Tr	(2)	7
207	mature	15	58	34	3	15	0.07	0.04	0.3	42	Tr	1	7
208	**Sheeps milk**, *raw*	44	120	170	18	150	0.03	0.10	0.7	82	Tr	N	N
209	**Soya milk**, plain	32	120	13	15	47	0.40	0.06	0.2	15	0.1	N	N
210	flavoured	61	110	14	18	53	0.40	0.02	0.2	130	0.1	N	N

Milk and milk products *continued*

No.	Food	Retinol µg	Carotene µg	Vitamin D µg	Vitamin E mg	Thiamin mg	Ribo-flavin mg	Niacin mg	Trypt 60 mg	Vitamin B6 mg	Vitamin B12 µg	Folate µg	Panto-thenate mg	Biotin µg	Vitamin C mg
198	**Condensed milk**, skimmed, *sweetened*	28	20	0.85	0.04	0.11	0.51	0.3	2.3	0.09	0.9	16	1.03	5.2	5
199	whole, *sweetened*	110	70	5.40	0.19	0.09	0.46	0.3	2.0	0.07	0.7	15	0.85	3.9	4
200	**Dried skimmed milk**[a]	350	5	2.10	0.27	0.38	1.63	1.0	8.5	0.60	2.6	51	3.28	20.1	13
201	*with vegetable fat*	395	15	10.50	1.32	0.23	1.20	0.6	5.5	0.35	2.3	36	2.15	15.0	11
202	**Evaporated milk**, whole	105	100	3.95[b]	0.19	0.07	0.42	0.2	2.0	0.07	0.1	11	0.75	4.0	1
203	**Flavoured milk**	20	8	0.01	0.03	0.03	0.17	0.1	0.8	0.03	0.1	2	0.30	2.2	Tr
204	**Goats milk**, pasteurised	44	Tr	0.11	0.03	0.04	0.13	0.3	0.7	0.06	0.1	1	0.41	3.0	1
205	**Human milk**, colostrum	155	(135)	N	1.30	Tr	0.03	0.1	0.7	Tr	0.1	2	0.12	Tr	7
206	transitional	85	(37)	N	0.48	0.01	0.03	0.1	0.5	Tr	Tr	3	0.20	0.2	6
207	mature	58	(24)	0.04	0.34	0.02	0.03	0.2	0.5	0.01	Tr	5	0.25	0.7	4
208	**Sheeps milk**, *raw*	83	Tr	0.18	0.11	0.08	0.32	0.4	1.3	0.08	0.6	5	0.45	2.5	5
209	**Soya milk**, plain	0	Tr	0	0.74	0.06	0.27	0.1	0.5	0.07	0	19	N	N	0
210	flavoured	0	N	0	0.66	0.06	0.03	0.1	0.5	0.07	0	15	N	N	Tr

[a] Unfortified skimmed milk powder contains approximately 8µg retinol, 3µg carotene, Tr vitamin D, and 0.01mg vitamin E per 100g. Some brands contain as much as 755µg retinol, 10µg carotene and 4.6µg vitamin D per 100g

[b] This is for fortified product. Unfortified evaporated milk contains approximately 0.09µg vitamin D per 100g

Milk and milk products *continued*

Composition of food per 100g

No.	Food	Description and main data sources	Edible proportion	Water g	Protein g	Fat g	Carbo-hydrate g	Energy value kcal	kJ
	Fresh creams (pasteurised)								
211	**Half**	10 samples, 5 brands	1.00	78.9	3.0	13.3	4.3	148	613
212	**Single**	10 samples, 5 brands	1.00	73.7	2.6	19.1	4.1	198	817
213	**Soured**	8 samples, 4 brands	1.00	72.5	2.9	19.9	3.8	205	845
214	**Whipping**	10 samples, 6 brands	1.00	55.4	2.0	39.3	3.1	373	1539
215	**Double**	12 samples, 5 brands	1.00	47.5	1.7	48.0	2.7	449	1849
216	**Clotted**	17 samples, 3 brands	1.00	32.2	1.6	63.5	2.3	586	2413
	Sterilised creams								
217	**Sterilised**, canned	13 cans, 6 brands	1.00	69.2	2.5	23.9	3.7	239	985
	UHT creams								
218	**Canned spray**	8 samples (Anchor)	1.00	58.4	1.9	32.0	3.5	309	1273
	Imitation creams								
219	**Dessert Top**	Manufacturer's data (Nestlés)	1.00	N	2.4	28.8	6.0	291	1201
220	**Dream Topping**, *made up with whole milk*	Recipe	1.00	69.9	3.8	13.5	12.1	182	757
221	*made up with semi-skimmed milk*	Recipe	1.00	71.5	3.9	11.7	12.2	166	694
222	**Elmlea**, single	Analysis and manufacturer's data (Van den Berghs)	1.00	N	3.2	18.0	4.1	190	786
223	whipping	Analysis and manufacturer's data (Van den Berghs)	1.00	N	2.5	33.0	3.2	319	1315
224	double	Analysis and manufacturer's data (Van den Berghs)	1.00	N	2.5	48.0	3.2	454	1870
225	**Tip Top**	Manufacturer's data (Nestlés)	1.00	N	5.0	6.5	8.5	110	460

Milk and milk products *continued*

Composition of food per 100g

No.	Food	Total nitrogen g	Fatty acids			Cholest-erol mg	Starch g	Total sugars g	Dietary fibre	
			Satd g	Mono unsatd g	Poly unsatd g				Southgate method g	Englyst method g
Fresh creams (pasteurised)										
211	**Half**	0.47	8.3	3.9	0.4	40	0	4.3	0	0
212	**Single**	0.41	11.9	5.5	0.5	55	0	4.1	0	0
213	**Soured**	0.45	12.5	5.8	0.6	60	0	3.8	0	0
214	**Whipping**	0.31	24.6	11.4	1.1	105	0	3.1	0	0
215	**Double**	0.27	30.0	13.9	1.4	130	0	2.7	0	0
216	**Clotted**	0.25	39.7	18.4	1.8	170	0	2.3	0	0
Sterilised creams										
217	**Sterilised**, canned	0.39	14.9	6.9	0.7	65	0	3.7	0	0
UHT creams										
218	**Canned spray**	0.30	20.0	9.3	0.9	85	0	3.5	0	0
Imitation creams										
219	**Dessert Top**	0.38	27.0	N	N	Tr	Tr	N	Tr[a]	Tr[a]
220	**Dream Topping**, *made up with whole milk*	0.59	11.7	1.0	0.2	11	2.1	10.0	Tr	Tr
221	*made up with semi-skimmed milk*	0.61	10.5	0.5	0.1	6	2.1	10.2	Tr	Tr
222	**Elmlea**, single	0.50	14.0	3.0	0.2	30	0	4.1	0.3[a]	0.3[a]
223	whipping	0.39	28.0	2.5	1.0	31	0	3.2	0.1[a]	0.1[a]
224	double	0.39	29.0	13.9	3.0	33	0	3.2	0.1[a]	0.1[a]
225	**Tip Top**	0.78	5.8	N	N	Tr	N	N	Tr[a]	Tr[a]

[a] Carob and guar gums are added as thickeners

Milk and milk products continued

No.	Food	Na	K	Ca	Mg	P	Fe	Cu	Zn	Cl	Mn	Se	I
						mg						μg	
Fresh creams (pasteurised)													
211	**Half**	49	120	99	11	82	0.1	Tr	0.3	77	Tr	Tr	N
212	**Single**	49	120	91	9	76	0.1	Tr	0.5	80	Tr	Tr	N
213	**Soured**	41	110	93	10	81	0.4	Tr	0.5	81	Tr	Tr	N
214	**Whipping**	40	80	62	7	58	Tr	Tr	0.3	59	Tr	Tr	N
215	**Double**	37	65	50	6	50	0.2	Tr	0.2	51	Tr	Tr	Tr
216	**Clotted**	18	55	37	5	40	0.1	0.09	0.2	40	Tr	Tr	Tr
Sterilised creams													
217	**Sterilised**, canned	53	110	86	10	73	0.8	Tr	1.1	78	Tr	Tr	N
UHT creams													
218	**Canned spray**	33	92	66	7	57	1.0	Tr	0.4	62	Tr	Tr	N
Imitation creams													
219	**Dessert Top**	50	100	80	8	70	N	N	N	N	N	N	N
220	**Dream Topping**, *made up with whole milk*	70	120	95	9	92	0.1	0.03	0.3	82	Tr	N	12
221	*made up with semi-skimmed milk*	70	130	99	9	94	0.1	0.03	0.4	82	Tr	N	12
222	**Elmlea**, single	40	120	110	N	95	N	N	N	N	N	N	N
223	whipping	31	93	89	N	74	N	N	N	N	N	N	N
224	double	31	93	89	N	74	N	N	N	N	N	N	N
225	**Tip Top**	90	220	160	16	150	N	N	N	N	N	N	N

Milk and milk products *continued*

No.	Food	Retinol µg	Carotene µg	Vitamin D µg	Vitamin E mg	Thiamin mg	Ribo-flavin mg	Niacin mg	Trypt 60 mg	Vitamin B6 mg	Vitamin B12 µg	Folate µg	Panto-thenate mg	Biotin µg	Vitamin C mg
Fresh creams (pasteurised)															
211	Half	190	54	0.10	0.29	0.03	0.18	0.1	0.7	0.05	0.3	6	0.26	1.7	1
212	Single	315	125	0.14	0.40	0.04	0.17	0.1	0.6	0.05	0.3	7	0.28	1.8	1
213	Soured	330	105	0.15	0.44	0.03	0.17	0.1	0.7	0.04	0.2	12	0.24	1.5	Tr
214	Whipping	565	265	0.22	0.86	0.02	0.17	Tr	0.5	0.04	0.2	7	0.22	1.4	1
215	Double	600	325	0.27	1.10	0.02	0.16	Tr	0.4	0.03	0.2	7	0.19	1.1	1
216	Clotted	705	685	0.28	1.48	0.02	0.16	Tr	0.4	0.03	0.1	6	0.14	1.0	Tr
Sterilised creams															
217	Sterilised, canned	240	215	Tr	0.48	0.02	0.16	0.1	0.6	0.02	0.1	1	0.25	2.1	Tr
UHT creams															
218	Canned spray	370	340	(0.18)	0.74	0.03	0.17	0.1	0.5	0.03	0.2	1	0.19	1.7	0
Imitation creams															
219	Dessert Top	Tr	400	Tr	N	N	N	N	0.6	N	N	N	N	N	Tr
220	Dream Topping, *made up with whole*														
	milk	41	N[a]	0.02	N	0.04	0.19	0.1	0.8	0.05	0.5	4	N	N	1
221	*made up with semi-skimmed milk*	16	N[a]	0.01	N	0.04	0.19	0.1	0.9	0.05	0.5	4	N	N	1
222	Elmlea, single	N	N	N	N	N	N	N	0.7	N	N	N	N	N	N
223	whipping	N	N	N	N	N	N	N	0.6	N	N	N	N	N	N
224	double	N	N	N	N	N	N	N	0.6	N	N	N	N	N	N
225	Tip Top	Tr	6	Tr	N	N	N	N	1.2	N	N	N	N	N	Tr

[a] β-Carotene is added as a colouring agent

Milk and milk products *continued*

Composition of food per 100g

No.	Food	Description and main data sources	Edible proportion	Water g	Protein g	Fat g	Carbo-hydrate g	Energy value kcal	kJ
	Cheeses								
226	**Brie**	10 samples	1.00	48.6	19.3	26.9	Tr	319	1323
227	**Camembert**	10 samples	1.00	50.7	20.9	23.7	Tr	297	1232
228	**Cheddar**, *average*	Weighted average from 5 countries	1.00	36.0	25.5	34.4	0.1	412	1708
229	vegetarian	10 samples	1.00	33.9	25.8	35.7	Tr	425	1759
230	**Cheddar-type**, *reduced fat*	10 samples, Tendale	1.00	47.1	31.5	15.0	Tr	261	1091
231	**Cheese spread**, plain	10 samples, 3 brands	1.00	53.3	13.5	22.8[a]	4.4	276	1143
232	**Cottage cheese**, plain	10-19 samples	1.00	79.1	13.8	3.9	2.1	98	413
233	*with additions*	10 samples, mixed, e.g. with pineapple, Cheddar cheese	1.00	76.9	12.8	3.8	2.6	95	400
234	reduced fat	6 samples, different brands	1.00	80.2	13.3	1.4	3.3	78	331
235	**Cream cheese**	3 samples	1.00	45.5	3.1	47.4	Tr	439	1807
236	**Danish blue**	10 samples	1.00	45.3	20.1	29.6	Tr	347	1437
237	**Edam**	10 samples	1.00	43.8	26.0	25.4	Tr	333	1382
238	**Feta**	18 samples, made from sheeps and goats milk	1.00	56.5	15.6	20.2	1.5	250	1037
239	**Fromage frais**, fruit	11 samples, 4 brands, mixed flavours	1.00	71.9	6.8	5.8	13.8	131	551
240	plain	12 samples, 3 brands	1.00	77.9	6.8	7.1	5.7	113	469
241	very low fat	10 samples, 4 brands, plain and fruit	1.00	83.7	7.7	0.2	6.8	58	247

[a] Reduced fat varieties contain approximately 9.0g fat per 100g

Milk and milk products *continued*

Composition of food per 100g

No.	Food	Total nitrogen g	Fatty acids			Cholest-erol mg	Starch g	Total sugars g	Dietary fibre	
			Satd g	Mono unsatd g	Poly unsatd g				Southgate method g	Englyst method g
Cheeses										
226	**Brie**	3.02	(16.8)	(7.8)	(0.8)	100	0	Tr	0	0
227	**Camembert**	3.27	(14.8)	(6.9)	(0.7)	75	0	Tr	0	0
228	**Cheddar**, *average*	4.00	21.7	9.4	1.4	100	0	0.1	0	0
229	*vegetarian*	4.04	22.5	9.8	1.5	105	0	Tr	0	0
230	**Cheddar-type**, *reduced fat*	4.94	9.4	4.4	0.4	43	0	Tr	0	0
231	**Cheese spread**, plain	2.11	14.3	6.6	0.7	(65)	0	4.4	0	0
232	**Cottage cheese**, plain	2.16	2.4	1.1	0.1	13	0	2.1	0	0
233	*with additions*	2.00	2.4	1.1	0.1	13	0	2.6	Tr	Tr
234	*reduced fat*	2.08	0.9	0.4	Tr	5	0	3.3	0	0
235	**Cream cheese**	0.49	29.7	13.7	1.4	95	0	Tr	0	0
236	**Danish blue**	3.15	(18.5)	(8.6)	(0.9)	75	0	Tr	0	0
237	**Edam**	4.08	(15.9)	(7.4)	(0.7)	80	0	Tr	0	0
238	**Feta**	2.45	(13.7)	(4.1)	(0.6)	70	0	1.5	0	0
239	**Fromage frais**, fruit	1.06	3.6	1.7	0.2	21	0	13.8	Tr	Tr
240	plain	1.06	4.4	2.1	0.2	25	0	5.7	0	0
241	very low fat	1.21	0.1	0.1	Tr	1	Tr	6.8	Tr	Tr

No.	Food	Na	K	Ca	Mg	P	Fe	Cu	Zn	Cl	Mn	Se	I
						mg						μg	
	Cheeses												
226	**Brie**	700	100	540	27	390	0.8	Tr	2.2	1060	Tr	N	N
227	**Camembert**	650	100	350	21	310	0.2	0.07	2.7	1120	Tr	N	N
228	**Cheddar**, *average*	670	77	720	25	490	0.3	0.03	2.3	1030	Tr	12	39
229	vegetarian	670	67	690	31	490	0.2	Tr	1.9	990	0.1	(12)	(46)
230	**Cheddar-type**, *reduced fat*	670	110	840	39	620	0.2	0.05	2.8	1110	Tr	(15)	N
231	**Cheese spread**, plain	1060	240	420	25	790	0.2	0.12	2.1	820	Tr	(6)	29
232	**Cottage cheese**, plain	380	89	73	9	160	0.1	0.04	0.6	550	Tr	(4)	N
233	with additions	360	130	110	12	160	0.1	0.05	0.5	590	Tr	(4)	N
234	reduced fat	(380)	(89)	(73)	(9)	(160)	(0.1)	(0.04)	(0.6)	(550)	Tr	(4)	N
235	**Cream cheese**	300	160	98	10	100	0.1	(0.04)	0.5	480	Tr	(1)	N
236	**Danish blue**	1260	89	500	27	370	0.2	0.08	2.0	1950	Tr	2	9
237	**Edam**	1020	97	770	39	530	0.4	0.05	2.2	1570	Tr	N	N
238	**Feta**	1440	95	360	20	280	0.2	0.07	0.9	2350	Tr	N	N
239	**Fromage frais**, fruit	35	110	86	8	110	0.1	0.02	0.4	78	Tr	(2)	N
240	plain	31	110	89	8	110	0.1	Tr	0.3	100	Tr	(2)	N
241	very low fat	(33)	(110)	(87)	(8)	(110)	(0.1)	(0.01)	(0.3)	(89)	Tr	(2)	N

Milk and milk products *continued*

No.	Food	Retinol μg	Carotene μg	Vitamin D μg	Vitamin E mg	Thiamin mg	Ribo-flavin mg	Niacin mg	Trypt 60 mg	Vitamin B6 mg	Vitamin B12 μg	Folate μg	Panto-thenate mg	Biotin μg	Vitamin C mg
Cheeses															
226	**Brie**	285	210	0.20	0.84	0.04[a]	0.43	0.4	4.5	0.15	1.2	58	0.35	5.6	Tr
227	**Camembert**	230	315	(0.18)	0.65	0.05[b]	0.52	1.0	4.9	0.22	1.1	102	0.36	7.6	Tr
228	**Cheddar**, *average*	325	225	0.26	0.53	0.03	0.40	0.1	6.0	0.10	1.1	33	0.36	3.0	Tr
229	vegetarian	385	460	0.27	0.80	0.03	0.45	Tr	6.1	0.11	1.2	25	0.46	2.6	Tr
230	**Cheddar-type**, *reduced fat*	165	100	0.11	0.39	0.03	0.53	0.1	7.4	0.13	1.3	56	0.51	3.8	Tr
231	**Cheese spread**, plain	275	105	0.17	0.24	0.05	0.36	0.1	3.2	0.08	0.6	19	0.51	3.6	Tr
232	**Cottage cheese**, plain	44	10	0.03	0.08	0.03	0.26	0.1	3.2	0.08	0.7	27	0.40	3.0	Tr
233	with additions	43	10	0.03	0.08	0.06	0.21	0.2	3.0	0.08	0.6	13	0.31	3.0	1
234	reduced fat	16	4	0.01	0.03	(0.03)	(0.26)	(0.1)	3.1	(0.08)	(0.7)	(27)	(0.40)	(3.0)	Tr
235	**Cream cheese**	385	220	0.27	1.00	0.03	0.13	0.1	0.7	0.04	0.3	11	0.27	1.6	Tr
236	**Danish blue**	280	250	(0.23)	0.76	0.03	0.41	0.5	4.7	0.12	1.0	50	0.53	2.7	Tr
237	**Edam**	175	150	(0.19)	0.48	0.03	0.35	0.1	6.1	0.09	2.1	40	0.38	1.8	Tr
238	**Feta**	220	33	0.50	0.37	0.04	0.21	0.2	3.5	0.07	1.1	23	0.36	2.4	Tr
239	**Fromage frais**, fruit	82	N	0.04	(0.01)	0.02	0.35	0.1	1.6	0.04	1.4	15	N	N	Tr
240	plain	100	Tr	0.05	0.02	0.04	0.40	0.1	1.6	0.10	1.4	15	N	N	Tr
241	very low fat	3	N	Tr	Tr	(0.03)	(0.37)	(0.1)	1.8	(0.07)	(1.4)	(15)	N	N	Tr

[a] The rind alone contains 0.5mg thiamin per 100g   [b] The rind alone contains 0.4mg thiamin per 100g

Milk and milk products *continued*

No.	Food	Description and main data sources	Edible proportion	Water g	Protein g	Fat g	Carbo-hydrate g	Energy value kcal	kJ
	Cheeses								
242	**Full fat soft cheese**	e.g. Philadelphia-type. Manufacturer's data plus							
		calculation	1.00	58.0	8.6	31.0	Tr	313	1293
243	**Gouda**	10 samples	1.00	40.1	24.0	31.0	Tr	375	1555
244	**Hard cheese**, *average*	Average of Cheddar, Derby, Double Gloucester and							
		Leicester	1.00	37.2	24.7	34.0	0.1	405	1679
245	**Lymeswold**	Mild blue full fat soft cheese, 10 samples	1.00	41.0	15.6	40.3	Tr	425	1756
246	**Medium fat soft cheese**	e.g. Philadelphia light, 5 samples, 3 brands	1.00	69.5	9.2	14.5	3.1	179	743
247	**Parmesan**	10 samples, block and powdered	1.00	18.4	39.4	32.7	Tr	452	1880
248	**Processed cheese**, plain	10 samples, blocks and slices	1.00	45.7	20.8	27.0[a]	0.9	330	1367
249	**Stilton**, blue	10-13 samples	1.00	38.6	22.7	35.5	0.1	411	1701
250	**White cheese**, *average*	Average of Caerphilly, Cheshire, Lancashire,							
		Wensleydale	1.00	41.4	23.4	31.3	0.1	376	1557

[a] Reduced fat varieties contain approximately 9.5g fat per 100g

Milk and milk products *continued*

Composition of food per 100g

No.	Food	Total nitrogen	Fatty acids			Cholest- erol	Starch	Total sugars	Dietary fibre	
			Satd	Mono unsatd	Poly unsatd				Southgate method	Englyst method
		g	g	g	g	mg	g	g	g	g
Cheeses										
242	Full fat soft cheese	1.35	19.4	9.0	0.9	90	0	Tr	0	0
243	Gouda	3.76	(19.4)	(9.0)	(0.9)	100	0	Tr	0	0
244	Hard cheese, *average*	3.87	21.3	9.9	1.0	100	0	0.1	0	0
245	Lymeswold	2.44	25.2	11.7	1.2	115	0	Tr	0	0
246	Medium fat soft cheese	1.45	9.1	4.2	0.4	42	0	3.1	0	0
247	Parmesan	6.17	(20.5)	(9.5)	(0.9)	100	0	Tr	0	0
248	Processed cheese, plain	3.26	16.6	7.7	1.2	85	0	0.9	0	0
249	Stilton, blue	3.56	22.2	10.3	1.0	105	0	0.1	0	0
250	White cheese, *average*	3.67	19.6	9.1	0.9	90	0	0.1	0	0

Milk and milk products *continued*

No.	Food	mg										µg	
		Na	K	Ca	Mg	P	Fe	Cu	Zn	Cl	Mn	Se	I
Cheeses													
242	**Full fat soft cheese**	(330)	(150)	(110)	(9)	(130)	(0.1)	(0.10)	(0.7)	(600)	Tr	(3)	N
243	**Gouda**	910	91	740	38	490	0.1	Tr	1.8	1440	Tr	N	N
244	**Hard cheese**, *average*	620	82	670	24	470	0.4	0.03	2.3	980	Tr	(12)	(44)
245	**Lymeswold**	560	85	270	19	260	0.3	0.03	1.8	910	Tr	(7)	(46)
246	**Medium fat soft cheese**	N	N	N	N	N	N	N	N	N	Tr	(3)	N
247	**Parmesan**	1090	110	1200	45	810	1.1	0.33	5.3	1820	0.1	11	N
248	**Processed cheese**, plain	1320	130	600	22	800	0.5	0.17	3.2	1100	Tr	(10)	(29)
249	**Stilton**, blue	930	130	320	20	310	0.3	0.18	2.5	1410	Tr	(11)	(46)
250	**White cheese**, *average*	530	88	560	19	400	0.4	0.12	3.1	840	Tr	(12)	(46)

Milk and milk products *continued*

No.	Food	Retinol µg	Carotene µg	Vitamin D µg	Vitamin E mg	Thiamin mg	Riboflavin mg	Niacin mg	Trypt 60 mg	Vitamin B6 mg	Vitamin B12 µg	Folate µg	Pantothenate mg	Biotin µg	Vitamin C mg
Cheeses															
242	**Full fat soft cheese**	N	N	N	N	(0.03)	(0.17)	(0.1)	2.0	(0.05)	(0.3)	(13)	(0.31)	(2.0)	Tr
243	**Gouda**	245	145	(0.24)	0.53	0.03	0.30	0.1	5.6	0.08	1.7	43	0.32	1.4	Tr
244	**Hard cheese**, *average*	335	225	0.26	0.50	0.03	0.43	0.1	5.8	0.11	1.3	28	0.34	3.0	Tr
245	**Lymeswold**	440	330	0.31	0.93	0.04	0.43	0.6	3.7	0.14	0.9	56	0.39	6.3	Tr
246	**Medium fat soft cheese**	195	175	0.11	0.78	N	N	N	2.2	N	N	N	N	N	Tr
247	**Parmesan**	345	210	(0.25)	0.70	0.03	0.44	0.1	9.3	0.13	1.9	12	0.43	3.3	Tr
248	**Processed cheese**, plain	270	95	0.21	0.55	0.03	0.28	0.1	4.9	0.08	0.9	18	0.31	2.3	Tr
249	**Stilton**, blue	355	185	0.27	0.61	0.03	0.43	0.5	5.3	0.16	1.0	77	0.71	3.6	Tr
250	**White cheese**, *average*	315	225	0.24	0.65	0.03	0.47	0.1	5.5	0.09	1.1	44	0.29	3.9	Tr

Milk and milk products continued

Composition of food per 100g

No.	Food	Description and main data sources	Edible proportion	Water g	Protein g	Fat g	Carbo-hydrate g	Energy value kcal	Energy value kJ
251	**Drinking yogurt**	5 samples (Ambrosia), UHT	1.00	84.4	3.1	Tr[a]	13.1	62	263
252	**Greek yogurt**, cows	5 samples, 3 brands, 'strained' variety	1.00	78.5	6.4	9.1[b]	2.0	115	477
253	sheep	3 samples (Total), 'set' variety	1.00	80.9	4.4	7.5	5.6	106	442
254	**Low calorie yogurt**	13 samples, 5 brands, assorted flavours	1.00	87.9	4.3	0.2	6.0	41	177
255	**Low fat yogurt**, plain	10 samples, 5 brands	1.00	84.9	5.1	0.8	7.5	56	236
256	flavoured	24 samples, 4 brands, assorted flavours	1.00	77.9	3.8	0.9	17.9	90	384
257	fruit	26 samples, 9 brands, assorted flavours	1.00	77.0	4.1	0.7	17.9	90	382
258	**Soya yogurt**	5 samples sweetened (Sojal)	1.00	82.4	5.0	4.2	3.9	72	305
259	**Tzatziki**	Yogurt-based Greek starter. Recipe	1.00	85.8	3.7	4.9	2.0	66	275
260	**Whole milk yogurt**, plain	22 samples, 2 brands	1.00	81.9	5.7	3.0	7.8	79	333
261	fruit	10 samples, assorted flavours, 'thick and creamy' type	1.00	73.1	5.1	2.8	15.7	105	441

[a] The fat content is variable. Non-UHT varieties contain 0.3 - 2g fat per 100g

[b] 'Set' varieties contain approximately 4g fat per 100g

Milk and milk products *continued*

No.	Food	Total nitrogen g	Fatty acids			Cholest-erol mg	Starch g	Total sugars g	Dietary fibre	
			Satd g	Mono unsatd g	Poly unsatd g				Southgate method g	Englyst method g
251	**Drinking yogurt**	0.48	Tr	Tr	Tr	Tr	0	13.1	Tr	Tr
252	**Greek yogurt**, cows	1.01	5.2	2.7	0.5	N	0	2.0	0	0
253	sheep	0.69	4.8	1.9	0.4	(14)	0	5.6	N	N
254	**Low calorie yogurt**	0.67	0.1	0.1	Tr	1	0	6.0	N	N
255	**Low fat yogurt**, plain	0.80	0.5	0.2	Tr	4	0	7.5	N	N
256	flavoured	0.59	0.5	0.3	0.1	4	0	17.9	N	N
257	fruit	0.64	0.4	0.2	Tr	4	0	17.9	N	N
258	**Soya yogurt**	0.88	0.6	0.9	2.4	0	Tr	3.9	N	N
259	**Tzatziki**	0.59	2.9	1.4	0.2	N	0.3	1.7	0.2	0.2
260	**Whole milk yogurt**, plain	0.89	1.7	0.9	0.2	11	0	7.8	N	N
261	fruit	0.80	1.5	0.8	0.2	10	0	15.7[a]	N	N

[a] 'Real' fruit yogurts contain 12.1g total sugars per 100g

Milk and milk products *continued*

Inorganic constituents per 100g

No.	Food	Na	K	Ca	Mg	P	Fe	Cu	Zn	Cl	Mn	Se	I
						mg						µg	
251	**Drinking yogurt**	47	130	100	11	81	0.1	0.01	0.3	75	Tr	(1)	N
252	**Greek yogurt**, cows	71	150	150	12	130	0.3	Tr	0.5	100	Tr	2	N
253	sheep	150	190	150	16	140	Tr	Tr	0.5	220	Tr	1	N
254	**Low calorie yogurt**	73	180	130	13	110	0.1	Tr	0.4	120	Tr	(1)	N
255	**Low fat yogurt**, plain	83	250	190	19	160	0.1	Tr	0.6	150	Tr	1	63
256	flavoured	65	190	150	15	120	0.1	Tr	0.5	130	Tr	(1)	N
257	fruit	64	210	150	15	120	0.1	Tr	0.5	130	Tr	(1)	48
258	**Soya yogurt**	N	N	N	N	N	N	N	N	N	N	N	N
259	**Tzatziki**	370	150	88	12	91	0.3	0.01	0.3	570	N	1	N
260	**Whole milk yogurt**, plain	80	280	200	19	170	0.1	Tr	0.7	170	Tr	(2)	(63)
261	fruit	82	210	160	16	130	Tr	Tr	0.5	150	Tr	(1)	(48)

Milk and milk products *continued*

No.	Food	Retinol µg	Carotene µg	Vitamin D µg	Vitamin E mg	Thiamin mg	Ribo-flavin mg	Niacin mg	Trypt 60 mg	Vitamin B6 mg	Vitamin B12 µg	Folate µg	Panto-thenate mg	Biotin µg	Vitamin C mg
251	**Drinking yogurt**	Tr	Tr	Tr	Tr	0.03	0.16	0.1	0.7	0.05	0.2	12	0.19	0.9	0
252	**Greek yogurt**, cows	115	(36)	0.05	0.38	0.03	0.36	0.1	1.5	0.05	0.2	6	N	N	Tr
253	sheep	86	(11)	0.24	0.73	0.05	0.33	0.2	1.0	0.08	0.2	3	N	N	Tr
254	**Low calorie yogurt**	Tr	Tr	Tr	0.03	0.04	0.29	0.1	1.0	0.07	(0.2)	8	N	N	1
255	**Low fat yogurt**, plain	8	5	0.01	0.01	0.05	0.25	0.1	1.2	0.09	0.2	17	0.45	2.9	1
256	flavoured	9	6	0.01	0.01	0.05	0.21	0.1	0.9	0.07	0.2	19	0.30	2.2	1
257	fruit	10	4	(0.01)	(0.01)	0.05	0.21	0.1	1.0	0.08	0.2	16	0.33	2.3	1
258	**Soya yogurt**	23	(3)	0	1.49	N	N	N	0.9	N	0	N	N	N	0
259	**Tzatziki**	60	(46)	0.03	0.20	0.03	0.20	0.10	0.90	0.04	0.1	7	N	N	1
260	**Whole milk yogurt**, plain	28	21	0.04	0.05	0.06	0.27	0.2	1.3	0.10	0.2	18	0.50	2.6	1
261	fruit	39	16	(0.04)	(0.05)	0.06	0.30	0.1	1.3	0.07	0.1	10	0.30	2.0	1

Milk and milk products *continued*

Composition of food per 100g

No.	Food	Description and main data sources	Edible proportion	Water g	Protein g	Fat g	Carbo-hydrate g	Energy value kcal	kJ
262	**Arctic roll**	10 samples, 2 brands	1.00	51.3	4.1	6.6	33.3	200	847
263	**Choc ice**	Plain and milk varieties; analysis and manufacturer's data (Birds Eye Wall's)	1.00	N	3.5	17.5	28.1	277	1157
264	**Chocolate nut sundae**	Recipe	1.00	46.0	3.0	15.3	34.2[a]	278	1165
265	**Cornetto**	Analysis and manufacturer's data (Birds Eye Wall's)	1.00	N	3.7	12.9	34.5	260	1092
266	**Frozen ice cream desserts**	6 samples, different types eg Sonata, Viennetta	1.00	61.7	3.3	14.2	22.8	227	946
267	**Ice cream**, dairy, vanilla	17 samples	1.00	61.9	3.6	9.8	24.4[a]	194	814
268	flavoured	17 samples, assorted flavours	1.00	59.8	3.5	8.0	24.7[a]	179	751
269	non-dairy, vanilla	11 samples, hard and soft scoop	1.00	65.3	3.2	8.7	23.1[a]	178	746
270	flavoured	14 samples, hard and soft scoop assorted flavours	1.00	64.9	3.1	7.4	23.2[a]	166	698
271	mixes	Prepared mix from ice cream parlour	1.00	63.4	4.1	7.9	25.1[a]	182	764
272	**Ice cream wafers**	6 samples, 2 brands	1.00	2.8	10.1	0.7	78.8	342	1458
273	**Sorbet**, lemon	Recipe	1.00	64.9	0.9	Tr	34.2	131	562

[a] Including oligosaccharides from the glucose syrup/maltodextrins in the product

Milk and milk products *continued*

Composition of food per 100g

No.	Food	Total nitrogen g	Fatty acids Satd g	Mono unsatd g	Poly unsatd g	Cholest- erol mg	Starch g	Total sugars g	Dietary fibre Southgate method g	Englyst method g
262	**Arctic roll**	0.66	3.1	2.5	0.8	30	8.0	25.3	0.8	Tr
263	**Choc ice**	0.55	10.8	4.8	1.1	7	N	N	Tr	Tr
264	**Chocolate nut sundae**	0.49	8.3	4.9	1.2	28	0.4	31.9[b]	0.2	0.1
265	**Cornetto**	0.59	6.7[a]	4.2[a]	1.3[a]	2[a]	9.0	25.5	N	N
266	**Frozen ice cream desserts**	0.53	11.2	1.8	0.4	3	Tr	22.8	Tr	Tr
267	**Ice cream, dairy, vanilla**	0.56	6.4	2.4	0.3	31	Tr	22.1[b]	Tr[c]	Tr[c]
268	flavoured	0.54	5.2	2.0	0.3	26	Tr	23.7[b]	Tr[c]	Tr[c]
269	non-dairy, vanilla	0.50	4.4	3.2	0.8	7	Tr	19.2[b]	Tr[c]	Tr[c]
270	flavoured	0.49	3.7	2.7	0.6	6	Tr	21.3[b]	Tr[c]	Tr[c]
271	mixes	0.65	4.0	2.9	0.7	(7)	Tr	21.7[b]	Tr[c]	Tr[c]
272	**Ice cream wafers**	1.77	N	N	N	0	77.7	1.1	N	N
273	**Sorbet**, lemon	0.14	Tr	Tr	Tr	0	0	34.2	0	0

[a] Strawberry variety only

[b] Not including oligosaccharides from the glucose syrup/maltodextrins in the product

[c] Gums and cellulose derivatives are added as stabilisers

No.	Food	mg										µg	
		Na	K	Ca	Mg	P	Fe	Cu	Zn	Cl	Mn	Se	I
262	**Arctic roll**	150	140	90	11	120	0.7	0.12	0.4	140	0.1	N	23
263	**Choc ice**	91	200	130	27	N	0.1	Tr	0.2	N	N	N	N
264	**Chocolate nut sundae**	150	170	80	N	N	N	N	N	N	N	N	N
265	**Cornetto**	91	170	120	21	N	N	N	N	N	N	N	N
266	**Frozen ice cream desserts**	84	200	110	19	99	0.5	0.04	0.4	110	0.1	N	N
267	**Ice cream**, dairy, vanilla	69	160	130	13	110	0.1	0.02	0.3	110	Tr	N	N
268	flavoured	61	180	110	19	99	0.5	0.05	0.4	100	Tr	N	N
269	non-dairy, vanilla	76	170	120	13	100	0.1	Tr	0.3	130	Tr	N	N
270	flavoured	72	160	120	13	99	0.1	Tr	0.4	120	Tr	N	N
271	mixes	59	180	140	15	120	0.1	Tr	0.5	110	Tr	N	N
272	**Ice cream wafers**	93	190	170	46	130	2.0	0.11	0.7	130	0.7	N	N
273	**Sorbet**, lemon	18	42	2	2	5	Tr	0.03	1.0	16	Tr	1	Tr

No.	Food	Retinol µg	Carotene µg	Vitamin D µg	Vitamin E mg	Thiamin mg	Ribo-flavin mg	Niacin mg	Trypt 60 mg	Vitamin B6 mg	Vitamin B12 µg	Folate µg	Panto-thenate mg	Biotin µg	Vitamin C mg
262	**Arctic roll**	N	N	N	N	0.07	0.11	0.2	0.9	0.06	0.3	12	N	N	0
263	**Choc ice**	1	5	Tr	N	N	N	N	0.8	N	N	N	N	N	N
264	**Chocolate nut sundae**	115	65	0.05	N	0.05	0.15	0.3	0.6	0.05	0.3	7	0.29	1.7	1
265	**Cornetto**	N	N	Tr	N	N	N	N	N	N	N	N	N	N	N
266	**Frozen ice cream desserts**	2	5	Tr	0.51	0.04	0.30	0.2	0.8	0.06	0.6	3	N	N	0
267	**Ice cream**, dairy, vanilla	115	195	0.12	0.21	0.04	0.25	0.1	0.8	0.08	0.4	7	0.44	2.5	1
268	flavoured	94	160	0.10	0.17	0.04	0.26	0.2	0.8	0.07	0.3	9	0.33	2.4	1
269	non-dairy, vanilla	1	6	Tr	0.84	0.04	0.24	0.1	0.7	0.07	0.5	8	0.43	3.0	1
270	flavoured	1	5	Tr	0.72	0.04	0.24	0.1	0.7	0.07	0.4	8	0.40	2.8	1
271	mixes	2	4	Tr	0.68	0.05	0.23	0.1	1.0	0.08	0.3	6	0.44	3.1	1
272	**Ice cream wafers**	0	0	0	N	0.20	0.04	2.3	2.1	0.15	0	15	N	N	0
273	**Sorbet**, lemon	0	0	0	Tr	Tr	0.04	Tr	0.3	0.01	Tr	1	0.04	0.7	Tr

Milk and milk products *continued*

Composition of food per 100g

Puddings and chilled desserts

No.	Food	Description and main data sources	Edible proportion	Water g	Protein g	Fat g	Carbo-hydrate g	Energy value kcal	kJ
274	**Cheesecake,** *frozen*	10 samples, assorted flavours, fruit topping	1.00	44.0	5.7	10.6	33.0	242	1017
275	**Creme caramel**	9 samples, 4 brands	1.00	72.0	3.0	2.2	20.6	109	462
276	**Custard,** *made up with whole milk*	Recipe	1.00	75.5	3.7	4.5	16.6	117	495
277	*made up with skimmed milk*	Recipe	1.00	79.3	3.8	0.1	16.8	79	339
278	*canned*	10 samples, 3 brands	1.00	77.2	2.6	3.0	15.4	95	401
279	**Instant dessert powder**	10 samples, 2 types, assorted flavours	1.00	1.0	2.4	17.3	60.1	391	1643
280	*made up with whole milk*	Recipe	1.00	72.1	3.1	6.3	14.8	125	523
281	*made up with skimmed milk*	Recipe	1.00	74.9	3.1	3.2	14.9	97	410
282	**Jelly,** *made with water*	Recipe	1.00	84.0	1.2	0	15.1	61	260
283	**Milk pudding,** *made with whole milk*	e.g. rice, sago, semolina, tapioca; recipe	1.00	72.4	3.9	4.3	19.9	129	543
284	*made with skimmed milk*	e.g. rice, sago, semolina, tapioca; recipe	1.00	76.0	4.0	0.2	20.1	93	398
285	**Mousse,** *chocolate*	10 samples, 4 brands, fresh	1.00	67.3	4.0	5.4	19.9	139	586
286	*fruit*	8 samples, assorted flavours, fresh	1.00	71.7	4.5	5.7	18.0	137	575
287	**Rice pudding,** *canned*	10 cans, 4 brands	1.00	77.6	3.4	2.5	14.0	89	374
288	**Trifle**	Recipe	1.00	67.2	3.6	6.3	22.3	160	674
289	*with fresh cream*	10 samples, individual and large	1.00	68.1	2.4	9.2	19.5	166	693

Milk and milk products *continued*

Composition of food per 100g

No.	Food	Total nitrogen g	Fatty acids Satd g	Mono unsatd g	Poly unsatd g	Cholest- erol mg	Starch g	Total sugars g	Dietary fibre Southgate method g	Englyst method g
	Puddings and chilled desserts									
274	**Cheesecake**, *frozen*	0.91	5.6	3.6	0.8	60	10.8	22.2	0.9	(0.9)
275	**Creme caramel**	0.47	N	N	N	N	2.6	18.0	N	N
276	**Custard**, *made up with whole milk*	0.58	2.8	1.3	0.2	16	5.1	11.4	Tr	Tr
277	*made up with skimmed milk*	0.60	0.1	Tr	Tr	2	5.1	11.6	Tr	Tr
278	*canned*	0.42	1.7	0.9	0.1	(11)	3.1	12.3	0.1	(0.1)
279	**Instant dessert powder**	0.39	15.9	0.3	0.2	1	19.4	40.7	1.0	(1.0)
280	*made up with whole milk*	0.48	4.9	1.0	0.1	12	3.5	11.3	0.2	(0.2)
281	*made up with skimmed milk*	0.50	2.9	0.1	Tr	2	3.5	11.4	0.2	(0.2)
282	**Jelly**, *made with water*	0.21	0	0	0	0	0	15.1	0	0
283	**Milk pudding**, *made with whole milk*	0.62	2.7	1.2	0.2	15	9.3	10.7	0.2	0.1
284	*made with skimmed milk*	0.64	0.1	Tr	Tr	2	9.3	10.9	0.2	0.1
285	**Mousse**, *chocolate*	0.63	N	N	N	N	2.4	17.5	N	N
286	*fruit*	0.71	N	N	N	N	Tr	18.0	N	N
287	**Rice pudding**, *canned*	0.53	1.6	0.7	0.1	(9)	5.8	8.2[a]	N	0.2
288	**Trifle**	0.59	3.1	2.0	0.7	44	5.5	16.8	0.5	0.5
289	*with fresh cream*	0.38	5.2	3.0	0.5	33	4.5	15.0	0.5	(0.5)

[a] Low calorie varieties contain approximately 3.1g total sugars per 100g

Milk and milk products *continued*

Inorganic constituents per 100g

No.	Food	mg										µg	
		Na	K	Ca	Mg	P	Fe	Cu	Zn	Cl	Mn	Se	I
	Puddings and chilled desserts												
274	**Cheesecake**, *frozen*	160	130	68	10	93	0.5	0.06	0.4	220	0.2	N	25
275	**Creme caramel**	70	150	94	9	77	Tr	Tr	0.3	100	Tr	N	33
276	**Custard**, *made up with whole milk*	81	160	130	13	110	0.1	Tr	0.4	140	N	N	N
277	*made up with skimmed milk*	81	180	140	14	110	0.1	Tr	0.5	140	N	N	N
278	*canned*	67	130	100	8	87	0.2	0.02	0.3	75	Tr	N	Tr
279	**Instant dessert powder**	1100	64	20	11	650	0.5	0.20	0.4	45	0.1	N	12
280	*made up with whole milk*	240	130	97	10	190	0.1	0.04	0.4	90	Tr	N	12
281	*made up with skimmed milk*	240	130	100	11	190	0.1	0.04	0.4	90	Tr	N	12
282	**Jelly**, *made with water*	5	5	7	Tr	1	0.4	0.01	N	6	N	N	N
283	**Milk pudding**, *made with whole milk*	59	160	130	13	110	0.1	0.01	0.5	110	N	N	N
284	*made with skimmed milk*	59	170	130	14	110	0.1	0.01	0.5	110	N	N	N
285	**Mousse**, *chocolate*	67	220	97	28	100	1.6	0.12	0.6	86	0.2	N	N
286	*fruit*	62	150	120	12	96	Tr	Tr	0.4	110	Tr	N	N
287	**Rice pudding**, *canned*	50	140	93	11	80	0.2	0.03	0.4	95	N	N	N
288	**Trifle**	53	140	79	15	85	0.5	0.04	0.4	87	0.1	N	83
289	*with fresh cream*	63	84	68	6	63	0.3	0.09	0.3	55	0.1	N	17

Milk and milk products *continued*

Puddings and chilled desserts

No.	Food	Retinol µg	Carotene µg	Vitamin D µg	Vitamin E mg	Thiamin mg	Ribo-flavin mg	Niacin mg	Trypt 60 mg	Vitamin B6 mg	Vitamin B12 µg	Folate µg	Panto-thenate mg	Biotin µg	Vitamin C mg
274	**Cheesecake**, *frozen*	N	N	N	N	0.04	0.16	0.3	1.4	0.02	0.5	7	N	N	0
275	**Creme caramel**	37	8	0.07	0.16	0.03	0.20	0.1	0.7	0.03	0.3	8	N	N	0
276	**Custard**, *made up with whole milk*	59	24	0.03	0.10	0.04	0.18	0.1	0.9	0.06	0.5	5	0.36	2.2	1
277	*made up with skimmed milk*	1	Tr	Tr	Tr	0.04	0.19	0.1	0.9	0.06	0.5	5	0.33	2.3	1
278	*canned*	N	N	N	N	0.04	0.10	Tr	0.6	0.03	Tr	2	N	N	0
279	**Instant dessert powder**	N	N	N	N	Tr	0.01	Tr	0.5	Tr	0.3	Tr	N	N	0
280	*made up with whole milk*	N	N	N	N	0.03	0.14	0.1	0.7	0.05	0.3	4	N	N	1
281	*made up with skimmed milk*	N	N	N	N	0.03	0.15	0.1	0.7	0.05	0.3	4	N	N	1
282	**Jelly**, *made with water*	0	0	0	0	0	0	0	0	0	0	0	0	0	0
283	**Milk pudding**, *made with whole milk*	56	22	0.03	0.10	0.04	0.16	0.1	0.9	0.06	0.4	3	0.30	2.2	1
284	*made with skimmed milk*	1	Tr	Tr	Tr	0.04	0.17	0.1	0.9	0.06	0.4	3	0.28	2.3	1
285	**Mousse**, *chocolate*	46	11	Tr	0.58	0.04	0.21	0.2	0.9	0.04	0.2	6	N	N	0
286	*fruit*	36	16	0.07	0.78	0.04	0.23	0.2	1.1	0.05	0.2	6	N	N	Tr
287	**Rice pudding**, *canned*	N	N	N	N	0.03	0.14	0.2	0.7	0.02	Tr	N	N	N	0
288	**Trifle**	70	33	0.17	0.40	0.06	0.13	0.3	0.9	0.06	0.4	8	0.34	3.1	4
289	*with fresh cream*	(70)	(33)	(0.17)	(0.40)	0.06	0.10	0.1	0.5	Tr	0.2	(8)	(0.34)	(3.1)	(4)

105

Eggs and egg dishes

The eggs and egg dishes in this section of the Tables are taken from the *'Milk Products and Eggs'* supplement.

Although most of the nutrients in eggs have been analysed, a few of the values for cooked eggs were derived by calculation from the amounts in raw eggs. Allowances have been made for any water loss or fat uptake in cases where eggs were cooked with fat.

Losses of labile vitamins assigned on recipe calculation were estimated using the figures in Section 3.4.

Eggs and egg dishes

290 to 305

Composition of food per 100g

No.	Food	Description and main data sources	Edible proportion	Water g	Protein g	Fat g	Carbo-hydrate g	Energy value kcal	kJ
290	**Eggs,** chicken, whole, *raw*[a]	Analysis of battery, deep litter and free range	1.00	75.1	12.5	10.8	Tr	147	612
291	white, *raw*	34 eggs and literature sources	1.00	88.3	9.0	Tr	Tr	36	153
292	yolk, *raw*	34 eggs and literature sources	1.00	51.0	16.1	30.5	Tr	339	1402
293	chicken, *boiled*	10 eggs	1.00	75.1	12.5	10.8	Tr	147	612
294	*fried in vegetable oil*	12 eggs, shallow fried	1.00	70.1	13.6	13.9	Tr	179	745
295	*poached*[b]	10 eggs, no fat added	1.00	75.1	12.5	10.8	Tr	147	612
296	*scrambled, with milk*	Recipe	1.00	62.4	10.7	22.6	0.6	247	1025
297	duck, whole, *raw*	Analytical and literature sources. Ref. Posati and Orr (1976)	1.00	70.6	14.3	11.8	Tr	163	680
298	**Egg fried rice**	Recipe	1.00	60.7	4.2	10.6	25.7	208	873
299	**Meringue**	Recipe	1.00	2.2	5.3	Tr	95.4	379	1616
300	*with cream*	Recipe. Ref. Wiles *et al.* (1980)	1.00	34.1	3.3	23.6	40.0	376	1570
301	**Omelette,** plain	Recipe	1.00	69.0	10.9	16.4	Tr	191	792
302	cheese	Recipe. Ref. Wiles *et al.* (1980)	1.00	57.7	15.9	22.6	Tr	266	1106
303	**Quiche,** cheese and egg	Recipe. Ref. Wiles *et al.* (1980)	1.00	46.7	12.5	22.2	17.3	314	1310
304	cheese and egg, wholemeal	Recipe	1.00	46.7	13.2	22.4	14.5	308	1283
305	**Scotch eggs,** retail	10 samples, 8 brands	1.00	54.0	12.0	17.1	13.1	251	1046

[a] An average egg is composed of 11% shell, 58% white and 31% yolk

[b] Eggs poached with fat added contain 74.4g water, 12.4g protein, 11.7g fat, Tr carbohydrate, 155 kcals and 644 kJ per 100g

Eggs and egg dishes

Composition of food per 100g

No.	Food	Total nitrogen g	Fatty acids			Cholesterol mg	Starch g	Total sugars g	Dietary fibre	
			Satd g	Mono unsatd g	Poly unsatd g				Southgate method g	Englyst method g
290	**Eggs**, chicken, whole, *raw*	2.01	3.1	4.7	1.2	385	0	Tr	0	0
291	white, *raw*	1.44	Tr	Tr	Tr	0	0	Tr	0	0
292	yolk, *raw*	2.58	8.7	13.2	3.4	1120	0	Tr	0	0
293	chicken, *boiled*	2.01	3.1	4.7	1.2	385	0	Tr	0	0
294	*fried in vegetable oil*	2.18	4.0	6.0	1.5	435	0	Tr	0	0
295	*poached*	2.01	3.1	4.7	1.2	385	0	Tr	0	0
296	*scrambled, with milk*	1.71	11.6	7.2	1.4	350	0	0.6	0	0
297	duck, whole, *raw*	2.29	2.9	4.9	2.0	680	0	Tr	0	0
298	**Egg fried rice**	0.69	1.5	4.1	4.2	70	24.8	0.9	0.9	0.4
299	**Meringue**	0.85	Tr	Tr	Tr	0	0	95.4	0	0
300	*with cream*	0.53	14.7	6.8	0.1	65	0	40.0	0	0
301	**Omelette**, plain	1.75	7.4	5.8	1.3	355	0	Tr	0	0
302	cheese	2.52	12.2	7.2	1.2	265	0	Tr	0	0
303	**Quiche**, cheese and egg	2.01	10.3	7.9	2.3	140	15.7	1.6	0.7	0.6
304	cheese and egg, wholemeal	2.12	10.4	8.0	2.4	140	12.7	1.7	1.8	1.9
305	**Scotch eggs**, retail	1.92	4.3	6.6	3.3	165	13.1	Tr	1.6	N

Egg and egg dishes

Inorganic constituents per 100g

No.	Food	Na	K	Ca	Mg	P	Fe	Cu	Zn	Cl	Mn	Se	I
						mg						μg	
290	**Eggs**, chicken, whole, *raw*	140	130	57	12	200	1.9	0.08	1.3	160	Tr	11	53
291	white, *raw*	190	150	5	11	33	0.1	0.02	0.1	170	Tr	6	(3)
292	yolk, *raw*	50	120	130	15	500	6.1	0.15	3.9	140	0.1	20	(140)
293	chicken, *boiled*	140	130	57	12	200	1.9	0.08	1.3	160	Tr	11	53
294	*fried in vegetable oil*	160	150	65	14	230	2.2	0.09	1.5	180	Tr	12	60
295	*poached*	140	130	57	12	200	1.9	0.08	1.3	160	Tr	11	53
296	*scrambled, with milk*	1030	130	63	17	180	1.6	0.07	1.1	1550	Tr	9	52
297	duck, whole, *raw*	120	190	63	16	200	2.9	(0.50)	1.4	N	(0.1)	N	N
298	**Egg fried rice**	27	72	13	5	67	0.5	0.07	0.7	38	0.3	5	10
299	**Meringue**	110	90	4	6	19	0.1	0.03	0.2	100	Tr	3	2
300	*with cream*	70	84	39	6	42	Tr	0.01	0.3	75	Tr	1	1
301	**Omelette**, plain	1030	110	51	16	170	1.7	0.07	1.1	1530	Tr	9	50
302	cheese	900	100	280	19	280	1.2	0.06	1.5	1360	Tr	10	46
303	**Quiche**, cheese and egg	340	120	260	17	220	1.0	0.06	1.1	530	0.1	7	31
304	cheese and egg, wholemeal	340	160	240	37	270	1.4	0.12	1.6	520	0.7	17	N
305	**Scotch eggs**, retail	670	130	50	15	170	1.8	0.23	1.2	980	0.2	N	17

No.	Food	Retinol µg	Carotene µg	Vitamin D µg	Vitamin E mg	Thiamin mg	Ribo-flavin mg	Niacin mg	Trypt 60 mg	Vitamin B6 mg	Vitamin B12 µg	Folate µg	Panto-thenate mg	Biotin µg	Vitamin C mg
290	**Eggs**, chicken, whole, *raw*	190	Tr	1.75[a]	1.11	0.09	0.47	0.1	3.7	0.12	2.5	50	1.77	20.0	0
291	white, *raw*	0	0	0	0	0.01	0.43	0.1	2.6	0.02	0.1	13	0.30	7.0	0
292	yolk, *raw*	535	Tr	4.94[a]	3.11	0.30	0.54	0.1	4.7	0.30	6.9	130	4.60	50.0	0
293	chicken, *boiled*	190	Tr	1.75[a]	1.11	0.07	0.35	0.1	3.7	0.12	1.1	39	1.30	16.0	0
294	*fried in vegetable oil*	215	Tr	1.99[a]	N	0.07	0.31	0.1	4.0	0.14	1.6	(40)	1.30	18.0	0
295	*poached*	190	Tr	1.75[a]	1.11	0.07	0.36	0.1	3.7	0.12	1.0	45	1.30	15.0	0
296	*scrambled, with milk*	295	72	1.55[a]	1.23	0.07	0.33	0.1	3.1	0.09	2.1	28	1.29	16.5	Tr
297	duck, whole, *raw*	540	120	5.00	N	0.16	0.47	0.2	4.2	0.25	5.4	80	N	N	0
298	**Egg fried rice**	34	8	0.31	0.20	0.03	0.08	0.3	1.1	0.06	0.4	8	0.39	4.5	Tr
299	**Meringue**	0	0	0	0	Tr	0.24	0.1	1.6	0.01	0.1	6	0.16	4.0	0
300	*with cream*	340	160	0.13	0.54	0.01	0.20	Tr	0.9	0.03	0.1	6	0.20	2.4	Tr
301	**Omelette**, plain	235	37	1.58	1.13	0.07	0.33	0.1	3.2	0.09	2.2	30	1.33	17.3	0
302	cheese	265	100	1.13	0.92	0.06	0.32	0.1	4.2	0.09	1.8	27	0.98	12.4	Tr
303	**Quiche**, cheese and egg	185	100	0.93	0.91	0.08	0.23	0.40	3.10	0.08	1.0	13	0.53	6.6	Tr
304	cheese and egg, wholemeal	185	100	0.93	1.14	0.10	0.24	1.20	3.20	0.13	1.0	17	0.61	7.8	Tr
305	**Scotch eggs**, retail	30	Tr	0.73	N	0.08	0.21	1.0	2.9	0.13	0.5	42	(1.10)	(8.7)	N

a If the hens have been fed a supplement, values may be considerably higher

Fats and oils

The number of foods in this section has been extended and now includes entries for different types of margarine and oils as well as reduced fat spreads analysed during production of the *'Milk Products and Eggs'* supplement.

Most oils show a very wide range of fatty acid composition depending on the variety, growing conditions and maturity of the seed.

The blend of fats and oils used in many of the foods included in this section can frequently be adjusted by manufacturers and this will alter the fatty acid composition. If accurate fatty acid data are required for specific products, and analytical facilities are not available, it is advisable to contact the manufacturer directly.

The profile for fatty acids in 'vegetable oil' was calculated from the values for the component soya, rape and corn oils, the proportions of which may vary, and this profile has been included to aid recipe calculation and survey work where unidentified oil has been consumed.

Fats and oils

Composition of food per 100g

No.	Food	Description and main data sources	Edible proportion	Water g	Protein g	Fat g	Carbo-hydrate g	Energy value kcal	kJ
Spreading fats									
306	**Butter**	Analysis and literature sources	1.00	15.6[a]	0.5	81.7[a,b]	Tr	737	3031
307	**Dairy/fat spread**	6 samples, Krona, Clover and Golden Churn	1.00	22.0	0.4	73.4	Tr	662	2723
308	**Low-fat spread**	4 samples, Gold, Delight, Outline and own brand	1.00	49.9	5.8	40.5	0.5	390	1605
309	**Margarine**	Mixed sample	1.00	16.0	0.2	81.6	1.0	739	3039
310	hard, *animal and vegetable fat*	10 samples, 2 brands and estimation from No. 309	1.00	16.0	0.2	81.6	1.0	739	3039
311	hard, *vegetable fat only*	4 samples of the same brand and estimation from No. 309	1.00	16.0	0.2	81.6	1.0	739	3039
312	soft, *animal and vegetable fat*	16 samples, 3 brands and estimation from No. 309	1.00	16.0	0.2	81.6	1.0	739	3039
313	soft, *vegetable fat only*	8 samples, 3 brands and estimation from No. 309	1.00	16.0	0.2	81.6	1.0	739	3039
314	polyunsaturated	18 samples, 3 brands and estimation from No. 309	1.00	16.0	0.2	81.6	1.0	739	3039
315	**Very low fat spread**	i.e. Gold Lowest, manufacturer's data (St Ivel)	1.00	N	8.3	25.0	3.6	273	1128
Animal fats									
316	**Compound cooking fat**	7 samples	1.00	Tr	Tr	99.3	Tr	894	3674
317	**Dripping**, beef	Analysed as purchased	1.00	1.0	Tr	99.0	Tr	891	3663
318	**Lard**	Analysed as purchased	1.00	1.0	Tr	99.0	Tr	891	3663
319	**Suet**, *shredded*	6 samples of the same brand	1.00	1.5	Tr	86.7	12.1	826	3402

a Unsalted butter contains 15.7g water and 82.7g fat per 100g

b 'Half-fat' butter spreads, e.g. Half-fat Anchor, Kerrygold light, contain 39-40g fat per 100g

No.	Food	Total nitrogen g	Fatty acids			Cholest-erol mg	Starch g	Total sugars g	Dietary fibre	
			Satd g	Mono unsatd g	Poly unsatd g				Southgate method g	Englyst method g
Spreading fats										
306	**Butter**	0.08	54.0	19.8	2.6	230	0	Tr	0	0
307	**Dairy/fat spread**	0.06	28.1	29.9	11.3	105	0	Tr	0	0
308	**Low-fat spread**	0.91	11.2	17.6	9.9	6	0	0.5	0	0
309	**Margarine**	0.03	N	N	N	N	0	1.0	0	0
310	hard, *animal and vegetable fat*	0.03	30.4	36.5	10.8	285	0	1.0	0	0
311	hard, *vegetable fat only*	0.03	35.9	33.0	9.4	15	0	1.0	0	0
312	soft, *animal and vegetable fat*	0.03	26.9	37.2	13.8	225	0	1.0	0	0
313	soft, *vegetable fat only*	0.03	25.0	31.0	21.8	9	0	1.0	0	0
314	*polyunsaturated*	0.03	16.2	20.6	41.1	7	0	1.0	0	0
315	**Very low fat spread**	1.30	6.5	13.5	3.5	N	Tr	3.6	0	0
Animal fats										
316	**Compound cooking fat**	Tr	38.1	45.6	11.1	375	0	Tr	0	0
317	**Dripping, beef**	Tr	54.8	36.7	2.5	94	0	Tr	0	0
318	**Lard**	Tr	40.8	43.8	9.6	93	0	Tr	0	0
319	**Suet,** *shredded*	Tr	48.0	32.1	2.1	82	11.9	0.2	0.6	0.5

Inorganic constituents per 100g

No.	Food	Na	K	Ca	Mg	P	Fe	Cu	Zn	Cl	Mn	Se	I
						mg						µg	
Spreading fats													
306	**Butter**	750[a]	15	15	2	24	0.2	0.03	0.1	1150[a]	Tr	Tr	38
307	**Dairy/fat spread**	760	43	14	2	18	Tr	Tr	Tr	1270	Tr	N	N
308	**Low-fat spread**	650	110	39	4	82	Tr	0.12	0.2	800	Tr	N	N
309	**Margarine**	800	5	4	1	12	0.3	0.04	N	1200	Tr	Tr	26
310	hard, *animal and vegetable fat*	800	5	4	1	12	0.3	0.04	N	1200	Tr	Tr	23
311	hard, *vegetable fat only*	800	5	4	1	12	0.3	0.04	N	1200	Tr	Tr	23
312	soft, *animal and vegetable fat*	800	5	4	1	12	0.3	0.04	N	1200	Tr	Tr	27
313	soft, *vegetable fat only*	800	5	4	1	12	0.3	0.04	N	1200	Tr	Tr	27
314	*polyunsaturated*	800	5	4	1	12	0.3	0.04	N	1200	Tr	Tr	(27)
315	**Very low fat spread**	1050	630	N	N	N	N	N	N	N	N	N	N
Animal fats													
316	**Compound cooking fat**	Tr	Tr	Tr	Tr	Tr	Tr	Tr	Tr	Tr	Tr	Tr	Tr
317	**Dripping**, beef	5	4	1	Tr	13	0.2	N	N	2	Tr	Tr	(5)
318	**Lard**	2	1	1	1	3	0.1	0.02	N	4	Tr	Tr	Tr
319	**Suet**, *shredded*	Tr	Tr	Tr	Tr	Tr	Tr	Tr	Tr	Tr	Tr	N	5

[a] Unsalted butter contains 11mg Na and 17mg Cl per 100g

No.	Food	Retinol μg	Carotene μg	Vitamin D μg	Vitamin E mg	Thiamin mg	Ribo-flavin mg	Niacin mg	Trypt 60 mg	Vitamin B₆ mg	Vitamin B₁₂ μg	Folate μg	Panto-thenate mg	Biotin μg	Vitamin C mg
	Spreading fats														
306	**Butter**	815	430	0.76	2.00	Tr	(0.02)	(Tr)	0.1	Tr	Tr	Tr	(0.04)	Tr	Tr
307	**Dairy/fat spread**	800	845	5.80	5.06	Tr	Tr	Tr	0.1	Tr	Tr	Tr	Tr	Tr	0
308	**Low-fat spread**	920	985	8.00	6.33	Tr	Tr	Tr	1.4	Tr	Tr	Tr	Tr	Tr	0
309	**Margarine**	780	750[a]	7.94	8.00[b]	Tr	Tr	Tr	Tr	Tr	Tr	Tr	Tr	Tr	0
310	*hard, animal and vegetable fat*	(665)	(750)[a]	(7.94)	N[b]	Tr	Tr	Tr	Tr	Tr	Tr	Tr	Tr	Tr	0
311	*hard, vegetable fat only*	(665)	(750)[a]	(7.94)	N[b]	Tr	Tr	Tr	Tr	Tr	Tr	Tr	Tr	Tr	0
312	*soft, animal and vegetable fat*	(860)	(750)[a]	(7.94)	N[b]	Tr	Tr	Tr	Tr	Tr	Tr	Tr	Tr	Tr	0
313	*soft, vegetable fat only*	(860)	(750)[a]	(7.94)	N[b]	Tr	Tr	Tr	Tr	Tr	Tr	Tr	Tr	Tr	0
314	*polyunsaturated*	(775)	(750)[a]	(7.94)	N[b]	Tr	Tr	Tr	Tr	Tr	Tr	Tr	Tr	Tr	0
315	**Very low fat spread**	N	N	N	N	Tr	Tr	Tr	1.9	Tr	Tr	Tr	Tr	Tr	0
	Animal fats														
316	**Compound cooking fat**	0	0	0	Tr	0	0	0	0	0	0	0	0	0	0
317	**Dripping,** *beef*	N	N	Tr	(0.30)	Tr	Tr	Tr	Tr	Tr	Tr	Tr	Tr	0	Tr
318	**Lard**	Tr	0	N	Tr	Tr	Tr	Tr	Tr	Tr	Tr	Tr	Tr	Tr	Tr
319	**Suet,** *shredded*	52	73	Tr	1.50	Tr	Tr	Tr	Tr	Tr	Tr	Tr	Tr	Tr	Tr

[a] Some brands may not contain β-carotene

[b] The vitamin E content will vary according to the blend of oils used

Fats and oils *continued*

Composition of food per 100g

No.	Food	Description and main data sources	Edible proportion	Water g	Protein g	Fat g	Carbo-hydrate g	Energy value kcal	kJ
Oils									
320	Coconut oil	Literature sources and estimation from No. 333	1.00	Tr	Tr	99.9	0	899	3696
321	Cod liver oil	3 samples	1.00	Tr	Tr	99.9	0	899	3696
322	Corn oil	Literature sources and estimation from No. 333; maize oil	1.00	Tr	Tr	99.9	0	899	3696
323	Cottonseed oil	Literature sources and estimation from No. 333	1.00	Tr	Tr	99.9	0	899	3696
324	Olive oil	Ref. Pellet and Shadarevian (1970)	1.00	Tr	Tr	99.9	0	899	3696
325	Palm oil	Literature sources and estimation from No. 333; refined oil	1.00	Tr	Tr	99.9	0	899	3696
326	Peanut oil	Literature sources and estimation from No. 333	1.00	Tr	Tr	99.9	0	899	3696
327	Rapeseed oil, high erucic acid	Literature sources and estimation from No. 333	1.00	Tr	Tr	99.9	0	899	3696
328	low erucic acid	Literature sources and estimation from No. 333	1.00	Tr	Tr	99.9	0	899	3696
329	Safflower oil	Literature sources and estimation from No. 333	1.00	Tr	Tr	99.9	0	899	3696
330	Sesame oil	Ref. Wu Leung *et al.* (1972)	1.00	0.1	0.2	99.7	0	881	3686
331	Soya oil	Literature sources and estimation from No. 333	1.00	Tr	Tr	99.9	0	899	3696
332	Sunflowerseed oil	Literature sources and estimation from No. 333	1.00	Tr	Tr	99.9	0	899	3696
333	Vegetable oil, blended, *average*	All kinds	1.00	Tr	Tr	99.9	0	899	3696
334	Wheatgerm oil	Literature sources and estimation from No. 333	1.00	Tr	Tr	99.9	0	899	3696
Ghee									
335	Ghee, butter	5 assorted samples	1.00	0.1	Tr	99.8	Tr	898	3693
336	palm	5 samples of the same brand	1.00	0.1	Tr	99.7	Tr	897	3689
337	vegetable	5 samples, 2 different types	1.00	0.1	Tr	99.8	Tr	898	3693

Fats and oils *continued*

Composition of food per 100g

No.	Food	Total nitrogen	Fatty acids			Cholest-erol	Starch	Total sugars	Dietary fibre	
			Satd	Mono unsatd	Poly unsatd				Southgate method	Englyst method
		g	g	g	g	mg	g	g	g	g
Oils										
320	**Coconut oil**	Tr	85.2	6.6	1.7	0	0	0	0	0
321	**Cod liver oil**	Tr	N	N	N	N	0	0	0	0
322	**Corn oil**	Tr	12.7	24.7	57.8	0	0	0	0	0
323	**Cottonseed oil**	Tr	25.6	21.3	48.1	0	0	0	0	0
324	**Olive oil**	Tr	14.0	69.7	11.2	0	0	0	0	0
325	**Palm oil**	Tr	45.3	41.6	8.3	0	0	0	0	0
326	**Peanut oil**	Tr	18.8	47.8	28.5	0	0	0	0	0
327	**Rapeseed oil**, high erucic acid	Tr	5.3	64.3	24.8	0	0	0	0	0
328	low erucic acid	Tr	6.6	57.2	31.5	0	0	0	0	0
329	**Safflower oil**	Tr	10.2	12.6	72.1	0	0	0	0	0
330	**Sesame oil**	0.03	14.2	37.3	43.9	0	0	0	0	0
331	**Soya oil**	Tr	14.5	23.2	56.5	0	0	0	0	0
332	**Sunflowerseed oil**	Tr	11.9	20.2	63.0	0	0	0	0	0
333	**Vegetable oil, blended**, *average*	Tr	10.4[a]	35.5[a]	48.2[a]	0	0	0	0	0
334	**Wheatgerm oil**	Tr	18.8	15.9	60.7	0	0	0	0	0
Ghee										
335	**Ghee**, butter	Tr	66.0	24.1	3.4	280	0	Tr	0	0
336	palm	Tr	47.0	35.3	8.9	0	0	Tr	0	0
337	vegetable	Tr	N[a]	N[a]	N[a]	0	0	Tr	0	0

[a] The fatty acid profile will depend on the blend of oils used

Fats and oils continued

Inorganic constituents per 100g

No.	Food	Na	K	Ca	Mg	P	Fe	Cu	Zn	Cl	Mn	Se	I
						mg						µg	
Oils													
320	**Coconut oil**	Tr	Tr	Tr	Tr	Tr	Tr	Tr	Tr	Tr	Tr	Tr	N
321	**Cod liver oil**	Tr	Tr	Tr	Tr	Tr	Tr	Tr	Tr	Tr	Tr	Tr	N
322	**Corn oil**	Tr	Tr	Tr	Tr	Tr	Tr	Tr	Tr	Tr	Tr	Tr	N
323	**Cottonseed oil**	Tr	N	Tr	Tr	Tr	Tr	Tr	Tr	Tr	Tr	Tr	N
324	**Olive oil**	Tr	Tr	Tr	Tr	Tr	0.4	Tr	Tr	Tr	Tr	Tr	N
325	**Palm oil**	Tr	Tr	Tr	Tr	Tr	Tr	Tr	Tr	Tr	Tr	Tr	N
326	**Peanut oil**	Tr	Tr	Tr	Tr	Tr	Tr	Tr	Tr	Tr	Tr	Tr	N
327	**Rapeseed oil**, high erucic acid	Tr	Tr	Tr	Tr	Tr	Tr	Tr	Tr	Tr	Tr	Tr	N
328	low erucic acid	Tr	Tr	Tr	Tr	Tr	Tr	Tr	Tr	Tr	Tr	Tr	N
329	**Safflower oil**	Tr	Tr	Tr	Tr	Tr	Tr	Tr	Tr	Tr	Tr	Tr	N
330	**Sesame oil**	2	20	10	Tr	N	0.1	Tr	Tr	N	Tr	Tr	N
331	**Soya oil**	Tr	Tr	Tr	Tr	Tr	Tr	Tr	Tr	Tr	Tr	Tr	N
332	**Sunflowerseed oil**	Tr	Tr	Tr	Tr	Tr	Tr	Tr	Tr	Tr	Tr	Tr	N
333	**Vegetable oil**, blended, average	Tr	Tr	Tr	Tr	Tr	Tr	Tr	Tr	Tr	Tr	Tr	11
334	**Wheatgerm oil**	Tr	Tr	Tr	Tr	Tr	Tr	Tr	Tr	Tr	Tr	Tr	N
Ghee													
335	**Ghee**, butter	2	3	Tr	Tr	Tr	0.2	Tr	Tr	28	Tr	Tr	N
336	palm	1	1	Tr	Tr	Tr	0.1	0.21	Tr	N	Tr	N	Tr
337	vegetable	1	1	Tr	Tr	Tr	Tr	0.14	Tr	N	Tr	N	N

Fats and oils *continued*

No.	Food	Retinol µg	Carotene µg	Vitamin D µg	Vitamin E mg	Thiamin mg	Ribo-flavin mg	Niacin mg	Trypt 60 mg	Vitamin B6 mg	Vitamin B12 µg	Folate µg	Panto-thenate mg	Biotin µg	Vitamin C mg
Oils															
320	**Coconut oil**	0	Tr	0	0.66	Tr	Tr	Tr	Tr	Tr	0	Tr	Tr	Tr	0
321	**Cod liver oil**	18000	Tr	210.0	20.00	0	0	0	0	0	0	0	0	0	0
322	**Corn oil**	0	Tr	0	17.24	Tr	Tr	Tr	Tr	Tr	0	Tr	Tr	Tr	0
323	**Cottonseed oil**	0	Tr	0	42.77	Tr	Tr	Tr	Tr	Tr	0	Tr	Tr	Tr	0
324	**Olive oil**	0	N	0	5.10	Tr	Tr	Tr	Tr	Tr	0	Tr	Tr	Tr	0
325	**Palm oil**	0	Tr[a]	0	33.12	Tr	Tr	Tr	Tr	Tr	0	Tr	Tr	Tr	0
326	**Peanut oil**	0	Tr	0	15.16	Tr	Tr	Tr	Tr	Tr	0	Tr	Tr	Tr	0
327	**Rapeseed oil**, high erucic acid	0	Tr	0	22.21	Tr	Tr	Tr	Tr	Tr	0	Tr	Tr	Tr	0
328	low erucic acid	0	Tr	0	22.21	Tr	Tr	Tr	Tr	Tr	0	Tr	Tr	Tr	0
329	**Safflower oil**	0	Tr	0	40.68	Tr	Tr	Tr	Tr	Tr	0	Tr	Tr	Tr	0
330	**Sesame oil**	0	Tr	0	N	0.01	0.07	0.1	Tr	Tr	0	Tr	Tr	Tr	0
331	**Soya oil**	0	Tr	0	16.29	Tr	Tr	Tr	Tr	Tr	0	Tr	Tr	Tr	0
332	**Sunflowerseed oil**	0	Tr	0	49.22	Tr	Tr	Tr	Tr	Tr	0	Tr	Tr	Tr	0
333	**Vegetable oil**, blended, *average*	0	Tr	0	N[b]	Tr	Tr	Tr	Tr	Tr	0	Tr	Tr	Tr	0
334	**Wheatgerm oil**	0	Tr	0	136.65	Tr	Tr	Tr	Tr	Tr	0	Tr	Tr	Tr	0
Ghee															
335	**Ghee**, butter	675	500	1.90	3.31	0	Tr	Tr	Tr	Tr	Tr	0	Tr	Tr	0
336	palm	Tr	Tr	0	7.40	0	0	Tr	Tr	Tr	0	0	Tr	Tr	0
337	vegetable	Tr	Tr	0	10.27[b]	0	0	Tr	Tr	Tr	0	0	Tr	Tr	0

[a] Unrefined palm oil contains approximately 30000µg β- and 24000µg α-carotene per 100g

[b] The vitamin E content will vary according to the type of oil

Meat and meat products

Section 2.5

Meat and Meat Products

This section of the Tables has been extended by inclusion of new foods in the cooked dishes and meat products sections. New analytical data has been included making nutrient coverage more extensive.

The nutrient values were constructed from the separable fat and lean in meat analysed independently following dissection of bacon and carcase meat into lean meat, separable fat and inedible matter. Cuts of meat were selected to have the most representative proportions of fat and lean, and these were used to calculate the nutrient values. The major source of variation in meat composition however is the proportion of lean to fat and it is extremely difficult to define the level of average fatness for a particular joint. In the sampling scheme, care was taken to select typical cuts with regard to degree of fatness. For this reason values for 'lean only' and 'lean and fat' are given for virtually all the items in the bacon, beef, lamb and pork groups and should provide a reasonable guide to the average composition of an item (Paul and Southgate, 1977). The percentage of separable lean in the edible portion of the different cuts analysed is given in the description of the sample. It is more accurate however to measure the proportion of lean and fat in any meat consumed and if meat has a different percentage of lean, to compute the nutrients for meat from the 'lean only' and 'fat, average, raw (or cooked)' values.

For raw meats, edible proportion is given for items that are normally purchased with bones or other inedible material, e.g. rind. For cooked items that are normally served with waste, the edible proportion is given where they have been weighed with the waste.

For weight loss on cooking and calculation of cooked edible proportion obtainable from raw meat, see Section 3.3.

Vitamin E, vitamin B$_6$, vitamin B$_{12}$, folic acid, pantothenic acid and biotin were analysed in pooled samples of both raw and cooked lean, and in addition vitamin E was analysed in both raw and cooked fat. The values for individual cuts have been calculated from these results.

Where new analytical data are available for liver and certain meat products, the values for retinol and carotene have been revised.

New analytical data for fatty acid fraction totals and cholesterol values, derived from UK MAFF studies, have been used in this edition and the 1978 edition data used where new data are not available. Where foods are fried, in the majority of cases the fatty acids have (where no analysis is available) been calculated from the fatty acid profile of the food itself. In foods where a significant amount of fat is absorbed on frying the fatty acids will reflect the profile of the cooking fat specified in the food name.

New values for starch and sugar have been calculated using the proportions found in flour or breadcrumbs. Glycogen is present in liver and the contribution from this has been included in the values for starch and carbohydrate.

The composition of cooked sausages can be affected by the cooking procedure and whether or not they have skins. Losses of water and to a certain extent fat are reduced if the sausage is not pricked during cooking and is cooked slowly to prevent bursting or extrusion at the ends. The figures in these tables refer to sausages with skins that were pricked prior to cooking. Low fat sausages (with skins) were not pricked prior to cooking.

The fatty acid profile of margarine used in recipe calculations is an average of hard, soft and polyunsaturated varieties.

Losses of labile vitamins assigned to cooked dishes and foods were estimated using the figures in Section 3.4.

Taxonomic names for foods in this part of the Tables can be found in Section 3.6.

Meat and meat products

Composition of food per 100g

No.	Food	Description and main data sources	Edible proportion	Water g	Protein g	Fat g	Carbo-hydrate g	Energy value kcal	kJ
Bacon									
338	fat only, *raw, average*	Fat from five different cuts	1.00	12.8	4.8	80.9	0	747	3075
339	*cooked, average*	Fat from five different cuts	1.00	13.8	9.3	72.8	0	692	2852
340	lean only, *raw, average*	Lean from five different cuts	1.00	67.0	20.2	7.4	0	147	617
341	**Collar joint**, lean and fat, *raw*	12 samples, boneless, 70% lean	0.91	51.3	14.6	28.9	0	319	1318
342	*boiled*	12 samples, 73% lean; soaked for 16 hours before cooking	1.00	49.0	20.4	27.0	0	325	1346
343	lean only, *boiled*	12 samples; soaked for 16 hours before cooking	1.00	60.8	26.0	9.7	0	191	801
344	**Gammon joint**, lean and fat, *raw*	12 samples, boneless, 80% lean	0.93	60.8	17.6	18.3	0	236	978
345	*boiled*	12 samples 80% lean; soaked for 16 hours before cooking	1.00	53.9	24.7	18.9	0	269	1119
346	lean only, *boiled*	12 samples; soaked for 16 hours before cooking	1.00	62.7	29.4	5.5	0	167	703
347	**Gammon rasher**, lean and fat, *grilled*	24 samples, 88% lean; rind removed before cooking	1.00	52.1	29.5	12.2	0	228	953
348	lean only, *grilled*	24 samples; rind removed before cooking	1.00	57.0	31.4	5.2	0	172	726

Meat and meat products

No.	Food	Total nitrogen g	Fatty acids			Cholest-erol mg	Starch g	Total sugars g	Dietary fibre	
			Satd g	Mono unsatd g	Poly unsatd g				Southgate method g	Englyst method g

Bacon

No.	Food	Total nitrogen g	Satd g	Mono unsatd g	Poly unsatd g	Cholest-erol mg	Starch g	Total sugars g	Southgate method g	Englyst method g
338	fat only, *raw, average*	0.76	31.5	36.1	8.9	198	0	0	0	0
339	*cooked, average*	1.48	28.5	32.9	7.6	270	0	0	0	0
340	lean only, *raw, average*	3.23	2.7	3.1	0.8	18	0	0ᵃ	0	0
341	**Collar joint**, lean and fat, *raw*	2.34	11.2	12.9	3.2	71	0	0	0	0
342	*boiled*	3.26	10.6	12.2	2.8	100	0	0	0	0
343	lean only, *boiled*	4.16	3.6	4.2	1.0	36	0	0	0	0
344	**Gammon joint**, lean and fat, *raw*	2.82	7.1	8.1	2.0	45	0	0	0	0
345	*boiled*	3.95	7.4	8.5	2.0	70	0	0	0	0
346	lean only, *boiled*	4.70	2.1	2.4	0.5	20	0	0	0	0
347	**Gammon rasher**, lean and fat, *grilled*	4.72	4.8	5.5	1.3	45	0	0	0	0
348	lean only, *grilled*	5.02	1.9	2.2	0.5	19	0	0	0	0

ᵃ Sweetcure bacon contains 0.5g sugars per 100g

Meat and meat products

Inorganic constituents per 100g

No.	Food	Na	K	Ca	Mg	P	Fe	Cu	Zn	Cl	Mn	Se	I
						mg						µg	
												Se	I
Bacon													
338	fat only, *raw, average*	560	75	3	4	38[a]	0.7	0.06	0.6	810	Tr	(1)	(11)
339	*cooked, average*	990	130	7	10	90	0.8	0.09	0.8	1520	Tr	(2)	(11)
340	lean only, *raw, average*	1870[a]	350	9	22	180[a]	1.2	0.12	2.5	2800[a]	0.02	(4)	(11)
341	**Collar joint**, lean and fat, *raw*	1690	260	7	16	140	1.2	0.11	2.4	2560	0.01	(4)	12
342	*boiled*	1100	170	13	15	140	1.6	0.18	3.9	1630	0.01	(4)	(12)
343	lean only, *boiled*	1350	210	15	19	170	1.9	0.22	5.1	2000	0.02	(4)	(12)
344	**Gammon joint**, lean and fat, *raw*	1180	310	7	20	160	0.9	0.11	1.7	1800	0.02	(4)	6
345	*boiled*	960	210	9	18	150	1.3	0.15	2.7	1440	0.02	(4)	(6)
346	lean only, *boiled*	1110	250	10	21	180	1.5	0.17	3.3	1670	0.02	(4)	(6)
347	**Gammon rasher**, lean and fat, *grilled*	2140	480	9	31	260	1.4	0.17	3.2	3290	0.02	(4)	(6)
348	lean only, *grilled*	2210	520	10	33	270	1.5	0.18	3.5	3410	0.02	(4)	(6)

[a] Sweetcure bacon contains 140mg P per 100g fat and 1200mg Na, 280mg P and 1500mg Cl per 100g lean

No.	Food	Retinol µg	Carotene µg	Vitamin D µg	Vitamin E mg	Thiamin mg	Ribo-flavin mg	Niacin mg	Trypt 60 mg	Vitamin B6 mg	Vitamin B12 µg	Folate µg	Panto-thenate mg	Biotin µg	Vitamin C mg
Bacon															
338	fat only, *raw, average* [a]	Tr	Tr	Tr	0.11	N	N	N	0.9	N	Tr	Tr	N	Tr	0
339	*cooked, average* [a]	Tr	Tr	Tr	0.36	N	N	N	1.7	N	Tr	Tr	N	Tr	0
340	lean only, *raw, average*	Tr	Tr	Tr	0.06	0.65	0.25	4.7	3.8	0.45	Tr	3	0.60	2	0
341	**Collar joint**, lean and fat, *raw*	Tr	Tr	Tr	0.07	0.41	0.21	2.7	2.7	0.32	Tr	2	0.40	1	0
342	*boiled*	Tr	Tr	Tr	(0.14)	0.27	0.22	2.6	3.8	(0.24)	Tr	Tr	(0.40)	(2)	0
343	lean only, *boiled*	Tr	Tr	Tr	(0.05)	0.37	0.30	3.6	4.9	(0.33)	Tr	(1)	(0.50)	(3)	0
344	**Gammon joint**, lean and fat, *raw*	Tr	Tr	Tr	0.07	0.62	0.17	4.1	3.3	0.36	Tr	3	0.50	2	0
345	*boiled*	Tr	Tr	Tr	(0.11)	0.44	0.15	3.4	4.6	(0.26)	Tr	Tr	(0.40)	(2)	0
346	lean only, *boiled*	Tr	Tr	Tr	(0.05)	0.55	0.19	4.2	5.5	(0.33)	Tr	(1)	(0.50)	(3)	0
347	**Gammon rasher**, lean and fat, *grilled*	Tr	Tr	Tr	(0.07)	0.88	0.24	6.3	5.5	(0.33)	Tr	(2)	(0.60)	(3)	0
348	lean only, *grilled*	Tr	Tr	Tr	(0.04)	1.00	0.27	7.1	5.9	(0.37)	Tr	(2)	(0.70)	(3)	0

[a] Bacon fat may contain small amounts of vitamins. However, it is extremely difficult to obtain satisfactory analytical values for inclusion in the tables

Meat and meat products *continued*

Composition of food per 100g

No.	Food	Description and main data sources	Edible proportion	Water g	Protein g	Fat g	Carbo-hydrate g	Energy value kcal	kJ
	Bacon								
349	**Rasher**, lean and fat, *raw, back*	36 samples, 59% lean; rind removed	0.94	40.5	14.2	41.2	0	428	1766
350	*raw, middle*	36 samples, 59% lean; rind removed	0.91	40.8	14.3	40.9	0	425	1756
351	*-, streaky*	36 samples, 61% lean; rind removed	0.86	41.8	14.6	39.5	0	414	1710
352	lean only, *fried, average*	Average of back, middle and streaky	1.00	39.2	32.8	22.3	0	332	1383
353	lean and fat, *fried, back*	36 samples, 68% lean; rind removed before cooking	1.00	29.7	24.9	40.6	0	465	1926
354	*fried, middle*	36 samples, 64% lean; rind removed before cooking	1.00	28.7	24.1	42.3	0	477	1975
355	*-, streaky*	36 samples, 60% lean; rind removed before cooking	1.00	27.5	23.1	44.8	0	496	2050
356	lean only, *grilled, average*	Average of back, middle and streaky	1.00	46.0	30.5	18.9	0	292	1218
357	lean and fat, *grilled, back*	36 samples, 72% lean; rind removed before cooking	1.00	36.0	25.3	33.8	0	405	1681
358	*grilled, middle*	36 samples, 70% lean; rind removed before cooking	1.00	35.2	24.9	35.1	0	416	1722
359	*-, streaky*	36 samples, 69% lean; rind removed before cooking	1.00	34.6	24.5	36.0	0	422	1749
360	**Ham**, canned	12 samples, 10 brands	1.00	72.5	18.4	5.1	0	120	502

Meat and meat products *continued*

No.	Food	Total nitrogen g	Fatty acids Satd g	Fatty acids Mono unsatd g	Fatty acids Poly unsatd g	Cholest- erol mg	Starch g	Total sugars g	Dietary fibre Southgate method g	Dietary fibre Englyst method g
	Bacon									
349	**Rasher**, lean and fat, *raw, back*	2.27	16.2	18.4	4.5	101	0	0	0	0
350	*raw, middle*	2.28	15.9	18.3	4.5	100	0	0	0	0
351	*-, streaky*	2.34	15.4	17.6	4.3	97	0	0	0	0
352	lean only, *fried, average*	5.24	8.3	9.6	2.2	75	0	0	0	0
353	lean and fat, *fried, back*	3.99	15.9	18.3	4.3	143	0	0	0	0
354	*fried, middle*	3.86	16.5	19.1	4.4	149	0	0	0	0
355	*-, streaky*	3.69	17.5	20.2	4.7	158	0	0	0	0
356	lean only, *grilled, average*	4.88	7.4	8.5	2.0	70	0	0	0	0
357	lean and fat, *grilled, back*	4.04	13.2	15.3	3.5	125	0	0	0	0
358	*grilled, middle*	3.98	13.8	15.9	3.7	130	0	0	0	0
359	*-, streaky*	3.92	14.1	16.3	3.8	133	0	0	0	0
360	**Ham**, canned	2.95	1.9	2.1	0.6	68	0	0	0	0

Bacon

No.	Food	Na	K	Ca	Mg	P	Fe	Cu	Zn	Cl	Mn	Se	I
						mg						µg	
349	**Rasher**, lean and fat, *raw, back*	1470	230	7	15	120	1.0	0.10	1.6	2140	0.01	(4)	13
350	*raw, middle*	1470	230	7	15	120	1.0	0.10	1.6	2150	0.01	(4)	(13)
351	*-, streaky*	1500	240	8	15	120	1.0	0.10	1.7	2190	0.01	(4)	13
352	lean only, *fried, average*	2280	380	16	25	210	1.5	0.14	3.6	3510	0.01	(4)	(10)
353	lean and fat, *fried, back*	1910	300	13	20	170	1.3	0.12	2.6	2970	0.01	(4)	11
354	*fried, middle*	1870	300	13	20	170	1.3	0.12	2.5	2910	0.01	(4)	(10)
355	*-, streaky*	1820	290	12	19	160	1.2	0.12	2.4	2840	0.01	(4)	9
356	lean only, *grilled, average*	2240	350	13	18	190	1.6	0.17	3.7	3250	0.01	(4)	(10)
357	lean and fat, *grilled, back*	2020	290	12	16	160	1.5	0.16	3.0	2970	0.01	(4)	11
358	*grilled, middle*	2000	290	12	16	160	1.5	0.16	2.9	2940	0.01	(4)	(10)
359	*-, streaky*	1990	290	12	16	160	1.5	0.15	2.9	2930	0.01	(4)	9
360	**Ham**, canned	1250	280	9	18	280	1.2	0.22	2.3	1670	0.02	(8)	11

No.	Food	Retinol µg	Carotene µg	Vitamin D µg	Vitamin E mg	Thiamin mg	Ribo-flavin mg	Niacin mg	Trypt 60 mg	Vitamin B6 mg	Vitamin B12 µg	Folate µg	Panto-thenate mg	Biotin µg	Vitamin C mg
	Bacon														
349	**Rasher**, lean and fat, *raw, back*	Tr	Tr	Tr	0.07	0.35	0.14	3.1	2.7	0.26	Tr	2	0.40	1	0
350	*raw, middle*	Tr	Tr	Tr	0.08	0.36	0.14	3.1	2.7	0.27	Tr	2	0.40	1	0
351	*-, streaky*	Tr	Tr	Tr	0.08	0.37	0.15	3.2	2.7	0.28	Tr	2	0.40	1	0
352	lean only, *fried, average*	Tr	Tr	Tr	0.06	0.61	0.31	7.7	6.1	0.45	Tr	2	0.50	4	0
353	lean and fat, *fried, back*	Tr	Tr	Tr	0.18	0.41	0.21	5.2	4.7	0.30	Tr	1	0.30	3	0
354	*fried, middle*	Tr	Tr	Tr	0.20	0.39	0.20	5.0	4.5	0.29	Tr	1	0.30	2	0
355	*-, streaky*	Tr	Tr	Tr	0.21	0.37	0.19	4.6	4.3	0.27	Tr	1	0.30	2	0
356	lean only, *grilled, average*	Tr	Tr	Tr	0.04	0.59	0.23	6.2	5.7	0.37	Tr	2	0.70	3	0
357	lean and fat, *grilled, back*	Tr	Tr	Tr	0.11	0.43	0.17	4.5	4.7	0.27	Tr	1	0.50	2	0
358	*grilled, middle*	Tr	Tr	Tr	0.11	0.41	0.16	4.4	4.6	0.26	Tr	1	0.50	2	0
359	*-, streaky*	Tr	Tr	Tr	0.12	0.40	0.16	4.2	4.6	0.25	Tr	1	0.50	2	0
360	**Ham**, canned	Tr	Tr	Tr	0.08	0.52	0.25	3.9	3.0	0.22	Tr	Tr	0.60	1	0[a]

a Some brands have ascorbic acid added and may contain from 12 to 60mg per 100g

Meat and meat products *continued*

Composition of food per 100g

No.	Food	Description and main data sources	Edible proportion	Water g	Protein g	Fat g	Carbo-hydrate g	Energy value kcal	Energy value kJ
Beef									
361	fat only, *raw, average*	Fat from six different cuts	1.00	24.0	8.8	66.9	0	637	2625
362	*cooked, average*	Fat from six different cuts	1.00	25.2	11.9	62.8	0	613	2526
363	lean only, *raw, average*	Lean from six different cuts	1.00	74.0	20.3	4.6	0	123	517
364	**Brisket,** lean and fat, *raw*	18 samples, boned and rolled, 77% lean	0.96	62.2	16.8	20.5	0	252	1044
365	*boiled*	18 samples, boned and rolled, 77% lean; salt added	1.00	48.4	27.6	23.9	0	326	1354
366	**Forerib,** lean and fat, *raw*	18 samples, with bone, 72% lean	0.76	57.4	16.0	25.1	0	290	1201
367	*roast*	18 samples, 72% lean; cooked on the bone	1.00	48.4	22.4	28.8	0	349	1446
368	lean only, *roast*	18 samples; cooked on the bone	1.00	59.1	27.9	12.6	0	225	941
369	**Mince,** *raw*	18 samples	1.00	64.5	18.8	16.2	0	221	919
370	*stewed*	18 samples; salt added	1.00	59.1	23.1	15.2	0	229	955
371	**Rump steak,** lean and fat, *raw*	18 samples, 86% lean	0.97	66.7	18.9	13.5	0	197	821
372	*fried*	18 samples, 87% lean	1.00	56.2	28.6	14.6	0	246	1026
373	*grilled*	18 samples, 89% lean	1.00	59.3	27.3	12.1	0	218	912
374	lean only, *fried*	18 samples	1.00	61.1	30.8	7.4	0	190	797
375	*grilled*	18 samples	1.00	63.8	28.6	6.0	0	168	708

Meat and meat products *continued*

Composition of food per 100g

No.	Food	Total nitrogen g	Fatty acids Satd g	Mono unsatd g	Poly unsatd g	Cholesterol mg	Starch g	Total sugars g	Dietary fibre Southgate method g	Englyst method g
Beef										
361	fat only, *raw, average*	1.40	28.6	32.4	2.7	88	0	0	0	0
362	*cooked, average*	1.91	26.9	30.4	2.6	84	0	0	0	0
363	lean only, *raw, average*	3.25	1.9	2.1	0.2	59	0	0	0	0
364	**Brisket**, lean and fat, *raw*	2.68	8.1	10.1	0.4	74	0	0	0	0
365	*boiled*	4.41	9.5	11.8	0.5	91	0	0	0	0
366	**Forerib**, lean and fat, *raw*	2.56	10.7	12.1	1.0	67	0	0	0	0
367	*roast*	3.59	12.3	13.9	1.2	82	0	0	0	0
368	lean only, *roast*	4.46	5.2	5.8	0.5	82	0	0	0	0
369	**Mince**, *raw*	3.01	6.9	7.8	0.7	(66)	0	0	0	0
370	*stewed*	3.69	6.5	7.4	0.6	(83)	0	0	0	0
371	**Rump steak**, lean and fat, *raw*	3.02	5.8	6.5	0.5	63	0	0	0	0
372	*fried*	4.58	6.2	7.1	0.6	82	0	0	0	0
373	*grilled*	4.36	5.2	5.8	0.5	82	0	0	0	0
374	lean only, *fried*	4.92	3.1	3.5	0.3	82	0	0	0	0
375	*grilled*	4.58	2.5	2.8	0.2	82	0	0	0	0

Meat and meat products *continued*

Inorganic constituents per 100g

No.	Food	Na	K	Ca	Mg	P	Fe	Cu	Zn	Cl	Mn	Se	I
						mg						µg	
Beef													
361	fat only, *raw, average*	33	100	10	7	60	1.0	0.11	1.0	39	0.03	(1)	(6)
362	*cooked, average*	50	160	14	11	90	1.4	0.14	1.4	56	0.04	(2)	(6)
363	lean only, *raw, average*	61	350	7	20	180	2.1	0.14	4.3	59	0.04	3	(6)
364	**Brisket,** lean and fat, *raw*	68	270	7	16	140	1.6	0.12	3.5	69	0.04	3	(6)
365	*boiled*	73	200	12	18	150	2.8	0.13	6.3	92	0.04	3	(6)
366	**Forerib,** lean and fat, *raw*	48	270	10	15	130	1.5	0.12	3.4	45	0.04	3	(6)
367	*roast*	51	260	14	18	150	1.9	0.16	5.2	56	0.04	3	(6)
368	lean only, *roast*	56	310	13	22	180	2.3	0.17	6.8	61	0.04	3	(6)
369	**Mince,** *raw*	86	290	15	17	160	2.7	0.15	4.3	86	0.04	3	(6)
370	*stewed*	320	290	18	20	170	3.1	0.24	5.8	470	0.04	3	(6)
371	**Rump steak,** lean and fat, *raw*	51	330	6	20	210	2.3	0.14	4.6	49	0.04	3	(6)
372	*fried*	54	360	7	24	220	3.2	0.15	5.3	56	0.04	3	(6)
373	*grilled*	55	380	7	25	220	3.4	0.18	4.9	61	0.04	3	(6)
374	lean only, *fried*	57	390	6	25	240	3.4	0.15	5.9	58	0.04	3	(6)
375	*grilled*	56	400	7	26	230	3.5	0.18	5.3	62	0.04	3	(6)

No.	Food	Retinol µg	Carotene µg	Vitamin D µg	Vitamin E mg	Thiamin mg	Ribo-flavin mg	Niacin mg	Trypt 60 mg	Vitamin B6 mg	Vitamin B12 µg	Folate µg	Panto-thenate mg	Biotin µg	Vitamin C mg
Beef															
361	fat only, *raw, average* [a]	N	N	N	0.32	N	N	N	1.9	N	Tr	N	N	Tr	0
362	*cooked, average* [a]	N	N	N	0.55	N	N	N	2.6	N	Tr	N	N	Tr	0
363	lean only, *raw, average*	Tr	Tr	Tr	0.15	0.07	0.24	5.2	4.3	0.32	2	10	0.70	Tr	0
364	**Brisket**, lean and fat, *raw*	Tr	Tr	Tr	0.19	0.05	0.16	3.7	3.6	0.25	1	8	0.50	Tr	0
365	*boiled*	Tr	Tr	Tr	0.35	0.04	0.30	4.3	5.9	0.25	2.0	13	0.70	Tr	0
366	**Forerib**, lean and fat, *raw*	Tr	Tr	Tr	0.20	0.04	0.14	3.6	3.4	0.23	1.0	7	0.50	Tr	0
367	*roast*	Tr	Tr	Tr	0.36	0.04	0.24	3.9	4.8	0.24	1.0	13	0.60	Tr	0
368	lean only, *roast*	Tr	Tr	Tr	0.29	0.05	0.33	5.5	6.0	0.33	2.0	17	0.90	Tr	0
369	**Mince**, *raw*	Tr	Tr	Tr	(0.18)	0.06	0.31	4.0	4.0	(0.27)	(2)	(9)	(0.60)	Tr	0
370	*stewed*	Tr	Tr	Tr	(0.31)	0.05	0.33	4.4	4.9	(0.30)	(2.0)	(16)	(0.80)	Tr	0
371	**Rump steak**, lean and fat, *raw*	Tr	Tr	Tr	0.17	0.08	0.26	4.2	4.0	0.27	2.0	9	0.60	Tr	0
372	*fried*	Tr	Tr	Tr	0.33	0.08	0.35	5.5	6.1	0.29	2.0	15	0.80	Tr	0
373	*grilled*	Tr	Tr	Tr	0.32	0.08	0.32	5.7	5.8	0.29	2.0	15	0.80	Tr	0
374	lean only, *fried*	Tr	Tr	Tr	0.29	0.09	0.40	6.3	6.6	0.33	2.0	17	0.90	Tr	0
375	*grilled*	Tr	Tr	Tr	0.29	0.09	0.36	6.4	6.1	0.33	2.0	17	0.90	Tr	0

[a] Beef fat may contain small amounts of some vitamins. However, it is extremely difficult to obtain satisfactory analytical values for inclusion in the tables

Meat and meat products *continued*

Composition of food per 100g

No.	Food	Description and main data sources	Edible proportion	Water g	Protein g	Fat g	Carbo-hydrate g	Energy value kcal	kJ
Beef									
376	**Salted**, *fat removed, raw*	Ref. Wu Leung *et al.* (1968)	1.00	50.6	27.1	0.4	0	119	498
377	*dried, raw*	Ref. Wu Leung *et al.* (1968)	1.00	29.4	55.4	1.5	0	250	1046
378	**Silverside**, lean and fat, *salted, boiled*	17 samples, 85% lean; soaked for 18 hours before cooking	1.00	54.5	28.6	14.2	0	242	1012
379	lean only, *salted, boiled*	17 samples; soaked for 18 hours before cooking	1.00	59.7	32.3	4.9	0	173	730
380	**Sirloin**, lean and fat, *raw*	18 samples, boneless, 72% lean	0.92	59.4	16.6	22.8	0	272	1126
381	*roast*	18 samples, boneless, 80% lean	1.00	54.3	23.6	21.1	0	284	1182
382	lean only, *roast*	18 samples, boneless	1.00	62.0	27.6	9.1	0	192	806
383	**Stewing steak**, lean and fat, *raw*	18 samples, 85% lean	0.96	68.7	20.2	10.6	0	176	736
384	*stewed*	18 samples, 92% lean; salt added	1.00	57.1	30.9	11.0	0	223	932
385	**Topside**, lean and fat, *raw*	18 samples, 87% lean	0.95	68.4	19.6	11.2	0	179	748
386	*roast*	18 samples, 88% lean	1.00	60.2	26.6	12.0	0	214	896
387	lean only, *roast*	18 samples	1.00	65.1	29.2	4.4	0	156	659

Meat and meat products *continued*

Composition of food per 100g

No.	Food	Total nitrogen g	Fatty acids			Cholest-erol mg	Starch g	Total sugars g	Dietary fibre	
			Satd g	Mono unsatd g	Poly unsatd g				Southgate method g	Englyst method g
Beef										
376	**Salted**, *fat removed, raw*	4.33	N	N	N	N	0	0	0	0
377	*dried, raw*	8.86	N	N	N	N	0	0	0	0
378	**Silverside**, lean and fat, *salted, boiled*	4.58	6.2	7.0	0.6	83	0	0	0	0
379	lean only, *salted, boiled*	5.16	2.0	2.3	0.2	82	0	0	0	0
380	**Sirloin**, lean and fat, *raw*	2.65	9.7	11.0	0.9	67	0	0	0	0
381	*roast*	3.77	9.0	10.2	0.9	83	0	0	0	0
382	lean only, *roast*	4.41	3.7	4.2	0.4	82	0	0	0	0
383	**Stewing steak**, lean and fat, *raw*	3.23	4.5	5.1	0.4	63	0	0	0	0
384	*stewed*	4.94	4.7	5.3	0.5	82	0	0	0	0
385	**Topside**, lean and fat, *raw*	3.13	4.1	5.5	0.7	62	0	0	0	0
386	*roast*	4.26	4.1	6.4	0.7	82	0	0	0	0
387	lean only, *roast*	4.67	1.4	2.2	0.2	82	0	0	0	0

Meat and meat products continued

Inorganic constituents per 100g

No.	Food	Na	K	Ca	Mg	P	Fe	Cu	Zn	Cl	Mn	Se	I
						mg						µg	
Beef													
376	**Salted**, *fat removed, raw*	N	N	91	N	N	5.4	N	N	N	N	N	N
377	*dried, raw*	N	N	49	N	N	4.9	N	N	N	N	N	N
378	**Silverside**, lean and fat, salted, boiled	910	200	11	18	140	2.8	0.25	5.5	1420	0.04	3	(6)
379	lean only, salted, boiled	1000	230	10	20	150	3.2	0.27	6.2	1560	0.04	3	(6)
380	**Sirloin**, lean and fat, raw	49	260	9	16	150	1.6	0.13	3.1	52	0.04	3	(6)
381	roast	54	300	10	19	170	1.9	0.18	4.6	64	0.04	3	(6)
382	lean only, roast	59	350	10	22	190	2.1	0.19	5.5	65	0.04	3	(6)
383	**Stewing steak**, lean and fat, raw	72	320	8	18	140	2.1	0.15	3.8	73	0.04	3	6
384	stewed	360	230	15	21	160	3.0	0.25	8.7	550	0.04	3	6
385	**Topside**, lean and fat, raw	43	340	5	19	170	1.9	0.13	3.3	44	0.04	3	(6)
386	roast	48	350	6	23	200	2.6	0.13	4.9	51	0.04	3	(6)
387	lean only, roast	49	370	6	24	210	2.8	0.14	5.5	51	0.04	3	(6)

Meat and meat products continued

No.	Food	Retinol μg	Carotene μg	Vitamin D μg	Vitamin E mg	Thiamin mg	Ribo-flavin mg	Niacin mg	Trypt 60 mg	Vitamin B6 mg	Vitamin B12 μg	Folate μg	Panto-thenate mg	Biotin μg	Vitamin C mg
Beef															
376	**Salted**, *fat removed, raw*	Tr	Tr	Tr	N	0.02	0.21	3.4	5.8	N	N	N	N	N	0
377	*dried, raw*	Tr	Tr	Tr	N	0.02	0.18	6.3	11.8	N	N	N	N	N	0
378	**Silverside**, lean and fat, salted, *boiled*	Tr	Tr	Tr	0.33	0.03	0.27	3.3	6.1	0.28	2.0	15	0.80	Tr	0
379	lean only, salted, *boiled*	Tr	Tr	Tr	0.29	0.04	0.32	3.9	6.9	0.33	2.0	17	0.90	Tr	0
380	**Sirloin**, lean and fat, *raw*	Tr	Tr	Tr	0.20	0.04	0.17	4.2	3.5	0.23	1.0	7	0.50	Tr	0
381	*roast*	Tr	Tr	Tr	0.34	0.06	0.25	4.8	5.0	0.26	2.0	14	0.70	Tr	0
382	lean only, *roast*	Tr	Tr	Tr	0.29	0.07	0.31	6.0	5.9	0.33	2.0	17	0.90	Tr	0
383	**Stewing steak**, lean and fat, *raw*	Tr	Tr	Tr	0.18	0.06	0.23	4.2	4.3	0.27	2	9	0.60	Tr	0
384	*stewed*	Tr	Tr	Tr	0.31	0.03	0.33	3.6	6.6	0.30	2.0	16	0.80	Tr	0
385	**Topside**, lean and fat, *raw*	Tr	Tr	Tr	0.17	0.05	0.21	4.8	4.2	0.28	2.0	9	0.60	Tr	0
386	*roast*	Tr	Tr	Tr	0.32	0.07	0.31	5.7	5.7	0.29	2.0	15	0.80	Tr	0
387	lean only, *roast*	Tr	Tr	Tr	0.29	0.08	0.35	6.5	6.2	0.33	2.0	17	0.90	Tr	0

Meat and meat products *continued*

Composition of food per 100g

No.	Food	Description and main data sources	Edible proportion	Water g	Protein g	Fat g	Carbo-hydrate g	Energy value kcal	kJ
	Lamb								
388	fat only, *raw, average*	Fat from six different cuts	1.00	21.2	6.2	71.8	0	671	2762
389	*cooked, average*	Fat from six different cuts	1.00	24.6	11.3	63.4	0	616	2538
390	lean only, *raw, average*	Lean from six different cuts	1.00	70.1	20.8	8.8	0	162	679
391	**Breast**, lean and fat, *raw*	15 samples, boneless, 59% lean	0.84	48.3	16.7	34.6	0	378	1564
392	*roast*	15 samples, boneless, 60% lean	1.00	43.6	19.1	37.1	0	410	1697
393	lean only, *roast*	15 samples, boneless	1.00	57.8	25.6	16.6	0	252	1049
394	**Chops**, loin, lean and fat, *raw*	15 samples, 60% lean	0.84	49.5	14.6	35.4	0	377	1558
395	*grilled*	15 samples, 70% lean	1.00	46.6	23.5	29.0	0	355	1473
396	*grilled, weighed with bone*	Calculated from lean and fat, grilled	0.79	36.3	18.3	22.6	0	277	1147
397	loin, lean only, *grilled*	15 samples	1.00	58.9	27.8	12.3	0	222	928
398	*grilled, weighed with fat and bone*	Calculated from lean only, grilled	0.55	32.4	15.3	6.8	0	122	512

Meat and meat products *continued*

Composition of food per 100g

No.	Food	Total nitrogen g	Fatty acids			Cholest-erol mg	Starch g	Total sugars g	Dietary fibre	
			Satd g	Mono unsatd g	Poly unsatd g				Southgate method g	Englyst method g

Lamb

No.	Food	Total nitrogen g	Satd g	Mono unsatd g	Poly unsatd g	Cholest-erol mg	Starch g	Total sugars g	Southgate method g	Englyst method g
388	fat only, *raw, average*	0.99	35.6	27.7	3.4	75	0	0	0	0
389	*cooked, average*	1.81	31.5	24.5	3.0	100	0	0	0	0
390	lean only, *raw, average*	3.33	4.2	3.3	0.4	79	0	0	0	0
391	**Breast**, lean and fat, *raw*	2.67	17.7	13.8	1.7	78	0	0	0	0
392	*roast*	3.05	18.4	14.3	1.8	106	0	0	0	0
393	lean only, *roast*	4.09	7.9	6.2	0.8	110	0	0	0	0
394	**Chops**, loin, lean and fat, *raw*	2.34	17.6	13.6	1.7	77	0	0	0	0
395	*grilled*	3.76	14.4	11.2	1.4	107	0	0	0	0
396	*grilled, weighed with bone*	2.93	11.2	8.7	1.1	84	0	0	0	0
397	loin, lean only, *grilled*	4.44	5.9	4.6	0.6	110	0	0	0	0
398	*grilled, weighed with fat and bone*	2.44	3.2	2.5	0.3	48	0	0	0	0

Meat and meat products *continued*

Inorganic constituents per 100g

No.	Food	Na	K	Ca	Mg	P	Fe	Cu	Zn	Cl	Mn	Se	I
						mg						µg	
Lamb													
388	fat only, *raw, average*	36	96	7	6	54	0.7	0.15	0.8	37	0.02	Tr	(5)
389	*cooked, average*	56	150	11	12	120	1.4	0.16	1.4	53	0.02	Tr	(5)
390	lean only, *raw, average*	88	350	7	24	190	1.6	0.17	4.0	76	0.02	1	(5)
391	**Breast**, lean and fat, *raw*	100	270	8	18	150	1.3	0.16	3.2	77	0.02	1	(5)
392	*roast*	73	250	10	18	150	1.5	0.19	3.6	74	0.02	1	(5)
393	lean only, *roast*	86	330	10	24	200	1.7	0.23	5.3	89	0.02	1	(5)
394	**Chops**, loin, lean and fat, *raw*	61	230	(7)[a]	17	140	1.2	0.16	2.1	60	0.02	1	(5)
395	*grilled*	72	320	(9)[a]	24	210	1.9	0.17	3.4	83	0.02	1	(5)
396	*grilled, weighed with bone*	56	250	(7)[a]	19	160	1.5	0.13	2.7	65	0.02	1	(4)
397	loin, lean only, *grilled*	75	380	(9)[a]	28	240	2.1	0.19	4.1	90	0.02	1	(5)
398	*grilled, weighed with fat and bone*	41	210	(5)[a]	15	130	1.2	0.10	2.3	50	0.01	1	(3)

[a] The calcium content is extremely variable as scrapings of bone may easily be included in the edible portion. This is a minimum value

Meat and meat products *continued*

Vitamins per 100g

No.	Food	Retinol µg	Carotene µg	Vitamin D µg	Vitamin E mg	Thiamin mg	Ribo-flavin mg	Niacin mg	Trypt 60 mg	Vitamin B6 mg	Vitamin B12 µg	Folate µg	Panto-thenate mg	Biotin µg	Vitamin C mg
	Lamb														
388	fat only, raw, average [a]	N	N	N	0.30	N	N	N	1.3	N	Tr	N	N	Tr	0
389	cooked, average [a]	N	N	N	0.18	N	N	N	2.4	N	Tr	N	N	Tr	0
390	lean only, raw, average	Tr	Tr	Tr	0.10	0.14	0.28	6.0	4.4	0.25	2	5	0.70	2	0
391	**Breast**, lean and fat, raw	Tr	Tr	Tr	0.18	0.08	0.17	3.8	3.6	0.15	1	3	0.40	1	0
392	roast	Tr	Tr	Tr	0.13	0.06	0.17	3.4	4.1	0.13	1.0	3	0.40	1	0
393	lean only, roast	Tr	Tr	Tr	0.10	0.10	0.29	5.6	5.5	0.22	2.0	4	0.70	2	0
394	**Chops**, loin, lean and fat, raw	Tr	Tr	Tr	0.18	0.09	0.16	4.0	3.1	0.15	1.0	3	0.40	1	0
395	grilled	Tr	Tr	Tr	0.12	0.11	0.21	5.1	5.0	0.15	2.0	3	0.50	1	0
396	grilled, weighed with bone	Tr	Tr	Tr	0.09	0.09	0.16	4.0	3.9	0.12	2.0	2	0.40	1	0
397	loin, lean only, grilled	Tr	Tr	Tr	0.10	0.15	0.30	7.2	5.9	0.22	2.0	4	0.70	2	0
398	grilled, weighed with fat and bone	Tr	Tr	Tr	0.06	0.08	0.17	4.0	3.3	0.12	1.0	2	0.40	1	0

[a] Lamb fat may contain small amounts of some vitamins. However, it is extremely difficult to obtain satisfactory analytical values for inclusion in the tables

143

Meat and meat products *continued*

Composition of food per 100g

Lamb

No.	Food	Description and main data sources	Edible proportion	Water g	Protein g	Fat g	Carbo-hydrate g	Energy value kcal	kJ
399	**Cutlets**, lean and fat, *raw*	15 samples, 59% lean	0.77	48.7	14.7	36.3	0	386	1593
400	*grilled*	15 samples, 67% lean	1.00	45.1	23.0	30.9	0	370	1534
401	*grilled, weighed with bone*	Calculated from lean and fat, grilled	0.68	29.8	15.2	20.4	0	244	1013
402	*lean only, grilled*	15 samples	1.00	58.9	27.8	12.3	0	222	928
403	*grilled, weighed with fat and bone*	Calculated from lean only, grilled	0.44	25.9	12.2	5.4	0	97	407
404	**Leg**, lean and fat, *raw*	15 samples, 80% lean	0.77	63.1	17.9	18.7	0	240	996
405	*roast*	15 samples, 82% lean	1.00	55.3	26.1	17.9	0	266	1106
406	*lean only, roast*	15 samples	1.00	61.8	29.4	8.1	0	191	800
407	**Scrag and neck**, lean and fat, *raw*	15 samples, 71% lean	0.61	55.7	15.6	28.2	0	316	1309
408	*stewed*	15 samples, 84% lean	1.00	52.6	25.6	21.1	0	292	1216
409	*lean only, stewed*	15 samples	1.00	55.9	27.8	15.7	0	253	1054
410	*stewed, weighed with fat and bone*	Calculated from lean only, stewed	0.51	28.5	14.1	8.0	0	128	536
411	**Shoulder**, lean and fat, *raw*	15 samples, 68% lean	0.79	56.1	15.6	28.0	0	314	1301
412	*roast*	15 samples, 73% lean	1.00	53.6	19.9	26.3	0	316	1311
413	*lean only, roast*	15 samples	1.00	64.8	23.8	11.2	0	196	819

Composition of food per 100g

No.	Food	Total nitrogen g	Fatty acids			Cholest-erol mg	Starch g	Total sugars g	Dietary fibre	
			Satd g	Mono unsatd g	Poly unsatd g				Southgate method g	Englyst method g
Lamb										
399	**Cutlets**, lean and fat, *raw*	2.35	18.0	14.0	1.7	78	0	0	0	0
400	*grilled*	3.68	15.3	11.9	1.5	107	0	0	0	0
401	*grilled, weighed with bone*	2.43	10.1	7.9	1.0	73	0	0	0	0
402	*lean only, grilled*	4.44	5.9	4.6	0.7	110	0	0	0	0
403	*grilled, weighed with fat and bone*	1.95	2.5	2.0	0.2	48	0	0	0	0
404	**Leg**, lean and fat, *raw*	2.86	9.3	7.2	0.9	78	0	0	0	0
405	*roast*	4.18	8.9	6.9	0.9	108	0	0	0	0
406	*lean only, roast*	4.71	3.9	3.0	0.4	110	0	0	0	0
407	**Scrag and neck**, lean and fat, *raw*	2.50	14.0	10.9	1.3	78	0	0	0	0
408	*stewed*	4.10	10.5	8.1	1.0	108	0	0	0	0
409	*lean only, stewed*	4.44	7.5	5.8	0.7	110	0	0	0	0
410	*stewed, weighed with fat and bone*	2.26	3.8	3.0	0.4	51	0	0	0	0
411	**Shoulder**, lean and fat, *raw*	2.50	13.9	10.8	1.3	68	0	0	0	0
412	*roast*	3.18	13.1	10.2	1.3	107	0	0	0	0
413	*lean only, roast*	3.80	5.4	4.2	0.5	110	0	0	0	0

Meat and meat products continued

Inorganic constituents per 100g

No.	Food	Na	K	Ca	Mg	P	Fe	Cu	Zn	Cl	Mn	Se	I
		mg										μg	
Lamb													
399	**Cutlets**, lean and fat, raw	60	230	(7)[a]	16	130	1.2	0.15	2.1	60	0.02	1	(5)
400	grilled	71	320	(9)[a]	23	200	1.9	0.18	3.3	82	0.02	1	(5)
401	grilled, weighed with bone	47	210	(6)[a]	15	130	1.3	0.12	2.2	54	0.01	1	(4)
402	lean only, grilled	75	380	(9)[a]	28	240	2.1	0.19	4.1	90	0.02	1	(5)
403	grilled, weighed with fat and bone	33	170	(4)[a]	12	110	0.9	0.08	1.8	40	0.01	1	(2)
404	**Leg**, lean and fat, raw	52	310	6	22	170	1.7	0.14	2.8	54	0.02	1	(5)
405	roast	65	310	8	25	200	2.5	0.28	4.6	62	0.02	1	(5)
406	lean only, roast	67	340	8	28	220	2.7	0.31	5.3	64	0.02	1	(5)
407	**Scrag and neck**, lean and fat, raw	71	260	(7)[a]	18	140	1.2	0.16	3.6	68	0.02	1	(5)
408	stewed	240	190	(9)[a]	18	190	2.2	0.22	6.1	340	0.02	1	(5)
409	lean only, stewed	250	200	(9)[a]	19	190	2.4	0.33	6.9	350	0.02	1	(5)
410	stewed, weighed with fat and bone	130	100	(5)[a]	10	100	1.2	0.17	3.5	180	0.02	1	(3)
411	**Shoulder**, lean and fat, raw	66	260	7	18	150	1.2	0.21	3.1	56	0.02	1	5
412	roast	61	260	9	19	150	1.6	0.15	4.3	60	0.02	1	5
413	lean only, roast	65	300	9	22	170	1.8	0.16	5.3	65	0.02	1	5

[a] The calcium content is extremely variable as scrapings of bone may easily be included in the edible portion. This is a minimum value

Meat and meat products continued

No.	Food	Retinol µg	Carotene µg	Vitamin D µg	Vitamin E mg	Thiamin mg	Ribo-flavin mg	Niacin mg	Trypt 60 mg	Vitamin B6 mg	Vitamin B12 µg	Folate µg	Panto-thenate mg	Biotin µg	Vitamin C mg
Lamb															
399	**Cutlets**, lean and fat, raw	Tr	Tr	Tr	0.18	0.09	0.16	3.9	3.1	0.15	1.0	3	0.40	1	0
400	grilled	Tr	Tr	Tr	0.13	0.10	0.20	4.8	4.9	0.15	2.0	3	0.50	1	0
401	grilled, weighed with bone	Tr	Tr	Tr	0.09	0.07	0.13	3.2	3.2	0.10	1.0	2	0.30	1	0
402	lean only, grilled	Tr	Tr	Tr	0.10	0.15	0.30	7.2	5.9	0.22	2.0	4	0.70	2	0
403	grilled, weighed with fat and bone	Tr	Tr	Tr	0.04	0.07	0.13	3.2	2.6	0.10	1.0	2	0.30	1	0
404	**Leg**, lean and fat, raw	Tr	Tr	Tr	0.14	0.14	0.25	5.7	3.8	0.20	2.0	4	0.60	1	0
405	roast	Tr	Tr	Tr	0.11	0.12	0.31	5.4	5.6	0.18	2.0	3	0.60	1	0
406	lean only, roast	Tr	Tr	Tr	0.10	0.14	0.38	6.6	6.3	0.22	2.0	4	0.70	2	0
407	**Scrag and neck**, lean and fat, raw	Tr	Tr	Tr	0.16	0.07	0.17	3.4	3.3	0.18	2.0	4	0.50	1	0
408	stewed	Tr	Tr	Tr	0.11	0.04	0.18	2.7	5.5	0.19	2.0	4	0.60	1	0
409	lean only, stewed	Tr	Tr	Tr	0.10	0.05	0.21	3.2	5.9	0.22	2.0	4	0.70	2	0
410	stewed, weighed with fat and bone	Tr	Tr	Tr	0.05	0.03	0.11	1.6	3.0	0.11	1.0	2	0.40	1	0
411	**Shoulder**, lean and fat, raw	Tr	Tr	Tr	0.17	0.10	0.18	3.6	3.3	0.17	2.0	3	0.50	1	0
412	roast	Tr	Tr	Tr	0.12	0.07	0.20	3.1	4.2	0.16	2.0	3	0.50	1	0
413	lean only, roast	Tr	Tr	Tr	0.10	0.10	0.27	4.3	5.1	0.22	2.0	4	0.70	2	0

Meat and meat products continued

Composition of food per 100g

Pork

No.	Food	Description and main data sources	Edible proportion	Water g	Protein g	Fat g	Carbo-hydrate g	Energy value kcal	kJ
414	fat only, *raw, average*	Fat from three different cuts	1.00	21.1	6.8	71.4	0	670	2757
415	*cooked, average*	Fat from three different cuts	1.00	20.9	14.8	62.2	0	619	2553
416	lean only, *raw, average*	Lean from three different cuts	1.00	71.5	20.7	7.1	0	147	615
417	**Belly rashers**, lean and fat, *raw*	15 samples, 56% lean	0.90	48.7	15.3	35.5	0	381	1574
418	*grilled*	15 samples, 56% lean	1.00	43.0	21.1	34.8	0	398	1646
419	**Chops,** loin, lean and fat, *raw*	15 samples, 65% lean; without kidney	0.84	54.3	15.9	29.5	0	329	1362
420	*grilled*	15 samples, 75% lean; without kidney	1.00	46.3	28.5	24.2	0	332	1380
421	*grilled, weighed with bone*	Calculated from lean and fat, grilled	0.78	36.1	22.2	18.8	0	258	1073
422	loin, lean only, *grilled*	15 samples	1.00	56.1	32.3	10.7	0	226	945
423	*grilled, weighed with fat and bone*	Calculated from lean only, grilled	0.59	33.1	19.1	6.3	0	133	558
424	**Leg,** lean and fat, *raw*	15 samples, fillet end, 73% lean	0.85	59.5	16.6	22.5	0	269	1115
425	*roast*	15 samples, fillet end, 76% lean	1.00	51.9	26.9	19.8	0	286	1190
426	lean only, *roast*	15 samples, fillet end	1.00	61.6	30.7	6.9	0	185	777
427	**Trotters and tails,** *salted, boiled*	23% trotters and 77% tails, boiled for 2 hours.	0.54	53.5	19.8	22.3	0	280	1162

Meat and meat products *continued*

Composition of food per 100g

No.	Food	Total nitrogen g	Fatty acids Satd g	Fatty acids Mono unsatd g	Fatty acids Poly unsatd g	Cholesterol mg	Starch g	Total sugars g	Dietary fibre Southgate method g	Dietary fibre Englyst method g
Pork										
414	fat only, *raw, average*	1.08	26.4	28.8	10.7	76	0	0	0	0
415	*cooked, average*	2.37	23.0	25.1	9.3	98	0	0	0	0
416	lean only, *raw, average*	3.29	2.5	2.8	1.0	69	0	0	0	0
417	**Belly rashers**, lean and fat, *raw*	2.44	13.1	14.3	5.3	72	0	0	0	0
418	*grilled*	3.38	12.9	14.1	5.2	105	0	0	0	0
419	**Chops**, loin, lean and fat, *raw*	2.55	10.9	11.9	4.4	72	0	0	0	0
420	*grilled*	4.55	9.0	9.8	3.6	108	0	0	0	0
421	*grilled, weighed with bone*	3.55	6.9	7.6	2.8	84	0	0	0	0
422	loin, lean only, *grilled*	5.17	3.8	4.1	1.5	110	0	0	0	0
423	*grilled, weighed with fat and bone*	3.05	2.2	2.4	0.9	65	0	0	0	0
424	**Leg**, lean and fat, *raw*	2.66	8.3	9.1	3.4	71	0	0	0	0
425	*roast*	4.30	7.3	8.0	3.0	107	0	0	0	0
426	lean only, *roast*	4.91	2.4	2.7	1.0	110	0	0	0	0
427	**Trotters and tails**, *salted, boiled*	3.17	N	N	N	N	0	0	0	0

Meat and meat products *continued*

Inorganic constituents per 100g

No.	Food	Na	K	Ca	Mg	P	Fe	Cu	Zn	Cl	Mn	Se	I
						mg						µg	
Pork													
414	fat only, *raw, average*	38	87	7	5	49	0.7	0.10	0.4	44	0.02	N	Tr
415	*cooked, average*	79	210	11	9	110	1.0	0.15	0.9	84	0.03	N	Tr
416	lean only, *raw, average*	76	370	8	22	200	0.9	0.15	2.4	71	0.04	14	(3)
417	**Belly rashers**, lean and fat, *raw*	73	220	8	14	120	0.8	0.15	1.8	71	0.03	14	(3)
418	*grilled*	95	310	11	19	170	1.0	0.16	2.6	92	0.03	14	(3)
419	**Chops**, loin, lean and fat, *raw*	56	290	(8)[a]	17	160	0.8	0.13	1.6	54	0.03	14	(3)
420	*grilled*	84	380	(11)[a]	26	230	1.2	0.17	2.9	79	0.03	14	(3)
421	*grilled, weighed with bone*	66	300	(9)[a]	20	180	0.9	0.13	2.3	62	0.02	11	(2)
422	loin, lean only, *grilled*	84	420	(9)[a]	29	260	1.2	0.16	3.5	78	0.04	14	(3)
423	*grilled, weighed with fat and bone*	50	250	(5)[a]	17	150	0.7	0.09	2.1	46	0.02	8	(2)
424	**Leg**, lean and fat, *raw*	59	300	7	18	160	0.8	0.12	1.8	59	0.03	14	3
425	*roast*	79	350	10	22	200	1.3	0.25	2.9	79	0.03	14	3
426	lean only, *roast*	79	390	9	25	230	1.3	0.29	3.5	76	0.04	14	3
427	**Trotters and tails**, salted, *boiled*	1615	30	129	8	110	0.7	0.07	2.4	2490	0.01	N	N

[a] The calcium content is extremely variable as scrapings of bone may easily be included in the edible portion. This is a minimum value

Meat and meat products *continued*

Pork

No.	Food	Retinol µg	Carotene µg	Vitamin D µg	Vitamin E mg	Thiamin mg	Ribo-flavin mg	Niacin mg	Trypt 60 mg	Vitamin B6 mg	Vitamin B12 µg	Folate µg	Panto-thenate mg	Biotin µg	Vitamin C mg
414	fat only, raw, average [a]	Tr	Tr	Tr	0.03	N	N	N	1.3	N	Tr	N	N	Tr	0
415	cooked, average [a]	Tr	Tr	Tr	0.12	N	N	N	2.8	N	Tr	N	N	Tr	0
416	lean only, raw, average	Tr	Tr	Tr	0	0.89	0.25	6.2	3.8	0.45	3	5	1.10	3	0
417	**Belly rashers**, lean and fat, raw	Tr	Tr	Tr	0.01	0.45	0.15	3.3	2.9	0.25	2.0	3	0.60	2	0
418	grilled	Tr	Tr	Tr	0.05	0.53	0.11	4.2	3.9	0.23	1.0	4	0.70	2	0
419	**Chops**, loin, lean and fat, raw	Tr	Tr	Tr	0.01	0.57	0.14	4.2	3.0	0.29	2.0	3	0.70	2	0
420	grilled	Tr	Tr	Tr	0.03	0.66	0.20	5.7	5.3	0.31	1.0	6	1.00	2	0
421	grilled, weighed with bone	Tr	Tr	Tr	0.02	0.51	0.16	4.4	4.1	0.24	1.0	5	0.80	2	0
422	loin, lean only, grilled	Tr	Tr	Tr	0	0.88	0.26	7.6	6.0	0.41	2.0	7	1.30	3	0
423	grilled, weighed with fat and bone	Tr	Tr	Tr	0	0.52	0.15	4.5	3.6	0.24	1.0	4	0.80	2	0
424	**Leg**, lean and fat, raw	Tr	Tr	Tr	0.01	0.73	0.20	4.5	3.1	0.33	2	4	0.80	2	0
425	roast	Tr	Tr	Tr	0.03	0.65	0.27	5.0	5.0	0.31	1.0	6	1.00	2	0
426	lean only, roast	Tr	Tr	Tr	0	0.85	0.35	6.6	5.7	0.41	2.0	7	1.30	3	0
427	**Trotters and tails**, salted, boiled	Tr	Tr	Tr	N	0.06	0.20	0.9	3.7	N	0.8	3	N	N	0

[a] Pork fat may contain small amounts of some vitamins. However, it is extremely difficult to obtain satisfactory analytical values for inclusion in the tables

Meat and meat products *continued*

Composition of food per 100g

No.	Food	Description and main data sources	Edible proportion	Water g	Protein g	Fat g	Carbo-hydrate g	Energy value kcal	Energy value kJ
	Veal								
428	**Cutlet,** *fried in vegetable oil*	Samples coated in egg and crumbs and fried	1.00	54.6	31.4	8.1	4.4	215	904
429	**Fillet,** *raw*	Lean samples	1.00	74.9	21.1	2.7	0	109	459
430	*roast*	Lean samples	1.00	55.1	31.6	11.5	0	230	963
431	**Chicken,** *meat only, raw*	15 samples, light and dark meat from dressed carcase	0.44	74.4	20.5	4.3	0	121	508
432	*meat and skin, raw*	15 samples, dressed carcase excluding waste	0.64	64.4	17.6	17.7	0	230	954
433	*light meat, raw*	15 samples	0.23	74.4	21.8	3.2	0	116	489
434	*dark meat, raw*	15 samples	0.21	74.5	19.1	5.5	0	126	528
435	*boiled, meat only*	15 samples, light and dark meat	1.00	63.4	29.2	7.3	0	183	767
436	*light meat*	15 samples	1.00	65.2	29.7	4.9	0	163	686
437	*dark meat*	15 samples	1.00	61.5	28.6	9.9	0	204	853
438	*roast, meat only*	15 samples, light and dark meat	1.00	68.4	24.8	5.4	0	148	621
439	*meat and skin*	15 samples	1.00	61.9	22.6	14.0	0	216	902
440	*light meat*	15 samples	1.00	68.5	26.5	4.0	0	142	599
441	*dark meat*	15 samples	1.00	68.2	23.1	6.9	0	155	648
442	*wing quarter, roast, meat only, weighed with bone*	Meat from whole wing quarter	0.50	34.2	12.4	2.7	0	74	311
443	*leg quarter, roast, meat only, weighed with bone*	Meat from whole leg quarter	0.38	42.4	15.4	3.4	0	92	388
444	*breaded, fried in vegetable oil*	4 samples, 4 brands; fried 8-12 minutes	1.00	53.2	18.0	12.7	14.8	242	1013

Meat and meat products continued

No.	Food	Total nitrogen g	Fatty acids			Cholest-erol mg	Starch g	Total sugars g	Dietary fibre	
			Satd g	Mono unsatd g	Poly unsatd g				Southgate method g	Englyst method g
428	**Cutlet,** *fried in vegetable oil*	5.02	0.8	2.7	3.7	109	(4.3)	(0.1)	(0.3)	(0.1)
	Veal									
429	**Fillet,** *raw*	3.37	0.9	1.2	0.4	84	0	0	0	0
430	*roast*	5.05	3.7	4.9	1.8	155	0	0	0	0
431	**Chicken,** *meat only, raw*	3.28	1.4	1.8	0.8	57	0	0	0	0
432	*meat and skin, raw*	2.82	5.9	7.5	3.3	99	0	0	0	0
433	*light meat, raw*	3.49	1.0	1.3	0.6	43	0	0	0	0
434	*dark meat, raw*	3.06	1.8	2.3	1.1	73	0	0	0	0
435	*boiled, meat only*	4.67	2.4	3.1	1.4	97	0	0	0	0
436	*light meat*	4.75	1.6	2.0	0.9	65	0	0	0	0
437	*dark meat*	4.58	3.2	4.1	1.9	132	0	0	0	0
438	*roast, meat only*	3.97	1.6	2.5	1.0	76	0	0	0	0
439	*meat and skin*	3.61	4.2	6.5	2.5	103	0	0	0	0
440	*light meat*	4.24	1.2	1.9	0.7	56	0	0	0	0
441	*dark meat*	3.69	2.1	3.2	1.2	97	0	0	0	0
442	*wing quarter, roast, meat only, weighed with bone*	1.99	0.8	1.2	0.5	38	0	0	0	0
443	*leg quarter, roast, meat only, weighed with bone*	2.46	1.0	1.6	0.6	48	0	0	0	0
444	*breaded, fried in vegetable oil*	2.88	2.1	5.3	4.7	36	14.0	0.8	N	0.7

Meat and meat products *continued*

Inorganic constituents per 100g

No.	Food	Na	K	Ca	Mg	P	Fe	Cu	Zn	Cl	Mn	Se	I
						mg						µg	
Veal													
428	**Cutlet**, *fried in vegetable oil*	110	420	10	33	280	1.6	N	N	120	N	N	N
429	**Fillet**, *raw*	110	360	8	25	260	1.2	0.06	2.8	68	0.01	(8)	N
430	*roast*	97	430	14	28	360	1.6	0.05	2.1	110	0.01	(8)	N
431	**Chicken**, *meat only, raw*	81	320	10	25	200	0.7	0.19	1.1	78	0.02	(7)	10
432	*meat and skin, raw*	70	260	10	20	160	0.7	0.16	1.0	69	0.02	(6)	10
433	*light meat, raw*	72	330	10	27	210[a]	0.5	0.14	0.7	70	0.02	(6)	10
434	*dark meat, raw*	89	300	11	22	180	0.9	0.25	1.6	86	0.02	(7)	10
435	*boiled, meat only*	82	300	11	25	190	1.2	0.20	2.0	90	0.02	(7)	10
436	*light meat*	70	370	9	26	200[a]	0.6	0.17	1.0	82	0.02	(6)	10
437	*dark meat*	95	230	12	22	180	1.9	0.23	3.1	99	0.02	(7)	10
438	*roast, meat only*	81	310	9	24	210	0.8	0.12	1.5	87	0.03	(7)	5
439	*meat and skin*	72	270	9	21	170	0.8	0.12	1.4	77	0.03	(6)	5
440	*light meat*	71	330	9	26	220[a]	0.5	0.11	1.0	79	0.03	(6)	5
441	*dark meat*	91	290	9	22	190	1.0	0.13	2.1	95	0.03	(7)	5
442	*wing quarter, roast, meat only, weighed with bone*	41	160	5	12	110	0.4	0.06	0.8	¯44	0.02	(7)	5
443	*leg quarter, roast, meat only, weighed with bone*	50	190	6	15	130	0.5	0.07	0.9	54	0.02	(7)	5
444	*breaded, fried in vegetable oil*	420	320	N	N	N	N	N	N	640	N	N	N

[a] In frozen chickens treated with polyphosphates the light meat may contain up to 250mg P per 100g

Meat and meat products *continued*

No.	Food	Retinol µg	Carotene µg	Vitamin D µg	Vitamin E mg	Thiamin mg	Ribo-flavin mg	Niacin mg	Trypt 60 mg	Vitamin B6 mg	Vitamin B12 µg	Folate µg	Panto-thenate mg	Biotin µg	Vitamin C mg
Veal															
428	**Cutlet**, *fried in vegetable oil*	Tr	Tr	Tr	N	N	N	N	6.7	N	1.0	N	N	Tr	0
429	**Fillet**, *raw*	Tr	Tr	Tr	N	0.10	0.25	7.0	4.5	0.30	1.0	5	0.60	Tr	0
430	*roast*	Tr	Tr	Tr	N	0.06	0.27	7.0	6.7	0.32	1.0	4	0.50	Tr	0
431	**Chicken**, *meat only, raw*	Tr	Tr	Tr	0.10	0.10	0.16	7.8	3.8	0.42	Tr	12	1.20	2	0
432	*meat and skin, raw*	Tr	Tr	Tr	N	0.08	0.14	6.0	3.3	0.30	Tr	8	0.90	2	0
433	*light meat, raw*	Tr	Tr	Tr	0.08	0.10	0.10	9.9	4.1	0.53	Tr	12	1.20	2	0
434	*dark meat, raw*	Tr	Tr	Tr	0.13	0.11	0.22	5.4	3.6	0.30	1.0	12	1.30	3	0
435	*boiled, meat only*	Tr	Tr	Tr	0.07	0.06	0.19	6.7	5.5	0.35	Tr	8	1.10	4	0
436	*light meat*	Tr	Tr	Tr	N	0.05	0.12	8.9	5.5	0.33	Tr	4	1.00	3	0
437	*dark meat*	Tr	Tr	Tr	N	0.07	0.28	4.3	5.3	0.37	1.0	13	1.10	4	0
438	*roast, meat only*	Tr	Tr	Tr	0.11	0.08	0.19	8.2	4.6	0.26	Tr	10	1.20	3	0
439	*meat and skin*	Tr	Tr	Tr	N	N	N	N	4.2	N	Tr	N	N	N	0
440	*light meat*	Tr	Tr	Tr	0.08	0.08	0.14	10.3	5.0	0.35	Tr	7	1.10	2	0
441	*dark meat*	Tr	Tr	Tr	0.15	0.09	0.24	6.1	4.3	0.16	1.0	13	1.30	3	0
442	*wing quarter, roast, meat only, weighed with bone*	Tr	Tr	Tr	0.06	0.04	0.10	4.1	2.3	0.13	Tr	5	0.60	2	0
443	*leg quarter, roast, meat only, weighed with bone*	Tr	Tr	Tr	0.07	0.05	0.12	5.1	2.9	0.16	Tr	6	0.70	2	0
444	*breaded, fried in vegetable oil*	Tr	N	Tr	N	0.12	0.06	8.2	3.4	0.53	Tr	N	1.20	2	0

Meat and meat products *continued*

Composition of food per 100g

No.	Food	Description and main data sources	Edible proportion	Water g	Protein g	Fat g	Carbo-hydrate g	Energy value kcal	kJ
445	**Duck**, *meat only, raw*	9 samples, meat from dressed carcase	0.28	75.0	19.7	4.8	0	122	513
446	*meat, fat and skin, raw*	9 samples, dressed carcase excluding waste	0.67	43.9	11.3	42.7	0	430	1772
447	*roast, meat only*	11 samples	1.00	64.2	25.3	9.7	0	189	789
448	*-, meat, fat and skin*	11 samples	1.00	49.6	19.6	29.0	0	339	1406
449	**Goose**, *roast, meat only*	Meat from carcase	1.00	46.7	29.3	22.4	0	319	1327
450	**Grouse**, *roast, meat only*	Meat from carcase	1.00	61.6	31.3	5.3	0	173	728
451	*roast, weighed with bone*	Calculated from roast, meat only	0.66	40.6	20.6	3.5	0	114	480
452	**Partridge**, *roast, meat only*	Meat from carcase	1.00	54.5	36.7	7.2	0	212	890
453	*roast, weighed with bone*	Calculated from roast, meat only	0.60	32.7	22.0	4.3	0	127	533
454	**Pheasant**, *roast, meat only*	Meat from carcase	1.00	56.9	32.2	9.3	0	213	892
455	*roast, weighed with bone*	Calculated from roast, meat only	0.63	35.8	20.3	5.9	0	134	563
456	**Pigeon**, *roast, meat only*	Meat from carcase	1.00	57.2	27.8	13.2	0	230	961
457	*roast, weighed with bone*	Calculated from roast, meat only	0.44	25.2	12.2	5.8	0	101	422

No.	Food	Total nitrogen g	Fatty acids Satd g	Mono unsatd g	Poly unsatd g	Cholest- erol mg	Starch g	Total sugars g	Dietary fibre Southgate method g	Englyst method g
445	**Duck**, *meat only, raw*	3.15	1.3	2.6	0.6	110	0	0	0	0
446	*meat, fat and skin, raw*	1.80	11.6	23.1	5.1	N	0	0	0	0
447	*roast, meat only*	4.05	2.7	5.3	1.2	160	0	0	0	0
448	*-, meat, fat and skin*	3.14	7.9	15.7	3.5	N	0	0	0	0
449	**Goose**, *roast, meat only*	4.69	N	N	N	N	0	0	0	0
450	**Grouse**, *roast, meat only*	5.00	1.2	0.7	3.1	N	0	0	0	0
451	*roast, weighed with bone*	3.30	0.8	0.4	2.1	N	0	0	0	0
452	**Partridge**, *roast, meat only*	5.87	1.9	3.3	1.7	N	0	0	0	0
453	*roast, weighed with bone*	3.52	1.1	2.0	1.0	N	0	0	0	0
454	**Pheasant**, *roast, meat only*	5.15	3.1	4.6	1.1	N	0	0	0	0
455	*roast, weighed with bone*	3.24	2.0	2.9	0.7	N	0	0	0	0
456	**Pigeon**, *roast, meat only*	4.44	N	N	N	N	0	0	0	0
457	*roast, weighed with bone*	1.95	N	N	N	N	0	0	0	0

Meat and meat products *continued*

No.	Food	Na	K	Ca	Mg	P	Fe	Cu	Zn	Cl	Mn	Se	I
						mg						µg	
445	**Duck**, *meat only, raw*	110	290	12	19	200	2.4	0.34	1.9	98	Tr	N	N
446	*meat, fat and skin, raw*	77	210	11	14	130	2.4	0.27	1.3	69	Tr	N	N
447	*roast, meat only*	96	270	13	20	200	2.7	0.31	2.6	96	Tr	N	N
448	*-, meat, fat and skin*	76	210	12	16	150	2.7	0.27	1.8	75	Tr	N	N
449	**Goose**, *roast, meat only*	150	410	10	31	270	4.6	0.15	2.6	160	0.01	N	N
450	**Grouse**, *roast, meat only*	96	470	30	41	340	7.6	0.10	1.5	130	0.06	N	N
451	*roast, weighed with bone*	63	310	20	27	220	5.0	0.05	0.7	88	0.03	N	N
452	**Partridge**, *roast, meat only*	100	410	46	36	310	7.7	N	N	99	N	N	N
453	*roast, weighed with bone*	60	240	28	22	190	4.6	N	N	59	N	N	N
454	**Pheasant**, *roast, meat only*	100	410	49	35	310	8.4	0.10	1.3	110	0.02	N	N
455	*roast, weighed with bone*	66	260	31	22	190	5.3	0.05	0.6	68	0.01	N	N
456	**Pigeon**, *roast, meat only*	110	410	16	34	400	19.4	0.33	1.7	99	0.05	N	N
457	*roast, weighed with bone*	46	180	7	15	180	8.5	0.15	0.7	44	0.02	N	N

Meat and meat products continued

No.	Food	Retinol µg	Carotene µg	Vitamin D µg	Vitamin E mg	Thiamin mg	Ribo-flavin mg	Niacin mg	Trypt 60 mg	Vitamin B6 mg	Vitamin B12 µg	Folate µg	Panto-thenate mg	Biotin µg	Vitamin C mg
445	**Duck**, meat only, raw	N	N	N	0.02	0.36	0.45	5.3	4.2	0.34	3	25	1.60	6	0
446	meat, fat and skin, raw	N	N	N	N	0.14	N	3.5	2.4	0.33	1.8	N	N	N	0
447	roast, meat only	N	N	N	0.02	0.26	0.47	5.1	5.4	0.25	3.0	10	1.50	4	0
448	-, meat, fat and skin	N	N	N	N	0.18	N	3.8	4.2	0.31	2.0	N	N	N	0
449	**Goose**, roast, meat only	N	N	N	N	0.12	0.51	4.6	5.5	0.42	1.9	12	1.40	3	0
450	**Grouse**, roast, meat only	N	N	N	N	0.19	0.80	7.0	5.8	0.64	0.9	37	N	N	0
451	roast, weighed with bone	N	N	N	N	0.10	0.41	3.6	3.9	0.33	0.5	19	N	N	0
452	**Partridge**, roast, meat only	N	N	N	N	N	N	N	6.9	N	N	N	N	N	0
453	roast, weighed with bone	N	N	N	N	N	N	N	4.1	N	N	N	N	N	0
454	**Pheasant**, roast, meat only	N	N	N	N	0.02	0.29	9.2	6.0	0.57	2.5	20	N	N	0
455	roast, weighed with bone	N	N	N	N	0.01	0.13	4.1	3.8	0.26	1.1	9	N	N	0
456	**Pigeon**, roast, meat only	N	N	N	Tr	0.27	N	7.0	5.2	0.82	N	8	N	N	0
457	roast, weighed with bone	N	N	N	Tr	0.12	N	3.1	2.3	0.36	N	3	N	N	0

Meat and meat products *continued*

Composition of food per 100g

No.	Food	Description and main data sources	Edible proportion	Water g	Protein g	Fat g	Carbohydrate g	Energy value kcal	kJ
458	**Turkey,** *meat only, raw*	5 samples, light and dark meat from dressed carcase	0.57	75.5	21.9	2.2	0	107	454
459	*meat and skin, raw*	5 samples, dressed carcase excluding waste	0.70	72.0	20.6	6.9	0	145	606
460	*light meat, raw*	5 samples	0.32	75.2	23.2	1.1	0	103	435
461	*dark meat, raw*	5 samples	0.25	75.9	20.3	3.6	0	114	478
462	*roast, meat only*	5 samples, light and dark meat	1.00	68.0	28.8	2.7	0	140	590
463	*meat and skin*	5 samples	1.00	65.0	28.0	6.5	0	171	717
464	*light meat*	5 samples	1.00	68.4	29.8	1.4	0	132	558
465	*dark meat*	5 samples	1.00	67.7	27.8	4.1	0	148	624
466	**Hare,** *stewed, meat only*	Meat from carcase	1.00	60.7	29.9	8.0	0	192	804
467	*stewed, weighed with bone*	Calculated from stewed, *meat only*	0.73	44.3	21.8	5.8	0	139	585
468	**Rabbit,** *meat only, raw*	9 samples, pieces of loin and leg	0.62	74.6	21.9	4.0	0	124	520
469	*stewed, meat only*	Pieces of loin and leg	1.00	63.9	27.3	7.7	0	179	749
470	*-, weighed with bone*	Calculated from stewed, *meat only*	0.64	32.5	13.9	3.9	0	91	381
471	**Venison,** *roast*	Haunch, meat only	1.00	56.8	35.0	6.4	0	198	832

Meat and meat products *continued*

Composition of food per 100g

No.	Food	Total nitrogen g	Fatty acids Satd g	Fatty acids Mono unsatd g	Fatty acids Poly unsatd g	Cholest- erol mg	Starch g	Total sugars g	Dietary fibre Southgate method g	Dietary fibre Englyst method g
458	**Turkey**, *meat only, raw*	3.51	0.7	0.9	0.4	61	0	0	0	0
459	*meat and skin, raw*	3.29	2.2	2.9	1.3	(191)	0	0	0	0
460	*light meat, raw*	3.71	0.3	0.5	0.2	49	0	0	0	0
461	*dark meat, raw*	3.24	1.2	1.5	0.7	81	0	0	0	0
462	*roast, meat only*	4.61	0.9	1.1	0.5	79	0	0	0	0
463	*meat and skin*	4.48	2.1	2.7	1.3	(191)	0	0	0	0
464	*light meat*	4.76	0.4	0.6	0.3	62	0	0	0	0
465	*dark meat*	4.44	1.3	1.7	0.8	100	0	0	0	0
466	**Hare**, *stewed, meat only*	4.78	N	N	N	N	0	0	0	0
467	*stewed, weighed with bone*	3.48	N	N	N	N	0	0	0	0
468	**Rabbit**, *meat only , raw*	3.50	1.6	0.8	1.3	71	0	0	0	0
469	*stewed, meat only*	4.37	3.2	1.5	2.5	137	0	0	0	0
470	*-, weighed with bone*	2.23	1.6	0.8	1.3	69	0	0	0	0
471	**Venison**, *roast*	5.60	N	N	N	N	0	0	0	0

Meat and meat products *continued*

Inorganic constituents per 100g

No.	Food	Na	K	Ca	Mg	P	Fe	Cu	Zn	Cl	Mn	Se	I
						mg						µg	
458	**Turkey**, *meat only, raw*	54	300	8	23	190	0.8	0.13	1.7	48	0.01	N	N
459	*meat and skin, raw*	49	270	9	19	170	0.8	0.12	1.6	43	0.01	N	N
460	*light meat, raw*	43	320	6	25	200	0.5	0.11	1.2	42	0.01	N	N
461	*dark meat, raw*	68	270	11	21	180	1.2	0.16	2.4	55	0.01	N	N
462	*roast, meat only*	57	310	9	27	220	0.9	0.15	2.4	52	0.01	N	N
463	*meat and skin*	52	280	9	24	200	0.9	0.14	2.1	47	0.01	N	N
464	*light meat*	45	340	7	29	230	0.5	0.14	1.5	42	0.01	N	N
465	*dark meat*	71	270	12	23	210	1.4	0.16	3.5	62	0.01	N	N
466	**Hare**, *stewed, meat only*	40	210	21	22	250	10.8	N	N	74	N	N	N
467	*stewed, weighed with bone*	29	150	15	16	180	7.9	N	N	54	N	N	N
468	**Rabbit**, *meat only , raw*	67	360	22	25	220	1.0	0.06	1.4	74	0.01	17	N
469	*stewed, meat only*	32	210	11	22	200	1.9	0.06	1.7	43	0.02	21	N
470	*-, weighed with bone*	16	110	6	11	100	1.0	0.03	0.9	22	0.01	11	N
471	**Venison**, *roast*	86	360	29	33	290	7.8	0.36	3.9	89	0.04	N	N

Meat and meat products *continued*

No.	Food	Retinol µg	Carotene µg	Vitamin D µg	Vitamin E mg	Thiamin mg	Ribo-flavin mg	Niacin mg	Trypt 60 mg	Vitamin B6 mg	Vitamin B12 µg	Folate µg	Panto-thenate mg	Biotin µg	Vitamin C mg
458	**Turkey**, *meat only, raw*	Tr	Tr	Tr	Tr	0.09	0.16	7.9	4.1	0.46	2.0	15	0.80	2	0
459	*meat and skin, raw*	N	N	N	N	N	N	N	3.8	N	N	N	N	N	0
460	*light meat, raw*	Tr	Tr	Tr	Tr	0.08	0.11	9.9	4.3	0.59	1.0	8	0.80	10	0
461	*dark meat, raw*	Tr	Tr	Tr	Tr	0.10	0.23	5.2	3.8	0.30	3.0	25	0.90	2	0
462	*roast, meat only*	Tr	Tr	Tr	Tr	0.07	0.21	8.5	5.4	0.32	2.0	15	0.80	2	0
463	*meat and skin*	N	N	N	N	N	N	N	5.2	N	N	N	N	N	0
464	*light meat*	Tr	Tr	Tr	Tr	0.07	0.14	10.0	5.6	0.31	1.0	13	0.70	1	0
465	*dark meat*	Tr	Tr	Tr	Tr	0.07	0.29	6.7	5.2	0.32	3.0	17	0.90	2	0
466	**Hare**, *stewed, meat only*	N	N	N	N	N	N	N	5.6	N	N	N	N	N	0
467	*stewed, weighed with bone*	N	N	N	N	N	N	N	4.1	N	N	N	N	N	0
468	**Rabbit**, *meat only , raw*	N	N	N	0.13	0.10	0.19	8.4	4.1	0.50	10.0	5	0.80	1	0
469	*stewed, meat only*	N	N	N	N	0.07	0.28	8.5	5.1	0.50	12.0	4	0.80	1	0
470	*-, weighed with bone*	N	N	N	N	0.04	0.14	4.3	2.6	0.26	6.0	2	0.40	Tr	0
471	**Venison**, *roast*	N	N	N	N	0.22	0.69	5.5	6.5	0.65	0.8	6	N	N	0

Meat and meat products *continued*

Composition of food per 100g

Offal

No.	Food	Description and main data sources	Edible proportion	Water g	Protein g	Fat g	Carbo-hydrate g	Energy value kcal	kJ
472	**Heart**, lamb, raw	12 samples; fat and valves removed	0.73	75.6	17.1	5.6	0	119	498
473	ox, *raw*	18 samples; fat and valves removed	0.81	76.3	18.9	3.6	0	108	455
474	*stewed*	18 samples; fat and valves removed before cooking	1.00	61.5	31.4	5.9	0	179	752
475	pig, *raw*	Fat removed	N	79.2	17.1	2.7	0	93	391
476	sheep, *roast*	Ventricles only	1.00	57.3	26.1	14.7	0	237	988
477	**Kidney**, lamb, *raw*	19 samples; core removed	0.93	78.9	16.5	2.7	0	90	380
478	*fried*	19 samples; core removed before cooking	1.00	66.5	24.6	6.3	0	155	651
479	ox, *raw*	18 samples; core removed	0.82	79.8	15.7	2.6	0	86	363
480	*stewed*	18 samples; core removed before cooking, salt added	1.00	64.1	25.6	7.7	0	172	720
481	pig, *raw*	20 samples; core removed	0.90	78.8	16.3	2.7	0	90	377
482	*stewed*	20 samples; core removed before cooking, salt added	1.00	66.3	24.4	6.1	0	153	641
483	**Liver**, calf, *raw*	12 samples	1.00	69.7	20.1	7.3	1.9[a]	153	642
484	*fried*	12 samples; coated in seasoned flour and fried	1.00	52.6	26.9	13.2	7.3[b]	254	1063
485	chicken, *raw*	16 samples	1.00	72.9	19.1	6.3	0.6[a]	135	567
486	*fried*	16 samples; coated in seasoned flour and fried	1.00	64.2	20.7	10.9	3.4[b]	194	810
487	lamb, *raw*	33 samples	1.00	67.3	20.1	10.3	1.6[a]	179	748
488	*fried*	18 samples; coated in seasoned flour and fried	1.00	58.4	22.9	14.0	3.9[b]	232	970
489	ox, *raw*	33 samples	1.00	68.6	21.1	7.8	2.2[a]	163	683
490	*stewed*	18 samples; coated in seasoned flour	1.00	62.6	24.8	9.5	3.6[b]	198	831
491	pig, *raw*	33 samples	1.00	69.5	21.3	6.8	2.1[a]	154	647
492	*stewed*	18 samples; coated in seasoned flour	1.00	62.1	25.6	8.1	3.6[b]	189	793

[a] As glycogen   [b] Including glycogen

Meat and meat products *continued*

Composition of food per 100g

No.	Food	Total nitrogen g	Fatty acids Satd g	Mono unsatd g	Poly unsatd g	Cholesterol mg	Starch g	Total sugars g	Dietary fibre Southgate method g	Englyst method g
Offal										
472	**Heart**, lamb, raw	2.73	2.1	1.7	0.5	140	0	0	0	0
473	ox, raw	3.03	1.7	0.9	0.1	140	0	0	0	0
474	stewed	5.02	2.9	1.6	0.2	230	0	0	0	0
475	pig, raw	2.74	N	N	N	N	0	0	0	0
476	sheep, roast	4.18	N	N	N	260	0	0	0	0
477	**Kidney**, lamb, raw	2.64	0.9	0.6	0.4	400	0	0	0	0
478	fried	3.94	2.1	1.5	0.9	610	0	0	0	0
479	ox, raw	2.51	1.1	0.7	0.1	400	0	0	0	0
480	stewed	4.09	3.2	2.0	0.5	690	0	0	0	0
481	pig, raw	2.61	0.9	0.7	0.4	410	0	0	0	0
482	stewed	3.91	2.0	1.6	0.9	700	0	0	0	0
483	**Liver**, calf, raw	3.21	2.2	1.3	1.9	370	1.9[a]	0	0	0
484	fried	4.31	4.0	2.3	3.4	330	7.3[b]	Tr	0.2	0.2
485	chicken, raw	3.05	2.0	1.4	1.3	380	0.6[a]	0	0	0
486	fried	3.31	3.4	2.4	2.2	350	3.4[b]	Tr	0.2	0.2
487	lamb, raw	3.22	2.9	3.0	1.5	430	1.6[a]	0	0	0
488	fried	3.67	4.0	4.1	2.1	400	3.9[b]	Tr	0.1	0.1
489	ox, raw	3.37	2.9	1.3	1.6	270	2.2[a]	0	0	0
490	stewed	3.96	3.5	1.5	2.0	240	3.6[b]	Tr	0	Tr
491	pig, raw	3.41	2.1	1.0	1.8	260	2.1[a]	0	0	0
492	stewed	4.09	2.5	1.3	2.2	290	3.6[b]	0	Tr	Tr

[a] As glycogen   [b] Including glycogen

Meat and meat products *continued*

Inorganic constituents per 100g

No.	Food	Na	K	Ca	Mg	P	Fe	Cu	Zn	Cl	Mn	Se	I
		mg										µg	
Offal													
472	**Heart**, lamb, *raw*	140	280	7	21	210	3.6	0.52	2.0	140	0.02	(2)	N
473	ox, *raw*	95	320	5	25	230	4.9	0.43	2.0	95	0.04	(3)	N
474	*stewed*	180	210	7	29	270	7.7	0.73	3.5	210	0.06	(3)	N
475	pig, *raw*	80	300	6	20	220	4.8	0.37	1.8	110	0.02	(5)	N
476	sheep, *roast*	150	370	10	35	390	8.1	N	N	130	N	N	N
477	**Kidney**, lamb, *raw*	220	270	10	17	260	7.4	0.42	2.4	270	0.16	93	N
478	*fried*	270	340	13	29	360	12.0	(0.65)	4.1	330	0.13	N	15
479	ox, *raw*	180	230	10	15	230	5.7	0.42	1.9	200	0.11	110	N
480	*stewed*	400	180	16	19	300	8.0	0.66	3.0	520	0.14	N	7
481	pig, *raw*	190	290	8	19	270	5.0	0.81	2.6	180	0.11	250	N
482	*stewed*	370	190	13	21	330	6.4	0.84	4.7	480	0.18	N	N
483	**Liver**, calf, *raw*	93	330	7	20	360	8.0	11.00	7.8	89	0.24	(22)	N
484	*fried*	170	410	15	26	470	7.5	12.00	6.2	210	0.29	N	N
485	chicken, *raw*	85	300	8	21	320	9.5	0.52	3.4	100	0.31	N	N
486	*fried*	240	290	15	23	350	9.1	0.53	3.4	350	0.35	N	5
487	lamb, *raw*	76	290	7	19	370	9.4	8.70	3.9	83	0.32	20	N
488	*fried*	190	300	12	22	400	10.0	9.90	4.4	250	0.45	N	13
489	ox, *raw*	81	320	6	19	360	7.0	2.50	4.0	90	0.37	14	N
490	*stewed*	110	250	11	19	380	7.8	2.30	4.3	120	0.44	N	N
491	pig, *raw*	87	320	6	21	370	21.0	2.70	6.9	95	0.32	(28)	N
492	*stewed*	130	250	11	22	390	17.0	2.50	8.2	150	0.40	N	N

Offal

No.	Food	Retinol µg	Carotene µg	Vitamin D µg	Vitamin E mg	Thiamin mg	Ribo-flavin mg	Niacin mg	Trypt 60 mg	Vitamin B6 mg	Vitamin B12 µg	Folate µg	Panto-thenate mg	Biotin µg	Vitamin C mg
472	**Heart**, lamb, *raw*	Tr	Tr	N	0.37	0.48	0.90	6.9	3.6	0.29	8.0	2	2.50	4	7
473	ox, *raw*	Tr	Tr	N	0.45	0.45	0.80	6.3	4.0	0.23	13.0	4	2.40	2	7
474	*stewed*	Tr	Tr	N	0.72	0.21	1.10	4.7	6.7	0.11	15.0	2	1.60	4	6
475	pig, *raw*	Tr	Tr	N	(0.37)	(0.48)	(0.90)	(6.9)	3.7	(0.29)	(8.0)	(2)	(2.50)	(4)	5
476	sheep, *roast*	Tr	Tr	N	(0.70)	(0.45)	(1.50)	(9.1)	5.6	(0.38)	(14)	(4)	(3.80)	(8)	(11)
477	**Kidney**, lamb, *raw*	105	0	N	0.45	0.49	1.80	8.3	3.5	0.30	55.0	31	4.30	37	7
478	*fried*	110	0	N	0.41	0.56	2.30	9.6	5.3	0.30	79.0	79	5.10	42	9
479	ox, *raw*	96	0	N	0.18	0.37	2.10	6.0	3.4	0.32	31.0	77	3.10	24	10
480	*stewed*	45	0	N	0.42	0.25	2.10	4.8	5.5	0.30	31.0	75	3.00	49	10
481	pig, *raw*	160	0	N	0.38	0.32	1.90	7.5	3.5	0.25	14.0	42	3.00	32	14
482	*stewed*	46	0	N	0.36	0.19	2.10	6.1	5.2	0.28	15.0	43	2.40	53	11
483	**Liver**, calf, *raw*	29730	100	0.3	0.24	0.21	3.10	12.4	4.3	0.54	100.0	240	8.40	39	18
484	*fried*	39780	100	0.3	0.50	0.27	4.20	15.6	5.8	0.73	87.0	320	8.80	53	13
485	chicken, *raw*	11325	0	0.2	0.25	0.36	2.70	10.2	4.1	0.40	56.0	590	6.10	210	23
486	*fried*	12230	0	N	0.34	0.37	1.70	10.5	4.4	0.45	49.0	500	5.50	170	13
487	lamb, *raw*	19895	60	0.5	0.46	0.27	3.30	14.2	4.3	0.42	84.0	220	8.20	41	10
488	*fried*	22680	60	0.5	0.32	0.26	4.40	15.2	4.9	0.49	81.0	240	7.60	41	12
489	ox, *raw*	16500	1540	1.1	0.42	0.23	3.10	13.4	4.5	0.83	110.0	330	8.10	33	23
490	*stewed*	20100	1540	1.1	0.44	0.18	3.60	10.3	5.3	0.52	110.0	290	5.70	50	15
491	pig, *raw*	17595	0	1.1	0.17	0.31	3.00	14.8	4.6	0.68	25.0	110	6.50	27	13
492	*stewed*	22820	0	1.1	0.16	0.21	3.10	11.5	5.5	0.64	26.0	110	4.60	34	9

Composition of food per 100g

No.	Food	Description and main data sources	Edible proportion	Water g	Protein g	Fat g	Carbo-hydrate g	Energy value kcal	kJ
Offal									
493	**Oxtail**, *stewed*	12 samples, lean only; salt added	1.00	53.9	30.5	13.4	0	243	1014
494	*stewed, weighed with fat and bones*	Calculated from oxtail stewed	0.37	20.5	11.6	5.1	0	92	386
495	**Sweetbread**, lamb, *raw*	12 samples	1.00	75.5	15.3	7.8	0	131	549
496	*fried*	12 samples; soaked 2 hours, boiled for 1 hour then coated with egg and breadcrumbs and fried	1.00	59.9	19.4	14.6	5.6	230	960
497	**Tongue**, lamb, *raw*	20 samples, fat and skin removed	0.57	67.9	15.3	14.6	0	193	800
498	ox, pickled, *raw*	6 samples, fat and skin removed	0.60	62.4	15.7	17.5	0	220	914
499	*boiled*	Fat and skin removed	1.00	48.6	19.5	23.9	0	293	1216
500	sheep, *stewed*	Fat and skin removed	1.00	56.9	18.2	24.0	0	289	1197
501	**Tripe**, *dressed*	18 samples; lime treated before purchase	1.00	88.1	9.4	2.5	0	60	252
502	*dressed, stewed*	18 samples; lime treated before purchase, stewed in milk	1.00	78.5	14.8	4.5	0	100	418
Meat products									
503	**Beefburgers**, *frozen, raw*	36 samples, 6 brands	1.00	56.3	15.2	20.5	5.3	265	1102
504	*frozen, fried*	36 samples, 6 brands	1.00	53.0	20.4	17.3	7.0	264	1099
505	**Black pudding**, *fried*	24 samples	1.00	44.0	12.9	21.9	15.0	305	1270
506	**Brawn**	10 samples	1.00	72.0	12.4	11.5	0	153	636
507	**Corned beef**, canned	18 samples	1.00	58.5	26.9	12.1	0	217	905
508	**Cornish pastie**	18 pasties, average 62% pastry, 38% filling	1.00	39.2	8.0	20.4	31.1	332	1388

Meat and meat products *continued*

Composition of food per 100g

No.	Food	Total nitrogen g	Fatty acids Satd g	Mono unsatd g	Poly unsatd g	Cholest-erol mg	Starch g	Total sugars g	Dietary fibre Southgate method g	Englyst method g
Offal										
493	**Oxtail**, *stewed*	4.88	N	N	N	110	0	0	0	0
494	*stewed, weighed with fat and bones*	1.85	N	N	N	N	0	0	0	0
495	**Sweetbread**, lamb, *raw*	2.44	3.0	2.4	0.3	260	0	0	0	0
496	*fried*	3.10	5.5	4.4	0.6	380	(5.4)	(0.2)	(0.4)	(0.1)
497	**Tongue**, lamb, *raw*	2.45	N	N	N	180	0	0	0	0
498	ox, pickled, *raw*	2.51	N	N	N	78	0	0	0	0
499	*boiled*	3.12	N	N	N	(100)	0	0	0	0
500	sheep, *stewed*	2.91	N	N	N	(270)	0	0	0	0
501	**Tripe**, *dressed*	1.50	1.1	0.7	Tr	95	0	0	0	0
502	*dressed, stewed*	2.37	2.4	1.3	0.1	160	0	Tr	0	0
Meat products										
503	**Beefburgers**, *frozen, raw*	2.43	9.5	9.2	0.8	96	4.2	1.1	1.3	N
504	*frozen, fried*	3.27	8.0	7.8	0.7	81	5.6	1.4	1.3	N
505	**Black pudding**, *fried*	2.06	(8.5)	(8.1)	(3.6)	68	(15.0)	Tr	0.5	N
506	**Brawn**	1.99	N	N	N	52	0	0	0	0
507	**Corned beef**, canned	4.30	6.3	4.8	0.3	93	0	0	0	0
508	**Cornish pastie**	1.28	(7.4)	(10.5)	(1.5)	49	(29.9)	(1.2)	1.2	0.9

Meat and meat products *continued*

No.	Food	mg										µg	
		Na	K	Ca	Mg	P	Fe	Cu	Zn	Cl	Mn	Se	I
Offal													
493	**Oxtail**, *stewed*	190	170	14	18	140	3.8	0.27	8.8	270	N	N	N
494	*stewed, weighed with fat and bones*	72	65	5	7	53	1.4	0.10	3.3	100	N	N	N
495	**Sweetbread**, lamb, *raw*	75	420	8	21	400	1.7	0.20	1.9	120	0.04	N	N
496	*fried*	210	260	34	23	420	1.8	0.22	2.1	260	(0.08)	N	N
497	**Tongue**, lamb, *raw*	420	250	6	33	170	2.2	0.64	2.7	550	N	N	N
498	*ox, pickled, raw*	1210	300	7	19	150	4.9	0.37	3.5	1750	0.01	N	N
499	*boiled*	1000	150	31	16	230	3.0	N	N	1450	0.01	N	N
500	*sheep, stewed*	80	110	11	13	200	3.4	N	N	80	N	N	N
501	**Tripe**, *dressed*	46	8	75	8	37	0.5	0.09	1.5	8	N	N	N
502	*dressed, stewed*	73	100	150	15	90	0.7	0.14	2.3	58	N	N	N
Meat products													
503	**Beefburgers**, *frozen, raw*	600	270	23	17	190	2.5	0.25	3.2	800	0.09	N	Tr
504	*frozen, fried*	880	340	33	23	250	3.1	0.28	4.2	1120	0.11	N	Tr
505	**Black pudding**, *fried*	1210	140	35	16	110	20.0	0.37	1.3	1770	N	N	N
506	**Brawn**	750	85	38	7	59	1.0	0.19	1.3	1110	N	N	N
507	**Corned beef**, canned	950	140	14	15	120	2.9	0.24	5.6	1430	0.02	(8)	14
508	**Cornish pastie**	590	190	60	18	110	1.5	0.35	1.0	860	0.20	N	3

No.	Food	Retinol µg	Carotene µg	Vitamin D µg	Vitamin E mg	Thiamin mg	Ribo-flavin mg	Niacin mg	Trypt 60 mg	Vitamin B₆ mg	Vitamin B₁₂ µg	Folate µg	Panto-thenate mg	Biotin µg	Vitamin C mg
Offal															
493	**Oxtail**, *stewed*	Tr	Tr	Tr	0.45	0.02	0.28	3.3	6.5	0.14	2.0	9	0.90	2	0
494	*stewed, weighed with fat and bones*	Tr	Tr	Tr	0.17	0.01	0.11	1.3	2.5	0.05	1.0	3	0.30	1	0
495	**Sweetbread**, lamb, raw	Tr	Tr	Tr	0.44	0.03	0.25	3.7	3.3	0.03	6.0	13	1.00	3	18
496	*fried*	Tr	Tr	Tr	1.20	0.03	0.24	2.1	4.1	0.02	4.0	14	0.80	5	18
497	**Tongue**, lamb, raw	Tr	Tr	Tr	0.21	0.17	0.49	4.9	3.3	0.17	7.0	4	1.00	1	7
498	ox, pickled, raw	Tr	Tr	Tr	0.28	0.10	0.38	6.4	3.4	0.18	5.0	6	0.80	2	3
499	boiled	Tr	Tr	Tr	(0.35)	(0.06)	(0.29)	(4.1)	4.2	(0.09)	(4.0)	(5)	(0.50)	(3)	(2)
500	sheep, stewed	Tr	Tr	Tr	(0.32)	(0.13)	(0.45)	(3.7)	3.9	(0.10)	(7.0)	(4)	(0.80)	(2)	(6)
501	**Tripe**, dressed	Tr	Tr	Tr	0.08	Tr	0.01	0.1	2.0	Tr	Tr	2	Tr	Tr	3
502	dressed, stewed	Tr	Tr	Tr	0.09	Tr	0.08	Tr	3.2	0.02	Tr	1	0.20	2	3
Meat products															
503	**Beefburgers**, *frozen, raw*	Tr	Tr	Tr	0.27	0.04	0.21	3.7	2.8	0.20	1.0	12	0.40	1	0
504	*frozen, fried*	Tr	Tr	Tr	0.58	0.02	0.23	4.2	3.8	0.20	2.0	15	0.50	2	0
505	**Black pudding**, *fried*	41	Tr	Tr	0.24	0.09	0.07	1.0	2.8	0.04	1.0	5	0.60	2	0
506	**Brawn**	Tr	Tr	Tr	0.06	0.05	0.08	1.0	2.3	0.05	Tr	3	0.90	Tr	0
507	**Corned beef**, canned	Tr	Tr	Tr	0.78	Tr	0.23	2.5	6.5	0.06	2.0	2	0.40	2	0
508	**Cornish pastie**	Tr	Tr	Tr	1.30	0.10	0.06	1.6	1.7	0.12	1.0	3	0.60	1	0

Meat and meat products *continued*

Composition of food per 100g

No.	Food	Description and main data sources	Edible proportion	Water g	Protein g	Fat g	Carbo-hydrate g	Energy value kcal	Energy value kJ
	Meat products								
509	**Faggots**	38 samples	1.00	47.1	11.1	18.5	15.3	268	1118
510	**Frankfurters**	12 samples, cans and packets, 6 brands	1.00	59.5	9.5	25.0	3.0	274	1135
511	**Grillsteaks**, *grilled*	10 samples; beef based; grilled 15-20 minutes	1.00	50.1	22.1	23.9	0.5	305	1268
512	**Haggis**, *boiled*	8 samples	1.00	46.2	10.7	21.7	19.2	310	1292
513	**Ham and pork**, chopped, canned	12 samples, 5 brands	1.00	58.5	14.4	23.6	1.4	275	1140
514	**Liver sausage**	24 samples	1.00	51.8	12.9	26.9	4.3	310	1283
515	**Luncheon meat**, canned	18 samples	1.00	51.5	12.6	26.9	5.5	313	1298
516	**Meat paste**	67 samples, beef, chicken, ham and tongue, liver and bacon	1.00	67.1	15.2	11.2	3.0	173	721
517	**Pate**, *liver*	20 samples, assorted types	1.00	50.6	13.1	28.9	1.0	316	1308
518	*low fat*	11 samples, assorted types; pork meat and liver based	1.00	65.0	18.0	12.0	2.8	191	795
519	**Polony**	24 samples	1.00	52.0	9.4	21.1	14.2	281	1168
520	**Pork pie**, *individual*	18 pies, average 55% pastry, 42% meat filling, 3% jelly	1.00	36.8	9.8	27.0	24.9	376	1564
521	**Salami**	24 samples, 8 different countries of origin	1.00	28.0	19.3	45.2	1.9	491	2031
522	**Sausage roll**, *flaky pastry*	Recipe	1.00	23.6	7.1	36.4	32.3	477	1985
523	*short pastry*	Recipe	1.00	22.3	8.0	31.9	37.5	459	1915

Meat and meat products *continued*

No.	Food	Total nitrogen g	Fatty acids			Cholesterol mg	Starch g	Total sugars g	Dietary fibre	
			Satd g	Mono unsatd g	Poly unsatd g				Southgate method g	Englyst method g
Meat products										
509	**Faggots**	1.78	N	N	N	79	(15.3)	Tr	0.5	N
510	**Frankfurters**	1.52	N	N	N	46	(3.0)	Tr	0.1	0.1
511	**Grillsteaks**, *grilled*	3.50	11.0	11.0	0.8	88	Tr	0.5	Tr	Tr
512	**Haggis**, *boiled*	1.71	7.6	6.9	1.4	91	(19.2)	Tr	N	N
513	**Ham and pork**, chopped, canned	2.30	8.8	11.2	2.3	60	1.2	0.2	0.3	0.3
514	**Liver sausage**	2.06	7.9	8.5	3.5	184	3.5	0.8	0.5	0.5
515	**Luncheon meat**, canned	2.02	9.8	12.4	3.3	73	5.5	Tr	0.4	0.3
516	**Meat paste**	2.43	N	N	N	68	(3.0)	Tr	0.1	0.1
517	**Pate**, liver	2.10	8.4	10.1	2.7	169	0.7	0.3	Tr	Tr
518	low fat	2.88	4.0	4.5	1.7	159	1.7	1.1	Tr	Tr
519	**Polony**	1.50	N	N	N	40	(14.2)	Tr	0.5	N
520	**Pork pie**, *individual*	1.56	10.2	12.5	2.7	52	(24.4)	(0.5)	0.9	0.9
521	**Salami**	3.09	N	N	N	79	1.9	Tr	0.1	0.1
522	**Sausage roll**, *flaky pastry*	1.18	13.4	15.6	5.3	49	31.1	1.2	1.5	1.2
523	*short pastry*	1.33	11.8	13.7	4.5	43	36.2	1.3	1.8	1.4

Meat and meat products *continued*

Inorganic constituents per 100g

No.	Food	Na	K	Ca	Mg	P	Fe	Cu	Zn	Cl	Mn	Se	I
						mg						μg	
	Meat products												
509	**Faggots**	820	170	55	18	120	8.3	0.60	1.6	1160	N	N	N
510	**Frankfurters**	980	98	34	9	130	1.5	0.24	1.4	1280	N	(8)	N
511	**Grillsteaks**, *grilled*	650	320	18	19	190	2.4	0.10	4.7	N	Tr	(3)	N
512	**Haggis**, *boiled*	770	170	29	36	160	4.8	0.44	1.9	1200	N	N	N
513	**Ham and pork**, chopped, canned	1090	230	14	13	250	1.2	0.24	2.9	1210	0.02	N	20
514	**Liver sausage**	860	170	26	12	230	6.4	0.63	2.3	1140	0.19	N	N[a]
515	**Luncheon meat**, canned	1050	140	15	8	200	1.1	0.33	2.2	1290	0.05	(7)	N[a]
516	**Meat paste**	740	160	86	15	170	2.3	0.31	2.3	1060	N	N	33
517	**Pate**, liver	790	160	15	11	490	7.1	0.59	2.9	950	0.10	N	N
518	low fat	710	190	14	14	230	6.3	0.46	2.7	N	0.20	N	N[a]
519	**Polony**	870	120	42	13	130	1.3	0.32	1.2	1160	N	N	7
520	**Pork pie**, *individual*	720	150	47	16	120	1.4	0.32	1.0	1030	N	N	15[a]
521	**Salami**	1850	160	10	10	160	1.0	0.24	1.7	2460	N	(7)	N
522	**Sausage roll**, *flaky pastry*	510	110	66	12	96	1.2	0.16	0.7	770	0.28	1	8
523	*short pastry*	530	120	76	14	110	1.3	0.17	0.7	800	0.3	2	8

[a] Iodine from erythrosine is present but largely unavailable

Meat and meat products *continued*

Meat products

No.	Food	Retinol µg	Carotene µg	Vitamin D µg	Vitamin E mg	Thiamin mg	Ribo-flavin mg	Niacin mg	Trypt 60 mg	Vitamin B6 mg	Vitamin B12 µg	Folate µg	Panto-thenate mg	Biotin µg	Vitamin C mg
509	**Faggots**	460	60	(0.2)	N	0.14	0.49	3.0	2.1	0.17	5.0	22	1.10	4	Tr
510	**Frankfurters**	Tr	Tr	Tr	0.25	0.08	0.12	1.5	1.5	0.03	1.0	1	0.40	2	0
511	**Grillsteaks**, *grilled*	Tr	Tr	Tr	0.15	0.13	0.14	4.2	4.7	0.18	2.5	N	0.63	2	Tr
512	**Haggis**, *boiled*	(1800)	Tr	(0.1)	0.41	0.16	0.35	1.5	2.0	0.07	2.0	8	0.50	12	Tr
513	**Ham and pork**, chopped, canned	Tr	Tr	Tr	0.11	0.19	0.21	3.2	2.7	0.05	1.0	1	0.40	2	0
514	**Liver sausage**	2605	Tr	(0.6)	0.10	0.17	1.58	4.3	2.4	0.14	8.0	19	1.50	7	Tr
515	**Luncheon meat**, canned	Tr	Tr	Tr	0.11	0.07	0.12	1.8	2.7	0.02	1.0	1	0.50	Tr	0[a]
516	**Meat paste**	Tr	Tr	Tr	0.17	0.03	0.26	3.8	2.8	0.08	3.0	9	0.30	3	0
517	**Pate**, liver	7330	130	N	N	0.13	1.10	2.9	2.8	0.25	7.2	89	2.15	10	N
518	low fat	5925	Tr	N	0.77	0.46	1.12	7.1	3.6	0.35	N	N	2.68	27	18
519	**Polony**	Tr	Tr	Tr	0.09	0.17	0.10	1.5	1.8	0.08	Tr	4	0.50	Tr	N
520	**Pork pie**, *individual*	Tr	Tr	Tr	0.43	0.16	0.09	1.8	2.1	0.06	1.0	3	0.60	1	0
521	**Salami**	Tr	Tr	Tr	0.28	0.21	0.23	4.6	3.6	0.15	1.0	3	0.80	3	N
522	**Sausage roll**, *flaky pastry*	110	105	1.1	1.31	0.10	0.04	1.7	1.5	0.06	0.3	4	0.24	1	Tr
523	*short pastry*	85	80	0.9	1.09	0.11	0.05	1.8	1.7	0.07	0.4	4	0.26	1	0

[a] Some brands have ascorbic acid added and may contain from 12 to 60mg per 100g

Meat and meat products continued

Composition of food per 100g

No.	Food	Description and main data sources	Edible proportion	Water g	Protein g	Fat g	Carbo-hydrate g	Energy value kcal	kJ
Meat products									
524	**Sausages**, beef, *raw*	20 samples	1.00	50.3	9.6	24.1	11.7	299	1242
525	*fried*	20 samples	1.00	47.7	12.9	18.0	14.9	269	1124
526	*grilled*	20 samples	1.00	47.9	13.0	17.3	15.2	265	1104
527	pork, *raw*	18 samples	1.00	45.4	10.6	32.1	9.5	367	1520
528	*fried*	18 samples	1.00	44.9	13.8	24.5	11.0	317	1317
529	*grilled*	18 samples	1.00	45.1	13.3	24.6	11.5	318	1320
530	low fat, *raw*	7 samples, 5 brands; pork sausages	1.00	61.9	12.5	9.5	8.1	166	694
531	*fried*	Samples as raw; fried 10-15 minutes	1.00	53.9	14.9	13.0	9.1	211	880
532	*grilled*	Samples as raw; grilled 10-15 minutes	1.00	50.1	16.2	13.8	10.8	229	959
533	**Saveloy**	60 samples	1.00	56.7	9.9	20.5	10.1	262	1088
534	**Steak and kidney pie**, *individual*	10 pies, purchased cooked; pastry top and bottom	1.00	42.6	9.1	21.2	25.6	323	1349
535	pastry top only	Recipe	1.00	49.0	15.2	18.4	15.9	286	1194
536	**Stewed steak**, canned, *with gravy*	12 samples, 8 brands	1.00	70.0	14.8	12.5	1.0	176	730
537	**Tongue**, canned	18 samples, lamb and ox	1.00	63.9	16.0	16.5	0	213	883
538	**White pudding**	6 samples	1.00	22.8	7.0	31.8	36.3	450	1876

Meat and meat products *continued*

Composition of food per 100g

No.	Food	Total nitrogen g	Fatty acids			Cholest-erol mg	Starch g	Total sugars g	Dietary fibre	
			Satd g	Mono unsatd g	Poly unsatd g				Southgate method g	Englyst method g
Meat products										
524	**Sausages**, beef, *raw*	1.54	10.0	11.2	1.4	40	9.9	1.8	0.7	0.5
525	*fried*	2.07	7.2	8.4	1.2	42	12.5	2.4	0.8	0.7
526	*grilled*	2.08	6.7	8.2	1.3	42	12.8	2.4	0.8	0.7
527	pork, *raw*	1.69	12.2	14.8	3.4	47	8.1	1.4	0.6	0.5
528	*fried*	2.20	9.4	11.1	2.6	53	9.2	1.7	0.7	0.7
529	*grilled*	2.13	9.5	11.0	2.7	53	9.7	1.8	0.7	0.7
530	low fat, *raw*	2.00	3.4	4.1	1.5	44	7.6	0.5	N	1.2
531	*fried*	2.38	4.3	5.7	2.3	49	8.4	0.7	N	1.4
532	*grilled*	2.59	5.0	6.0	2.2	55	9.9	0.9	N	1.5
533	**Saveloy**	1.59	N	N	N	45	(10.1)	Tr	0.4	N
534	**Steak and kidney pie**, *individual*	1.46	8.4	9.7	2.1	36	23.3	2.3	1.0	0.9
535	*pastry top only*	2.49	7.0	7.8	2.5	141	15.5	0.4	0.7	0.6
536	**Stewed steak**, canned, *with gravy*	2.37	5.9	5.5	0.4	57	1.0	Tr	Tr	Tr
537	**Tongue**, canned	2.56	6.4	7.9	1.2	110	0	0	0	0
538	**White pudding**	1.12	N	N	N	22	(36.3)	Tr	3.1	N

Meat and meat products *continued*

Inorganic constituents per 100g

No.	Food	Na	K	Ca	Mg	P	Fe	Cu	Zn	Cl	Mn	Se	I
						mg						µg	
Meat products													
524	**Sausages**, beef, *raw*	810	150	48	13	150	1.4	0.23	1.2	1100	(0.16)	N	7
525	*fried*	1090	180	64	16	210	1.6	0.36	1.6	1470	(0.21)	N	Tr
526	*grilled*	1100	190	73	17	210	1.7	0.30	1.7	1490	(0.21)	N	Tr
527	pork, *raw*	760	160	41	11	160	1.1	0.27	1.2	1030	(0.16)	N	Tr
528	*fried*	1050	200	55	15	210	1.5	0.37	1.7	1440	(0.21)	N	Tr
529	*grilled*	1000	200	53	15	220	1.5	0.34	1.6	1340	(0.21)	N	Tr
530	low fat, *raw*	910	160	90	14	170	0.9	0.11	1.2	N	0.20	N	N
531	*fried*	950	190	110	16	190	1.2	0.07	1.4	N	0.20	N	N
532	*grilled*	1190	220	130	19	230	1.3	0.08	1.7	N	0.24	N	N
533	**Saveloy**	890	160	23	9	210	1.5	0.27	1.4	1030	N	N	N
534	**Steak and kidney pie**, *individual*	510	140	53	18	110	2.5	0.10	1.2	720	N	N	11
535	*pastry top only*	660	240	35	19	140	2.8	0.21	2.4	1000	0.20	27	N
536	**Stewed steak**, canned, *with gravy*	380	240	14	14	98	2.1	0.19	3.3	550	0.02	N	N
537	**Tongue**, canned	1050	97	32	14	140	2.5	0.29	2.3	1430	N	N	N
538	**White pudding**	370	190	38	61	230	2.1	0.43	1.6	600	N	N	N

Meat products

No.	Food	Retinol µg	Carotene µg	Vitamin D µg	Vitamin E mg	Thiamin mg	Ribo-flavin mg	Niacin mg	Trypt 60 mg	Vitamin B6 mg	Vitamin B12 µg	Folate µg	Panto-thenate mg	Biotin µg	Vitamin C mg
524	**Sausages**, beef, *raw*	Tr	Tr	Tr	0.43	0.03	0.13	5.0	2.1	0.06	Tr	2	0.50	2	N[a]
525	*fried*	Tr	Tr	Tr	0.28	0	0.14	6.9	2.8	0.07	1.0	2	0.50	2	N[a]
526	*grilled*	Tr	Tr	Tr	0.22	0	0.14	5.4	2.8	0.07	1.0	4	0.50	2	N[a]
527	pork, *raw*	Tr	Tr	Tr	0.24	0.04	0.12	3.4	2.3	0.07	1.0	1	0.60	2	N[a]
528	*fried*	Tr	Tr	Tr	0.28	0.01	0.16	4.4	2.9	0.07	1.0	2	0.60	3	N[a]
529	*grilled*	Tr	Tr	Tr	0.22	0.02	0.15	4.0	2.8	0.06	1.0	3	0.60	3	N[a]
530	low fat, *raw*	Tr	Tr	Tr	0.25	0.03	0.11	2.1	2.3	(0.21)	(1.0)	(2)	(0.74)	3	N[a]
531	*fried*	Tr	Tr	Tr	0.28	Tr	0.12	2.2	2.8	0.12	(1.0)	(2)	0.96	3	N[a]
532	*grilled*	Tr	Tr	Tr	0.29	Tr	0.13	2.8	3.0	0.11	(1.0)	(3)	1.04	3	N[a]
533	**Saveloy**	Tr	Tr	Tr	0.08	0.14	0.09	1.9	1.9	0.06	Tr	1	0.40	Tr	N[a]
534	**Steak and kidney pie**, *individual*	N	Tr	N	N	0.12	0.15	1.7	1.7	0.06	2.0	8	(0.30)	(1)	0
535	*pastry top only*	80	50	0.5	0.74	0.13	0.52	3.6	3.2	0.18	8.3	13	0.81	4	1
536	**Stewed steak**, canned, *with gravy*	Tr	Tr	Tr	0.59	Tr	0.13	2.4	2.8	0.07	1.0	4	0.30	1	0
537	**Tongue**, canned	Tr	Tr	Tr	0.26	0.04	0.39	2.5	3.8	0.04	5.0	2	0.40	2	0
538	**White pudding**	Tr	Tr	Tr	1.00	0.26	0.08	0.5	1.3	0.06	1.0	6	0.80	18	0

[a] Ascorbic acid is added as an antioxidant. Measurable levels may be present

Meat and meat products *continued*

Composition of food per 100g

Meat dishes

No.	Food	Description and main data sources	Edible proportion	Water g	Protein g	Fat g	Carbo-hydrate g	Energy value kcal	kJ
539	**Beef chow mein**	12 samples from different shops. Noodles with beef and vegetables in sauce	1.00	71.7	6.7	6.0	14.7	136	571
540	**Beef curry,** *retail*	6 samples, 3 brands; cooked according to packet directions	1.00	69.5	13.5	6.6	6.3	137	575
541	*with rice*	Calculated from sample proportions as 57% curry and 43% rice	1.00	68.9	8.8	4.3	16.9	137	579
542	**Beef kheema**	Recipe	1.00	42.7	18.2	37.7	0.3	413	1709
543	**Beef koftas**	Recipe	1.00	40.8	23.3	27.6	3.4[a]	353	1469
544	**Beef steak pudding**	Recipe	1.00	57.5	10.8	12.3	18.8[a]	224	939
545	**Beef stew**	Recipe	1.00	77.3	9.7	7.2	4.6[a]	120	503
546	**Bolognese sauce**	Recipe	1.00	74.7	8.0	11.1	3.7[a]	145	602
547	**Chicken curry,** *with bone*	Recipe	0.75	49.5	7.7	12.7	2.3[a]	154	639
548	*without bone*	Recipe	1.00	66.0	10.2	17.0	3.1[a]	205	850
549	*retail*	7 samples, 5 brands; Korma and Masala varieties cooked according to packet directions	1.00	68.7	12.1	8.9	5.4	149	621
550	*with rice*	Calculated from sample proportions as 55% curry and 45% rice	1.00	68.4	7.8	5.5	16.9	144	607
551	**Chilli con carne**	Recipe	1.00	67.6	11.0	8.5	8.3[a]	151	632
552	**Curried meat**	Recipe	1.00	67.9	8.5	10.5	9.1[a]	162	677

[a] Including oligosaccharides

No.	Food	Total nitrogen g	Fatty acids			Cholest- erol mg	Starch g	Total sugars g	Dietary fibre	
			Satd g	Mono unsatd g	Poly unsatd g				Southgate method g	Englyst method g
Meat dishes										
539	**Beef chow mein**	1.07	1.3	3.1	1.4	N	12.3	2.4	N	N
540	**Beef curry,** *retail*	2.16	3.1	2.5	0.6	32	1.8	4.5	N	1.2
541	*with rice*	1.42	1.9	1.6	0.6	21	14.3	2.6	N	0.7
542	**Beef kheema**	2.92	13.0	14.5	8.1	83	Tr	0.3	0.2	0.1
543	**Beef koftas**	3.73	7.0	11.3	7.5	111	1.0[a]	1.4	0.7	0.4
544	**Beef steak pudding**	1.77	6.1	5.0	0.5	34	17.5[a]	0.9	1.0	0.8
545	**Beef stew**	1.55	3.3	3.2	0.3	30	2.4[a]	1.8	0.7	0.6
546	**Bolognese sauce**	1.28	3.1	4.7	2.6	25	0.1[a]	3.3	1.1	1.0
547	**Chicken curry,** *with bone*	1.22	1.7	4.5	5.6	28	0.4[a]	1.2	0.7	0.7
548	*without bone*	1.63	2.2	6.0	7.5	37	0.5[a]	1.6	1.0	0.9
549	*retail*	1.94	4.0	2.9	1.5	51	1.0	4.4	N	1.3
550	*with rice*	1.27	2.3	1.7	1.1	28	14.5	2.4	N	0.8
551	**Chilli con carne**	1.77	3.0	3.8	1.1	27	4.4[a]	2.8	3.2	2.3
552	**Curried meat**	1.37	2.9	3.6	2.9	23	1.7[a]	6.3	1.8	1.4

[a] Not including oligosaccharides

No.	Food	Na	K	Ca	Mg	P	Fe	Cu	Zn	Cl	Mn	Se	I
						mg						μg	
Meat dishes													
539	**Beef chow mein**	590	N	N	N	N	1.3	N	N	910	N	N	N
540	**Beef curry,** *retail*	450	340	N	N	N	N	N	N	690	N	N	N
541	*with rice*	260	260	N	N	N	N	N	N	400	N	N	N
542	**Beef kheema**	150	300	16	17	160	2.7	0.15	4.2	190	0.05	3	9
543	**Beef koftas**	990	480	24	33	230	2.8	0.17	3.8	1500	0.13	5	14
544	**Beef steak pudding**	330	190	41	14	120	1.4	0.12	1.8	430	0.20	2	6
545	**Beef stew**	330	200	15	11	72	1.1	0.09	1.8	500	0.10	2	4
546	**Bolognese sauce**	430	310	23	14	79	1.4	0.11	1.7	670	0.10	2	4
547	**Chicken curry,** *with bone*	460	200	18	16	86	1.5	0.09	0.6	700	0.12	2	4
548	*without bone*	620	270	24	24	110	1.9	0.13	0.8	930	0.16	3	6
549	*retail*	450	300	N	N	N	N	N	N	640	N	N	N
550	*with rice*	250	230	N	N	N	N	N	N	350	N	N	N
551	**Chilli con carne**	250	440	29	33	130	2.2	0.19	2.2	380	0.24	3	4
552	**Curried meat**	470	260	29	20	80	2.4	0.11	1.6	720	0.20	1	3

Meat and meat products *continued*

No.	Food	Retinol µg	Carotene µg	Vitamin D µg	Vitamin E mg	Thiamin mg	Ribo-flavin mg	Niacin mg	Trypt 60 mg	Vitamin B6 mg	Vitamin B12 µg	Folate µg	Panto-thenate mg	Biotin µg	Vitamin C mg
	Meat dishes														
539	**Beef chow mein**	Tr	Tr	Tr	(0.43)	0.03	0.03	N	1.1	N	Tr	N	N	N	Tr
540	**Beef curry**, *retail*	Tr	Tr	Tr	0.62	0.05	0.16	2.4	1.6	0.20	N	N	0.71	3	Tr
541	*with rice*	Tr	Tr	Tr	0.35	0.03	0.09	1.8	1.9	0.14	N	N	0.45	2	Tr
542	**Beef kheema**	72	95	0.1	4.04	0.05	0.24	3.2	3.9	0.22	1.9	5	0.48	Tr	2
543	**Beef koftas**	22	100	0.2	3.81	0.09	0.23	4.3	5.0	0.31	2.5	9	0.70	2	1
544	**Beef steak pudding**	4	7	Tr	0.18	0.09	0.09	1.8	2.3	0.14	0.9	1	0.27	Tr	Tr
545	**Beef stew**	1	1100	Tr	0.21	0.05	0.09	1.7	2.1	0.14	0.9	Tr	0.26	Tr	1
546	**Bolognese sauce**	Tr	1275	Tr	1.93	0.07	0.11	1.6	1.6	0.16	0.8	7	0.33	1	4
547	**Chicken curry**, *with bone*	290	51	Tr	2.82	0.06	0.11	2.4	1.4	0.16	1.4	11	0.47	5	2
548	*without bone*	385	68	Tr	3.76	0.08	0.14	3.3	1.9	0.21	1.9	15	0.62	7	2
549	*retail*	N	370	N	1.30	0.20	0.14	3.8	2.3	0.24	N	N	1.02	3	1
550	*with rice*	N	N	N	0.71	0.11	0.08	2.5	2.4	0.16	N	N	0.61	2	1
551	**Chilli con carne**	Tr	145	Tr	1.19	0.11	0.14	1.9	2.2	0.21	0.8	14	0.37	1	9
552	**Curried meat**	Tr	6	Tr	1.62	0.06	0.08	1.6	1.8	0.12	0.6	5	0.22	1	1

Meat and meat products *continued*

Composition of food per 100g

No.	Food	Description and main data sources	Edible proportion	Water g	Protein g	Fat g	Carbo-hydrate g	Energy value kcal	kJ
Meat dishes									
553	**Hot pot**	Recipe	1.00	73.5	9.4	4.5	10.1[a]	114	482
554	**Irish stew**	Recipe	1.00	76.2	5.3	7.6	9.1[a]	123	515
555	*weighed with bones*	Calculated from Irish stew	0.91	69.3	4.8	6.9	8.3[a]	112	470
556	**Lamb kheema**	Recipe	1.00	51.2	14.6	29.1	2.3[a]	328	1357
557	**Lasagne,** *frozen, cooked*	10 samples, 3 brands. Calculated from frozen using 5.3% weight loss	1.00	75.2	5.0	3.8	12.8	102	430
558	**Moussaka**	Recipe	1.00	68.0	9.1	13.6	7.0[a]	184	768
559	**Mutton biriani**	Recipe	1.00	50.2	7.5	16.9	25.1[a]	276	1154
560	**Mutton curry**	Recipe	1.00	44.0	14.9	33.4	3.9[a]	374	1547
561	**Pancake roll**	18 samples, 12 shops. Vegetable and beansprout filling	1.00	58.3	6.6	12.5	20.9	217	909
562	**Shepherd's pie**	Recipe	1.00	75.6	8.0	6.2	8.2[a]	118	496

[a] Including oligosaccharides

Composition of food per 100g

553 to 562

No.	Food	Total nitrogen g	Fatty acids			Cholest-erol mg	Starch g	Total sugars g	Dietary fibre	
			Satd g	Mono unsatd g	Poly unsatd g				Southgate method g	Englyst method g
	Meat dishes									
553	Hot pot	1.50	1.8	2.1	0.2	25	6.7[a]	2.8	1.4	1.2
554	Irish stew	0.85	3.5	2.8	0.3	20	7.2[a]	1.5	1.0	0.9
555	*weighed with bones*	0.77	3.2	2.5	0.3	18	6.5[a]	1.4	0.9	0.8
556	Lamb kheema	2.33	16.7	9.1	1.2	101	0.2[a]	1.6	0.6	0.5
557	Lasagne, *frozen, cooked*	0.79	1.9	1.4	0.2	11	10.7	2.1	0.5	N
558	Moussaka	1.46	4.5	5.2	3.0	40	3.9[a]	2.6	1.0	0.9
559	Mutton biriani	1.22	9.7	4.9	1.0	50	21.3[a]	3.3	1.4	0.7
560	Mutton curry	2.38	19.5	10.2	1.3	113	0.3[a]	2.7	1.0	0.8
561	Pancake roll	1.06	3.7	5.9	2.4	0	17.9	3.1	N	N
562	Shepherd's pie	1.28	2.4	2.7	0.5	28	7.0[a]	1.0	0.7	0.6

[a] Not including oligosaccharides

Meat dishes

No.	Food	Na	K	Ca	Mg	P	Fe	Cu	Zn	Cl	Mn	Se	I
						mg						μg	
553	Hot pot	660	350	18	18	81	1.2	0.12	1.7	1030	0.20	2	5
554	Irish stew	360	260	10	14	60	0.6	0.09	1.1	570	0.10	1	3
555	*weighed with bones*	330	240	9	13	55	0.5	0.08	1.0	520	0.10	1	3
556	Lamb kheema	650	330	16	24	150	1.9	0.13	2.3	990	0.20	1	12
557	Lasagne, *frozen, cooked*	430	150	71	17	83	0.7	0.17	0.7	860	0.20	N	N
558	Moussaka	320	270	81	16	120	1.1	0.09	1.7	500	0.10	3	10
559	Mutton biriani	270	250	41	22	110	0.9	0.16	1.3	420	0.40	3	12
560	Mutton curry	830	370	22	27	160	2.1	0.14	2.4	1270	0.30	1	15
561	Pancake roll	610	N	N	N	N	2.1	N	N	940	N	N	N
562	Shepherd's pie	450	230	16	14	71	1.2	0.11	1.9	700	0.10	1	5

Meat and meat products *continued*

Meat dishes

No.	Food	Retinol µg	Carotene µg	Vitamin D µg	Vitamin E mg	Thiamin mg	Ribo-flavin mg	Niacin mg	Trypt 60 mg	Vitamin B6 mg	Vitamin B12 µg	Folate µg	Panto-thenate mg	Biotin µg	Vitamin C mg
553	Hot pot	Tr	1305	Tr	0.26	0.13	0.08	1.7	2.0	0.28	0.8	11	0.36	Tr	3
554	Irish stew	Tr	2	Tr	0.14	0.10	0.04	0.9	1.1	0.20	0.5	1	0.22	Tr	3
555	weighed with bones	Tr	2	Tr	0.13	0.09	0.04	0.8	1.0	0.18	0.5	1	0.20	Tr	3
556	Lamb kheema	140	190	0.1	0.73	0.12	0.16	3.9	3.1	0.18	1.6	4	0.44	1	2
557	Lasagne, *frozen, cooked*	N	N	N	N	0.02	0.12	0.8	1.1	0.05	0.3	4	N	N	Tr
558	Moussaka	35	61	0.1	1.65	0.08	0.14	1.3	2.0	0.18	0.8	10	0.35	1	2
559	Mutton biriani	105	240	0.1	0.83	0.15	0.08	2.3	1.6	0.16	0.5	7	0.34	2	2
560	Mutton curry	185	250	0.2	0.94	0.14	0.17	4.0	3.1	0.21	1.6	6	0.47	1	3
561	Pancake roll	0	3	0	(0.77)	0.09	0.05	N	1.1	N	0	N	N	N	0
562	Shepherd's pie	15	15	0.1	0.29	0.09	0.11	1.6	1.7	0.25	0.6	13	0.43	Tr	3

Fish and fish products

This section of the Tables has been extended both in terms of foods and data included.

Fish are mainly drawn from a wild population which means that their composition is probably more variable than that of foods drawn from domesticated inbred stock whose nutrition has been closely controlled. There is considerable variation in composition within one species and this variation is probably greater than that between species.

In this edition cartilaginous fish (dogfish and skate) have been moved into the white fish group. However, it should be noted that there are major compositional differences between white fish and cartilaginous fish. Cartilaginous fish contain a very high proportion of non-protein nitrogen and appropriate adjustments were made before protein values were calculated.

For raw fish, edible proportion is given for items that are normally purchased with bones or other inedible material e.g. shells. For cooked items that are normally served with waste, the edible proportion is given where they have been weighed with the waste.

For weight loss on cooking and calculation of cooked edible portion obtainable from raw fish, see Section 3.3.

The fat contents of many fish show considerable seasonal changes and it is difficult to assign definite values. The actual fat content of fish normally landed and consumed shows less variation because the fish tend to be caught during a limited part of the cycle; the values used are therefore based on the fat content of the fish during the period when the major landings of the species are made.

New analytical data for fatty acids and cholesterol in fish and fish products have been added where available and where not, data from the 1978 edition applied. For fish commonly purchased fried from fish and chip shops, the fatty acid profile has been calculated as fried in blended oil or dripping. For other fried fish where a significant amount of fat is absorbed the fatty acids have been calculated as fried in average vegetable oil.

New analytical data have supplied values for copper, zinc, water soluble vitamins and some retinol values for fatty fish.

In fish with fine bones it is often difficult to remove the bones completely, whether before analysis or before consumption. The calcium and phosphorus content of these fish is more variable than in a fish which can be boned easily. The values in the tables are based on samples which have been prepared for consumption in the normal way.

The crustaceans and molluscs tend to accumulate many cations from their environment, and the concentration of iron, copper and zinc reported in these fish shows very wide variation, depending on the source of the samples and the levels of metallic contamination to which they have been exposed.

Values for starch and sugars have been calculated for this group using the revised data for flour and breadcrumbs. The data for this section are therefore more extensive and up-to-date than those in the 1978 edition.

Canned tuna has been recently analysed and a lower value found for fat than was quoted in the 1978 edition. The new analytical values have been included in this section.

Losses of labile vitamins assigned to cooked dishes or foods were estimated from figures in Section 3.4.

Taxonomic names for foods in this part of the Tables can be found in Section 3.6.

Fish and fish products

Composition of food per 100g

No.	Food	Description and main data sources	Edible proportion	Water g	Protein g	Fat g	Carbo-hydrate g	Energy value kcal	kJ
White fish									
563	**Cod**, *raw, fillets*	Samples from 3 different shops	0.89	82.1	17.4	0.7	0	76	322
564	*baked, fillets*	Samples baked in the oven with added butter	1.00	76.6	21.4	1.2	0	96	408
565	*–, fillets, weighed with bones and skin*	Samples baked in oven with butter added	0.85	65.1	18.3	1.0	0	82	348
566	*poached, fillets*	Poached in milk with butter and salt added	1.00	77.7	20.9	1.1	0	94	396
567	*–, fillets, weighed with bones and skin*	Calculated from poached fillets	0.87	67.6	18.2	1.0	0	82	346
568	*frozen, raw, steaks*	12 packets, 3 brands	1.00	83.9	15.6	0.6	0	68	287
569	*–, grilled, steaks*	12 samples; grilled with butter and salt added	1.00	78.0	20.8	1.3	0	95	402
570	*in batter, fried in blended oil*	Cooked samples from fish and chip shops	1.00	60.9	19.6	10.3	7.5	199	834
571	*–, fried in dripping*	Cooked samples from fish and chip shops	1.00	60.9	19.6	10.3	7.5	199	834
572	*dried, salted, boiled*	Soaked 24 hours and boiled	0.83	64.9	32.5	0.9	0	138	586
573	**Dogfish**, *in batter, fried in blended oil*	7 cooked samples from fish and chip shops	1.00	54.2	16.7[a]	18.8	7.7	265	1103
574	*in batter, fried in blended oil, weighed with waste*	Calculated from fried in blended oil	0.92	49.9	15.4[a]	17.3	7.1	244	1016
575	*–, fried in dripping*	7 cooked samples from fish and chip shops	1.00	54.2	16.7[a]	18.8	7.7	265	1103
576	*–, fried in dripping, weighed with waste*	Calculated from fried in dripping	0.92	49.9	15.4[a]	17.3	7.1	244	1016

[a] (Total N – non-protein N) × 6.25

No.	Food	Total nitrogen g	Fatty acids Satd g	Mono unsatd g	Poly unsatd g	Cholest- erol mg	Starch g	Total sugars g	Dietary fibre Southgate method g	Englyst method g
White fish										
563	**Cod**, *raw, fillets*	2.78	0.1	0.1	0.3	46	0	0	0	0
564	*baked, fillets*	3.43	0.5	0.2	0.2	48	0	0	0	0
565	*-, fillets, weighed with bones and skin*	2.92	0.4	0.2	0.2	41	0	0	0	0
566	*poached, fillets*	3.35	0.4	0.2	0.3	60	0	0	0	0
567	*-, fillets, weighed with bones and skin*	2.91	0.4	0.2	0.2	52	0	0	0	0
568	*frozen, raw, steaks*	2.49	0.1	0.1	0.2	46	0	0	0	0
569	*-, grilled, steaks*	3.32	0.5	0.2	0.3	60	0	0	0	0
570	*in batter, fried in blended oil*	3.14	0.9	5.1	3.7	N	(7.4)	(0.1)	0.3	(0.3)
571	*-, fried in dripping*	3.14	4.7	3.1	0.2	N	(7.4)	(0.1)	0.3	(0.3)
572	*dried, salted, boiled*	5.20	0.2	0.1	0.4	59	0	0	0	0
573	**Dogfish**, *in batter, fried in blended oil*	3.42	1.6	9.4	6.9	N	(7.5)	(0.1)	0.3	(0.3)
574	*in batter, fried in blended oil, weighed with waste*	3.15	1.5	8.6	6.3	N	(7.0)	(0.1)	0.3	(0.3)
575	*-, fried in dripping*	3.42	10.4	7.0	0.5	N	(7.5)	(0.1)	0.3	(0.3)
576	*-, fried in dripping, weighed with waste*	3.15	9.6	6.4	0.5	N	(7.0)	(0.1)	0.3	(0.3)

Fish and fish products

Inorganic constituents per 100g

No.	Food	Na	K	Ca	Mg	P	Fe	Cu	Zn	Cl	Mn	Se	I
						mg						µg	
White fish													
563	**Cod**, *raw, fillets*	77	320	16	23	170	0.3	0.06	0.4	110	0.01	28	110
564	*baked, fillets*	340	350	22	26	190	0.4	0.07	0.5	520	0.01	34	(110)
565	*-, fillets, weighed with bones and skin*	290	300	19	22	160	0.4	0.06	0.4	440	0.01	29	(90)
566	*poached, fillets*	110	330	29	26	180	0.3	0.09	0.5	150	0.01	33	(110)
567	*-, fillets, weighed with bones and skin*	96	290	25	23	160	0.3	0.08	0.4	130	0.01	29	(90)
568	*frozen, raw, steaks*	68	310	11	22	160	0.3	0.06	0.3	95	0.01	25	110
569	*-, grilled, steaks*	91	380	10	26	200	0.4	0.07	0.5	130	0.01	33	(110)
570	*in batter, fried in blended oil*	100	370	80	24	200	0.5	(0.07)	N	150	N	N	N
571	*-, fried in dripping*	100	370	80	24	200	0.5	(0.07)	N	150	N	N	N
572	*dried, salted, boiled*	400	31	22	35	160	1.8	N	N	670	0.01	52	N
573	**Dogfish**, *in batter, fried in blended oil*	290	310	42	23	220	1.1	0.13	0.5	340	N	N	N
574	*in batter, fried in blended oil, weighed with waste*	270	290	39	21	200	1.0	0.12	0.4	310	N	N	N
575	*-, fried in dripping*	290	310	42	23	220	1.1	0.13	0.5	340	N	N	N
576	*-, fried in dripping, weighed with waste*	270	290	39	21	200	1.0	0.12	0.4	310	N	N	N

Fish and fish products

No.	Food	Retinol µg	Carotene µg	Vitamin D µg	Vitamin E mg	Thiamin mg	Ribo-flavin mg	Niacin mg	Trypt 60 mg	Vitamin B₆ mg	Vitamin B₁₂ µg	Folate µg	Panto-thenate mg	Biotin µg	Vitamin C mg
	White fish														
563	**Cod**, *raw, fillets*	2	Tr	Tr	0.44	0.08	0.07	1.7	3.2	0.33	2	12	(0.20)	(3)	Tr
564	*baked, fillets*	(2)	Tr	Tr	0.59	0.07	0.07	1.7	4.0	0.38	2.0	12	(0.20)	(3)	Tr
565	*-, fillets, weighed with bones and skin*	(2)	Tr	Tr	0.50	0.06	0.06	1.3	3.4	0.32	2.0	10	(0.17)	(3)	Tr
566	*poached, fillets*	(2)	Tr	Tr	0.61	0.08	0.08	1.7	3.9	0.37	2.0	14	(0.19)	(3)	Tr
567	*-, fillets, weighed with bones and skin*	(2)	Tr	Tr	0.53	0.07	0.07	1.5	3.4	0.32	2.0	12	(0.16)	(3)	Tr
568	*frozen, raw, steaks*	2	Tr	Tr	N	0.06	0.05	1.5	2.9	0.34	1.0	6	(0.20)	(3)	Tr
569	*-, grilled, steaks*	(2)	Tr	Tr	N	0.08	0.06	1.9	3.9	0.41	2.0	10	(0.25)	(3)	Tr
570	*in batter, fried in blended oil*	N	Tr	Tr	N	(0.20)	(0.15)	(2.0)	3.7	N	N	N	N	N	Tr
571	*-, fried in dripping*	N	Tr	Tr	N	(0.20)	(0.15)	(2.0)	3.7	N	N	N	N	N	Tr
572	*dried, salted, boiled*	(2)	Tr	Tr	N	Tr	Tr	N	6.1	N	Tr	Tr	N	Tr	Tr
573	**Dogfish**, *in batter, fried in blended oil*	94	Tr	N	2.10	0.06	0.10	5.6	N	N	N	N	N	N	Tr
574	*in batter, fried in blended oil, weighed with waste*	86	Tr	N	1.90	0.06	0.09	5.2	N	N	N	N	N	N	Tr
575	*-, fried in dripping*	(94)	Tr	N	2.10	0.06	0.10	5.6	N	N	N	N	N	N	Tr
576	*-, fried in dripping, weighed with waste*	(86)	Tr	Tr	1.90	0.06	0.09	5.2	N	N	N	N	N	N	Tr

Fish and fish products continued

Composition of food per 100g

No.	Food	Description and main data sources	Edible proportion	Water	Protein	Fat	Carbo-hydrate	Energy value kcal	kJ
				g	g	g	g	kcal	kJ
White fish									
577	**Haddock**, *raw*	Fillets	N	81.3	16.8	0.6	0	73	308
578	*steamed*	Middle cut	1.00	75.1	22.8	0.8	0	98	417
579	*steamed, weighed with bones and skin*	Calculated from steamed	0.76	57.1	17.3	0.6	0	75	316
580	*in crumbs, fried in blended oil*	Samples coated in crumbs and fried	1.00	65.1	21.4	8.3	3.6	174	729
581	*-, fried in blended oil, weighed with bones*	Calculated from fried in blended oil	0.92	60.0	19.7	7.6	3.3	160	669
582	*-, fried in dripping*	Samples coated in crumbs and fried	1.00	65.1	21.4	8.3	3.6	174	729
583	*-, fried in dripping, weighed with bones*	Calculated from fried in dripping	0.92	60.0	19.7	7.6	3.3	160	669
584	*smoked, steamed*	Flesh only	1.00	71.6	23.3	0.9	0	101	429
585	*steamed, weighed with bones and skin*	Calculated from smoked, steamed	0.65	46.5	15.1	0.6	0	66	279
586	**Halibut**, *raw*	Literature sources	N	78.1	17.7	2.4	0	92	390
587	*steamed*	Middle cut	1.00	70.9	23.8	4.0	0	131	553
588	*steamed, weighed with bones and skin*	Calculated from steamed	0.76	53.8	18.0	3.0	0	99	417

Fish and fish products continued

Composition of food per 100g

No.	Food	Total nitrogen g	Fatty acids Satd g	Mono unsatd g	Poly unsatd g	Cholesterol mg	Starch g	Total sugars g	Dietary fibre Southgate method g	Englyst method g
	White fish									
577	**Haddock**, raw	2.68	0.1	0.1	0.2	36	0	0	0	0
578	steamed	3.65	0.2	0.1	0.3	48	0	0	0	0
579	steamed, weighed with bones and skin	2.77	0.1	0.1	0.2	44	0	0	0	0
580	in crumbs, fried in blended oil	3.42	0.7	4.1	3.0	N	(3.5)	(0.1)	0.2	(0.2)
581	-, fried in blended oil, weighed with bones	3.15	0.7	3.8	2.8	N	(3.2)	(0.1)	0.1	(0.1)
582	-, fried in dripping	3.42	3.8	2.5	0.2	N	(3.5)	(0.1)	0.2	(0.2)
583	-, fried in dripping, weighed with bones	3.15	3.5	2.3	0.1	N	(3.2)	(0.1)	0.1	(0.1)
584	smoked, steamed	3.73	0.2	0.1	0.3	54	0	0	0	0
585	steamed, weighed with bones and skin	2.42	0.1	0.1	0.2	35	0	0	0	0
586	**Halibut**, raw	2.83	0.3	0.5	0.8	50	0	0	0	0
587	steamed	3.80	0.5	0.9	1.3	60	0	0	0	0
588	steamed, weighed with bones and skin	2.88	0.4	0.7	1.0	46	0	0	0	0

Fish and fish products continued

Inorganic constituents per 100g

No.	Food	Na	K	Ca	Mg	P	Fe	Cu	Zn	Cl	Mn	Se	I
						mg						µg	
White fish													
577	**Haddock**, raw	120	300	18	23	170	0.6	0.05	0.3	160	0.01	22	250
578	steamed	120	320	55	28	230	0.7	0.02	(0.4)	140	0.01	30	(250)
579	steamed, weighed with bones and skin	92	250	41	21	180	0.5	0.02	(0.3)	110	0.01	22	(190)
580	in crumbs, fried in blended oil	180	350	110	31	250	1.2	0.01	0.4	180	0.03	30	330
581	-, fried in blended oil, weighed with bones	160	320	100	28	230	1.1	0.01	0.4	170	0.03	28	310
582	-, fried in dripping	180	350	110	31	250	1.2	0.01	0.4	180	0.03	30	330
583	-, fried in dripping, weighed with bones	160	320	100	28	230	1.1	0.01	0.4	170	0.03	28	310
584	smoked, steamed	1220	290	58	25	250	1.0	0.03	0.3	1900	0.01	30	(250)
585	steamed, weighed with bones and skin	790	190	37	17	160	0.7	0.02	0.2	1230	0.01	20	(160)
586	**Halibut**, raw	(84)	(260)	(10)	(17)	(190)	(0.5)	(0.05)	N	(60)	N	N	N
587	steamed	110	340	13	23	260	0.6	0.07	N	80	N	N	N
588	steamed, weighed with bones and skin	84	260	10	18	190	0.5	0.05	N	61	N	N	N

Fish and fish products *continued*

Vitamins per 100g

No.	Food	Retinol µg	Carotene µg	Vitamin D µg	Vitamin E mg	Thiamin mg	Riboflavin mg	Niacin mg	Trypt 60 mg	Vitamin B6 mg	Vitamin B12 µg	Folate µg	Pantothenate mg	Biotin µg	Vitamin C mg
White fish															
577	**Haddock**, *raw*	Tr	Tr	Tr	N	0.07	0.10	4.0	3.1	0.20	1.0	13	0.20	5	Tr
578	*steamed*	Tr	Tr	Tr	N	(0.08)	(0.13)	(5.1)	4.3	(0.25)	(1.0)	(16)	(0.20)	(6)	Tr
579	*steamed, weighed with bones and skin*	Tr	Tr	Tr	N	(0.06)	(0.10)	(3.9)	3.2	(0.19)	(1.0)	(12)	(0.15)	(5)	Tr
580	*in crumbs, fried in blended oil*	Tr	Tr	Tr	N	(0.08)	(0.08)	(2.8)	4.0	(0.24)	(1.3)	N	(0.26)	(2)	Tr
581	*-, fried in blended oil, weighed with bones*	Tr	Tr	Tr	N	(0.07)	(0.07)	(2.6)	3.7	(0.22)	(1.2)	N	(0.24)	(1)	Tr
582	*-, fried in dripping*	Tr	Tr	Tr	N	(0.08)	(0.08)	(2.8)	4.0	(0.24)	(1.3)	N	(0.26)	(2)	Tr
583	*-, fried in dripping, weighed with bones*	Tr	Tr	Tr	N	(0.07)	(0.07)	(2.6)	3.7	(0.22)	(1.2)	N	(0.24)	(1)	Tr
584	*smoked, steamed*	Tr	Tr	Tr	N	(0.10)	(0.11)	(1.7)	4.4	(0.35)	(3.0)	(5)	(0.20)	(3)	Tr
585	*steamed, weighed with bones and skin*	Tr	Tr	Tr	N	(0.07)	(0.07)	(1.1)	2.8	(0.28)	(2.0)	(3)	(0.13)	(2)	Tr
586	**Halibut**, *raw*	Tr[a]	Tr	Tr[a]	0.90	0.08	0.10	5.0	3.3	0.20	1.0	12	0.30	5	Tr
587	*steamed*	Tr[a]	Tr	Tr[a]	(1.00)	(0.08)	(0.11)	(5.2)	4.4	(0.23)	(1.0)	(14)	(0.28)	(5)	Tr
588	*steamed, weighed with bones and skin*	Tr[a]	Tr	Tr[a]	(0.76)	(0.06)	(0.08)	(4.0)	3.4	(0.17)	(1.0)	(11)	(0.21)	(4)	Tr

[a] These are values for Atlantic halibut. Pacific halibut have been reported to contain 120µg retinol and 1µg vitamin D per 100g

Fish and fish products continued

No.	Food	Description and main data sources	Edible proportion	Water g	Protein g	Fat g	Carbo-hydrate g	Energy value kcal	Energy value kJ
White fish									
589	**Lemon sole**, *raw*	Literature sources	N	81.2	17.1	1.4	0	81	343
590	*steamed*	Flesh only	1.00	77.2	20.6	0.9	0	91	384
591	*steamed, weighed with bones and skin*	Calculated from steamed	0.71	54.9	14.6	0.6	0	64	270
592	*in crumbs, fried*	Samples coated in crumbs and fried	1.00	60.4	16.1	13.0	9.3	216	904
593	*-, fried, weighed with bones*	Calculated from fried	0.79	47.7	12.7	10.3	7.4	171	715
594	**Plaice**, *raw*	8 fish, purchased whole	0.42	79.5	17.9	2.2	0	91	386
595	*steamed*	Flesh only	1.00	78.0	18.9	1.9	0	93	392
596	*steamed, weighed with bones and skin*	Calculated from steamed	0.54	42.1	10.2	1.0	0	50	210
597	*in batter, fried in blended oil*	6 cooked samples from fish and chip shops	1.00	52.4	15.8	18.0	14.4	279	1165
598	*-, fried in dripping*	6 cooked samples from fish and chip shops	1.00	52.4	15.8	18.0	14.4	279	1165
599	*in crumbs, fried, fillets*	8 samples coated in egg and crumbs, and fried	1.00	59.9	18.0	13.7	8.6	228	951

Fish and fish products *continued*

No.	Food	Total nitrogen g	Fatty acids Satd g	Mono unsatd g	Poly unsatd g	Cholest- erol mg	Starch g	Total sugars g	Dietary fibre Southgate method g	Englyst method g
White fish										
589	**Lemon sole**, *raw*	2.74	0.2	0.3	0.5	60	0	0	0	0
590	*steamed*	3.29	0.1	0.2	0.3	60	0	0	0	0
591	*steamed, weighed with bones and skin*	2.34	0.1	0.1	0.2	47	0	0	0	0
592	*in crumbs, fried*	2.57	1.3	4.4	6.0	N	(9.0)	(0.3)	0.4	(0.4)
593	*-, fried, weighed with bones*	2.03	1.0	3.5	4.7	N	(7.2)	(0.2)	0.3	(0.3)
594	**Plaice**, *raw*	2.86	0.3	0.6	0.5	42	0	0	0	0
595	*steamed*	3.02	0.3	0.5	0.5	50	0	0	0	0
596	*steamed, weighed with bones and skin*	1.63	0.2	0.3	0.2	27	0	0	0	0
597	*in batter, fried in blended oil*	2.52	1.5	8.9	6.6	N	(14.1)	(0.3)	0.5	N
598	*-, fried in dripping*	2.52	8.2	5.5	0.4	N	(14.1)	(0.3)	0.5	N
599	*in crumbs, fried, fillets*	2.88	1.4	4.7	6.3	N	(8.3)	(0.3)	0.4	N

Fish and fish products continued

Inorganic constituents per 100g

No.	Food	Na	K	Ca	Mg	P	Fe	Cu	Zn	Cl	Mn	Se	I
						mg						µg	
												Se	I

White fish

No.	Food	Na	K	Ca	Mg	P	Fe	Cu	Zn	Cl	Mn	Se	I
589	**Lemon sole**, raw	(95)	(230)	(17)	(17)	200	(0.5)	(0.10)	0.7	97	N	(44)	N
590	steamed	120	280	21	20	250	0.6	0.12	0.8	120	N	(44)	N
591	steamed, weighed with bones and skin	82	200	15	14	180	0.4	0.09	0.6	83	N	(35)	N
592	in crumbs, fried	140	250	95	22	240	1.1	0.16	N	120	N	N	N
593	-, fried, weighed with bones	110	200	75	16	190	0.9	0.13	N	98	N	N	N
594	**Plaice**, raw	120	280	51	22	180	0.3	0.05	0.5	170	0.01	36	28
595	steamed	120	280	38	24	250	0.6	0.02	0.3	110	0.01	38	(28)
596	steamed, weighed with bones and skin	65	150	20	13	130	0.3	0.01	0.2	61	0.01	20	(15)
597	in batter, fried in blended oil	220	230	93	21	170	1.0	0.17	1.0	280	(0.16)	N	(31)
598	-, fried in dripping	220	230	93	21	170	1.0	0.17	1.0	280	(0.16)	N	(31)
599	in crumbs, fried, fillets	220	280	67	24	180	0.8	0.20	0.7	310	(0.16)	(38)	31

200

Fish and fish products continued

Vitamins per 100g

No.	Food	Retinol µg	Carotene µg	Vitamin D µg	Vitamin E mg	Thiamin mg	Ribo-flavin mg	Niacin mg	Trypt 60 mg	Vitamin B6 mg	Vitamin B12 µg	Folate µg	Panto-thenate mg	Biotin µg	Vitamin C mg
White fish															
589	**Lemon sole**, *raw*	Tr	Tr	Tr	N	0.09	0.08	3.5	3.2	N	1	11	0.30	(5)	Tr
590	*steamed*	Tr	Tr	Tr	N	(0.09)	(0.09)	(3.6)	3.8	N	(1.0)	(13)	(0.31)	(5)	Tr
591	*steamed, weighed with bones and skin*	Tr	Tr	Tr	N	(0.06)	(0.06)	(2.6)	2.7	N	(1.0)	(9)	(0.22)	(4)	Tr
592	*in crumbs, fried*	Tr	Tr	Tr	N	N	N	N	3.0	N	N	N	N	N	Tr
593	*-, fried, weighed with bones*	Tr	Tr	Tr	N	N	N	N	2.4	N	N	N	N	N	Tr
594	**Plaice**, *raw*	Tr	Tr	Tr	N	0.30	0.10	3.2	3.3	0.43	2.0	10	0.80	N	Tr
595	*steamed*	Tr	Tr	Tr	N	(0.30)	(0.11)	(3.2)	3.5	(0.47)	(2.0)	(11)	(0.70)	N	Tr
596	*steamed, weighed with bones and skin*	Tr	Tr	Tr	N	(0.16)	(0.06)	(1.7)	1.9	(0.25)	(1.0)	(6)	(0.38)	N	Tr
597	*in batter, fried in blended oil*	Tr	Tr	Tr	N	0.20	0.15	2.0	2.9	N	N	N	N	N	Tr
598	*-, fried in dripping*	Tr	Tr	Tr	N	0.20	0.15	2.0	2.9	N	N	N	N	N	Tr
599	*in crumbs, fried, fillets*	Tr	Tr	Tr	N	0.23	0.18	2.9	3.4	0.36	1.0	17	N	N	Tr

Fish and fish products continued

Composition of food per 100g

No.	Food	Description and main data sources	Edible proportion	Water g	Protein g	Fat g	Carbo-hydrate g	Energy value kcal	kJ
White fish									
600	**Saithe**, *raw*	Literature sources	N	81.0	(17.0)	(0.5)	0	(73)	(308)
601	*steamed*	Pieces from tail end	1.00	74.8	23.3	0.6	0	99	418
602	*steamed, weighed with bones and skin*	Calculated from steamed	0.85	63.5	19.8	0.5	0	84	355
603	**Skate**, *in batter, fried*	6 cooked samples from fish and chip shops	1.00	61.8	17.9[a]	12.1	4.9	199	830
604	*in batter, fried, weighed with waste*	Calculated from in batter, fried	0.82	50.7	14.7[a]	9.9	4.0	163	680
605	**Whiting**, *steamed*	Flesh only	1.00	76.9	20.9	0.9	0	92	389
606	*steamed, weighed with bones*	Calculated from steamed	0.93	52.2	14.3	0.6	0	63	265
607	*in crumbs, fried*	Samples coated in crumbs and fried	1.00	63.0	18.1	10.3	7.0	191	801
608	*-, fried, weighed with bones*	Calculated from in crumbs, fried	0.90	56.8	16.3	9.3	6.3	173	722

[a] (Total N – non-protein N) $\times$ 6.25

Fish and fish products *continued*

600 to 608

Composition of food per 100g

No.	Food	Total nitrogen g	Fatty acids			Cholest-erol mg	Starch g	Total sugars g	Dietary fibre	
			Satd g	Mono unsatd g	Poly unsatd g				Southgate method g	Englyst method g
White fish										
600	**Saithe,** *raw*	(2.72)	0.1	0.1	0.2	40	0	0	0	0
601	*steamed*	3.73	0.1	0.1	0.2	53	0	0	0	0
602	*steamed, weighed with bones and skin*	3.17	0.1	0.1	0.2	45	0	0	0	0
603	**Skate,** *in batter, fried*	3.67	1.2	4.1	5.6	N	(4.8)	(0.1)	0.2	(0.2)
604	*in batter, fried, weighed with waste*	3.01	1.0	3.4	4.6	N	(3.9)	(0.1)	0.2	(0.2)
605	**Whiting,** *steamed*	3.35	0.1	0.3	0.2	46	0	0	0.2	0
606	*steamed, weighed with bones*	2.28	0.1	0.2	0.1	31	0	0	0	0
607	*in crumbs, fried*	2.90	1.0	3.5	4.7	N	(6.8)	(0.2)	0.3	(0.3)
608	*-, fried, weighed with bones*	2.61	0.9	3.2	4.3	N	(6.1)	(0.2)	0.3	(0.3)

203

Fish and fish products *continued*

Inorganic constituents per 100g

No.	Food	mg										µg	
		Na	K	Ca	Mg	P	Fe	Cu	Zn	Cl	Mn	Se	I
White fish													
600	**Saithe**, *raw*	(73)	(260)	(14)	(23)	(190)	(0.5)	0.05	0.5	(200)	0.01	20	36
601	*steamed*	97	350	19	31	250	0.6	0.07	0.7	83	0.01	28	(48)
602	*steamed, weighed with bones and skin*	83	300	16	26	210	0.5	0.06	0.6	71	0.01	24	(41)
603	**Skate**, *in batter, fried*	140	240	50	27	180	1.0	0.09	0.9	170	N	N	N
604	*in batter, fried, weighed with waste*	110	200	40	22	150	0.8	0.07	0.7	140	N	N	N
605	**Whiting**, *steamed*	130	300	42	28	190	1.0	0.02	0.3	93	0.01	22	(67)
606	*steamed, weighed with bones*	86	200	29	19	130	0.7	0.01	0.2	63	0.01	15	(46)
607	*in crumbs, fried*	200	320	48	33	260	0.7	N	N	190	N	N	N
608	*-, fried, weighed with bones*	180	290	43	29	230	0.6	N	N	180	N	N	N

Fish and fish products *continued*

No.	Food	Retinol μg	Carotene μg	Vitamin D μg	Vitamin E mg	Thiamin mg	Ribo-flavin mg	Niacin mg	Trypt 60 mg	Vitamin B6 mg	Vitamin B12 μg	Folate μg	Panto-thenate mg	Biotin μg	Vitamin C mg
White fish															
600	**Saithe,** raw	Tr	Tr	Tr	0.36	0.10	0.20	3.4	3.2	0.47	4.0	N	0.38	7	Tr
601	steamed	Tr	Tr	Tr	(0.47)	(0.12)	(0.26)	(4.0)	4.4	(0.62)	(5.0)	N	(0.40)	(8)	Tr
602	steamed, weighed with bones and skin	Tr	Tr	Tr	(0.40)	(0.10)	(0.22)	(3.4)	3.7	(0.53)	(4.0)	N	(0.34)	(7)	Tr
603	**Skate,** in batter, fried	9	Tr	N	1.20	0.03	0.10	2.4	N	N	N	N	N	N	Tr
604	in batter, fried, weighed with waste	7	Tr	N	1.00	0.02	0.08	2.0	N	N	N	N	N	N	Tr
605	**Whiting,** steamed	Tr	Tr	Tr	N	N	N	N	3.9	N	N	N	N	N	Tr
606	steamed, weighed with bones	Tr	Tr	Tr	N	N	N	N	2.7	N	N	N	N	N	Tr
607	in crumbs, fried	Tr	Tr	Tr	N	N	N	N	3.4	N	N	N	N	N	Tr
608	-, fried, weighed with bones	Tr	Tr	Tr	N	N	N	N	3.1	N	N	N	N	N	Tr

Fish and fish products *continued*

609 to 621

Composition of food per 100g

No.	Food	Description and main data sources	Edible proportion	Water g	Protein g	Fat g	Carbo-hydrate g	Energy value kcal	Energy value kJ
	Fatty fish								
609	**Anchovies**, canned in oil, *drained*	10 assorted brands	N	41.6	25.2	19.9	0	280	1165
610	**Herring**, *raw*	12 fish, sampled in November; flesh only	0.55	63.9[a]	16.8	18.5[b]	0	234	970
611	*fried*	Flesh, skin and roes; covered in oatmeal	1.00	58.7	23.1	15.1	1.5	234	975
612	*fried, weighed with bones*	Calculated from fried	0.88	51.6	20.3	13.3	1.3	206	858
613	*grilled*	12 fish, flesh only	1.00	65.5	20.4	13.0	0	199	828
614	*grilled, weighed with bones*	Calculated from grilled	0.68	44.5	13.9	8.8	0	135	562
615	**Kipper**, *baked*	Flesh only	1.00	58.7	25.5	11.4	0	205	855
616	*baked, weighed with bones*	Calculated from baked	0.54	31.6	13.8	6.2	0	111	464
617	**Mackerel**, *raw*	Literature sources	N	64.0	19.0	16.3	0	223	926
618	*fried*	Flesh only	1.00	65.6	21.5	11.3	0	188	784
619	*fried, weighed with bones*	Calculated from fried	0.73	47.8	15.7	8.3	0	138	574
620	*smoked*	10 samples, flesh and skin	0.98	47.1	18.9	30.9	0	354	1465
621	**Pilchards**, canned in tomato sauce	6 cans, 4 brands; whole contents	1.00	70.0	18.8	5.4	0.7	126	531

[a] The values for water content can vary throughout the year from about 75g per 100g in February – April to 60g per 100g in July – October

[b] The values for fat content can vary throughout the year from about 5g per 100g in February – April to 20g per 100g in July – October

Fish and fish products continued

Composition of food per 100g

No.	Food	Total nitrogen g	Fatty acids Satd g	Fatty acids Mono unsatd g	Fatty acids Poly unsatd g	Cholesterol mg	Starch g	Total sugars g	Dietary fibre Southgate method g	Dietary fibre Englyst method g
Fatty fish										
609	**Anchovies**, canned in oil, *drained*	4.03	N	N	N	N	0	0	0	0
610	**Herring**, *raw*	2.69	5.3	8.4	3.1	70	0	0	0	0
611	*fried*	3.69	4.3	6.9	2.5	(80)	1.5	Tr	1.3	N
612	*fried, weighed with bones*	3.24	3.8	6.0	2.2	(70)	1.3	Tr	1.2	N
613	*grilled*	3.26	3.7	5.9	2.1	50	0	0	0	0
614	*grilled, weighed with bones*	2.22	2.5	4.0	1.4	34	0	0	0	0
615	**Kipper**, *baked*	4.08	1.8	6.0	2.5	38	0	0	0	0
616	*baked, weighed with bones*	2.20	1.0	3.3	1.3	20	0	0	0	0
617	**Mackerel**, *raw*	3.04	3.3	8.0	3.3	55	0	0	0	0
618	*fried*	3.44	2.3	5.6	2.3	(62)	0	0	0	0
619	*fried, weighed with bones*	2.51	1.7	4.1	1.7	(45)	0	0	0	0
620	*smoked*	3.00	6.3	15.1	6.3	104	0	0	0	0
621	**Pilchards**, canned in tomato sauce	3.01	1.1	1.5	2.3	56	0.1	0.6	Tr	Tr

Fatty fish

No.	Food	mg										µg	
		Na	K	Ca	Mg	P	Fe	Cu	Zn	Cl	Mn	Se	I
609	**Anchovies**, canned in oil, *drained*	3930	230	300	56	300	4.1	0.17	3.2	6090	0.18	N	N
610	**Herring**, *raw*	67	340	33	29	210	0.8	0.12	0.5	76	0.04	34	29
611	*fried*	100	420	39	35	340	1.9	N	N	130	N	46	N
612	*fried, weighed with bones*	89	370	34	31	300	1.7	N	N	110	N	41	N
613	*grilled*	170	370	33	32	240	1.0	0.11	0.5	220	0.05	41	(32)
614	*grilled, weighed with bones*	120	250	22	22	160	0.7	0.07	0.4	150	0.03	28	(21)
615	**Kipper**, *baked*	990	520	65	48	430	1.4	0.14	1.3	1520	0.04	43	70
616	*baked, weighed with bones*	540	280	35	26	230	0.8	0.08	0.7	820	0.02	23	38
617	**Mackerel**, *raw*	(130)	(360)	(24)	(30)	240	(1.0)	0.19	0.5	(97)	0.02	30	170
618	*fried*	150	420	28	35	280	1.2	0.20	N	110	0.02	34	(190)
619	*fried, weighed with bones*	110	310	21	25	200	0.9	0.15	N	83	0.02	25	(130)
620	*smoked*	750	310	20	28	210	1.2	0.09	1.1	1130	0.02	27	(150)
621	**Pilchards**, canned in tomato sauce	370	420	300	39	350	2.7	0.19	1.6	580	0.11	30	64

Fish and fish products *continued*

No.	Food	Retinol µg	Carotene µg	Vitamin D µg	Vitamin E mg	Thiamin mg	Ribo-flavin mg	Niacin mg	Trypt 60 mg	Vitamin B_6 mg	Vitamin B_{12} µg	Folate µg	Panto-thenate mg	Biotin µg	Vitamin C mg
	Fatty fish														
609	**Anchovies**, canned in oil, *drained*	62	Tr	N	N	Tr	0.10	3.8	4.7	N	11	18	N	N	Tr
610	**Herring**, *raw*	45	Tr	22.5	0.21	Tr	0.18	4.1	3.1	0.45	6	5	1.00	10	Tr
611	*fried*	(49)	Tr	25.0	(0.30)	Tr	(0.18)	(4.0)	4.3	(0.57)	(11.0)	(10)	(0.88)	(10)	Tr
612	*fried, weighed with bones*	(43)	Tr	22.0	(0.26)	Tr	(0.16)	(3.5)	3.8	(0.50)	(10.0)	(9)	(0.77)	(9)	Tr
613	*grilled*	34	Tr	25.0	0.30	Tr	0.18	4.0	3.8	0.57	11.0	10	0.88	10	Tr
614	*grilled, weighed with bones*	23	Tr	17.0	0.20	Tr	0.12	2.7	2.6	0.39	8.0	7	(0.60)	(7)	Tr
615	**Kipper**, *baked*	33	Tr	25.0	(0.30)	Tr	(0.18)	(4.0)	(4.8)	(0.57)	(11.0)	(10)	(0.88)	(10)	Tr
616	*baked, weighed with bones*	18	Tr	13.5	(0.16)	Tr	(0.10)	(2.2)	2.6	(0.31)	(6.0)	(5)	(0.48)	(5)	Tr
617	**Mackerel**, *raw*	45	Tr	17.5	N	0.09	0.35	8.0	3.6	0.70	10	N	1.00	7	Tr
618	*fried*	43	Tr	(21.1)	N	(0.09)	(0.38)	(8.7)	4.0	(0.84)	(12.0)	N	(0.96)	(8)	Tr
619	*fried, weighed with bones*	32	Tr	(15.4)	N	(0.07)	(0.28)	(6.4)	2.9	(0.61)	(9.0)	N	(0.70)	(6)	Tr
620	*smoked*	25	Tr	8.0	0.25	0.26	0.52	9.7	3.5	1.03	5.6	N	1.03	3	Tr
621	**Pilchards**, canned in tomato sauce	8	(142)	8.0	0.70	0.02	0.29	7.6	3.5	0.27	12.0	N	0.85	11	Tr

Fish and fish products *continued*

622 to 633

Composition of food per 100g

No.	Food	Description and main data sources	Edible proportion	Water g	Protein g	Fat g	Carbohydrate g	Energy value kcal	kJ
Fatty fish									
622	**Salmon**, *raw*	Atlantic salmon; literature sources	N	68.0	(18.4)	(12.0)	0	(182)	(757)
623	*steamed*	Shoulder cut, flesh only	1.00	65.4	20.1	13.0	0	197	823
624	*steamed, weighed with bones and skin*	Calculated from steamed	0.81	53.0	16.3	10.5	0	160	666
625	*canned*	10 cans, red salmon; backbone and skin removed	0.98	70.4	20.3	8.2	0	155	649
626	*smoked*	4 samples	1.00	64.9	25.4	4.5	0	142	598
627	**Sardines**, canned in tomato sauce	10 cans, 4 brands; whole contents	1.00	65.0	17.8	11.6	0.5	177	740
628	*canned in oil, drained*	10 cans, 6 brands	0.83	58.4	23.7	13.6	0	217	906
629	**Trout**, brown, *steamed*	Flesh only	1.00	70.6	23.5	4.5	0	135	566
630	*steamed, weighed with bones*	Calculated from steamed	0.66	46.5	15.5	3.0	0	89	375
631	**Tuna**, canned in oil, *drained*	6 cans, 2 brands; skipjack tuna	0.79	63.3	27.1	9.0	0	189	794
632	*canned in brine, drained*	10 cans, 9 brands	0.81	74.6	23.5	0.6	0	99	422
633	**Whitebait**, *fried*	Whole fish; rolled in flour and fried	1.00	23.5	19.5	47.5	5.3	525	2174

Fish and fish products *continued*

No.	Food	Total nitrogen g	Fatty acids Satd g	Mono unsatd g	Poly unsatd g	Cholesterol mg	Starch g	Total sugars g	Dietary fibre Southgate method g	Englyst method g
	Fatty fish									
622	**Salmon**, *raw*	2.94	2.2	5.1	3.4	50	0	0	0	0
623	*steamed*	3.21	2.4	5.5	3.7	55	0	0	0	0
624	*steamed, weighed with bones and skin*	2.60	1.9	4.5	3.0	44	0	0	0	0
625	*canned*	3.24	1.5	3.5	2.4	34	0	0	0	0
626	*smoked*	4.06	0.8	1.9	1.3	(50)	0	0	0	0
627	**Sardines**, canned in tomato sauce	2.84	3.3	3.4	3.7	76	Tr	0.5	Tr	Tr
628	*canned in oil, drained*	3.79	2.8	4.7	4.8	65	0	0	0	0
629	**Trout**, brown, *steamed*	3.76	(1.0)	(1.6)	(1.5)	97	0	0	0	0
630	*steamed, weighed with bones*	2.48	(0.7)	(1.1)	(1.0)	64	0	0	0	0
631	**Tuna**, canned in oil, *drained*	4.30	1.4	2.2	4.5	50	0	0	0	0
632	*canned in brine, drained*	3.80	0.2	0.1	0.2	51	0	0	0	0
633	**Whitebait**, *fried*	3.12	4.4	15.1	20.6	N	(5.2)	(0.1)	0.2	0.2

Fish and fish products *continued*

Inorganic constituents per 100g

No.	Food	mg										µg	
		Na	K	Ca	Mg	P	Fe	Cu	Zn	Cl	Mn	Se	I
Fatty fish													
622	**Salmon**, *raw*	(98)	(310)	(27)	(26)	(280)	(0.7)	0.20	0.8	(59)	0.02	20	N
623	*steamed*	110	330	29	29	300	0.8	(0.22)	(0.9)	64	0.02	22	N
624	*steamed, weighed with bones and skin*	87	270	23	23	250	0.6	(0.18)	(0.7)	52	0.02	18	N
625	*canned*	570	300	93	30	240	1.4	0.09	0.9	880	Tr	25	59
626	*smoked*	1880	420	19	32	250	0.6	0.09	0.4	2850	(0.02)	(24)	N
627	**Sardines**, *canned in tomato sauce*	700	410	460	51	400	4.6	0.23	2.7	1110	0.24	37	N
628	*canned in oil, drained*	650	430	550	52	520	2.9	0.19	3.0	1000	0.19	50	23
629	**Trout**, *brown, steamed*	88[a]	370	36	31	270	1.0	(0.04)	(0.5)	70[a]	0.01	24	16
630	*steamed, weighed with bones*	58	250	24	20	180	0.7	(0.03)	(0.3)	46	0.01	16	11
631	**Tuna**, *canned in oil, drained*	290	260	12	33	190	1.6	0.20	1.1	530	0.05	90	14
632	*canned in brine, drained*	320	230	8	27	170	1.0	0.05	0.7	550	Tr	78	13
633	**Whitebait**, *fried*	230	110	860	50	860	5.1	N	N	330	N	N	N

[a] Sea trout contains 210mg Na and 260mg Cl per 100g

Fish and fish products *continued*

No.	Food	Retinol µg	Carotene µg	Vitamin D µg	Vitamin E mg	Thiamin mg	Ribo-flavin mg	Niacin mg	Trypt 60 mg	Vitamin B6 mg	Vitamin B12 µg	Folate µg	Panto-thenate mg	Biotin µg	Vitamin C mg
Fatty fish															
622	**Salmon**, raw	13[a]	Tr	Tr[a]	N	0.20	0.15	7.0	3.4	0.75	5	26	2.00	5	Tr
623	steamed	10[a]	Tr	Tr[a]	N	(0.20)	(0.11)	(7.0)	3.8	(0.83)	(6.0)	(29)	(1.80)	(4)	Tr
624	steamed, *weighed with bones and skin*	8[a]	Tr	Tr[a]	N	(0.16)	(0.09)	(5.7)	3.0	(0.67)	(5.0)	(23)	(1.50)	(3)	Tr
625	canned	35	Tr	12.5	1.50	0.04	0.18	7.0	3.8	0.45	4.0	12	0.50	5	Tr
626	smoked	(13)[a]	Tr	Tr[a]	N	0.16	0.17	8.8	4.7	N	N	N	N	N	Tr
627	**Sardines**, canned in tomato sauce	9	142	7.5	0.53	0.02	0.28	5.5	3.3	0.35	14	13	0.50	5	Tr
628	canned in oil, *drained*	(11)	Tr	7.5	0.30	0.04	0.36	8.2	4.4	0.48	28	8	0.50	5	Tr
629	**Trout**, brown, steamed	(39)	Tr	N	N	(0.20)	(0.12)	(4.2)	4.4	(0.35)	(4.9)	N	(1.41)	(2)	Tr
630	steamed, *weighed with bones*	(26)	Tr	N	N	(0.13)	(0.08)	(2.8)	2.9	(0.23)	(3.2)	N	(0.90)	(2)	Tr
631	**Tuna**, canned in oil, *drained*	N	Tr	5.8	N	0.02	0.12	16.1	5.0	0.51	4.8	N	0.32	3	Tr
632	canned in brine, *drained*	N	Tr	4.0	0.55	0.02	0.11	14.4	4.4	0.47	4.3	16	0.29	2	Tr
633	**Whitebait**, *fried*	N	Tr	N	N	N	N	N	3.6	N	N	N	N	N	Tr

[a] These are values for Atlantic salmon. Pacific salmon may contain 90 (20 –150) µg retinol and 12.5 (5 – 20) µg vitamin D per 100g

Fish and fish products continued

Composition of food per 100g

No.	Food	Description and main data sources	Edible proportion	Water g	Protein g	Fat g	Carbo-hydrate g	Energy value kcal	Energy value kJ
Crustacea									
634	**Crab**, *boiled*	Boiled in fresh water	1.00	72.5	20.1	5.2	0	127	534
635	*boiled, weighed with shell*	Calculated from boiled	0.20	14.5	4.0	1.0	0	25	105
636	*canned*	6 cans, 2 brands	1.00	79.2	18.1	0.9	0	81	341
637	**Lobster**, *boiled*	Boiled in fresh water	1.00	72.4	22.1	3.4	0	119	502
638	*boiled, weighed with shell*	Calculated from boiled	0.36	26.1	7.9	1.2	0	42	179
639	**Prawns**, *boiled*	Samples cooked in sea or salt water	1.00	70.0	22.6	1.8	0	107	451
640	*boiled, weighed with shell*	Calculated from boiled	0.38	26.6	8.6	0.7	0	41	172
641	**Scampi**, *in breadcrumbs, frozen, fried*	5 packets	1.00	39.4	12.2	17.6	28.9	316	1321
642	**Shrimps**, *frozen, shell removed*	10 assorted brands	1.00	81.2	16.5	0.8	0	73	310
643	*canned, drained*	10 cans, 3 brands	0.65	74.9	20.8	1.2	0	94	398
Molluscs									
644	**Cockles**, *boiled*	Samples cooked in sea or salt water	1.00	78.9	11.3	0.3	Tr	48	203
645	**Mussels**, *boiled*	Boiled in fresh water	1.00	79.0	17.2	2.0	Tr	87	366
646	*boiled, weighed with shell*	Calculated from boiled	0.30	23.7	5.2	0.6	Tr	26	111
647	**Squid**, *frozen, raw*	5 assorted brands	0.59	84.2	13.1	1.5	0	66	278
648	**Whelks**, *boiled, weighed with shell*	Samples cooked in sea or salt water	0.15	11.6	2.8	0.3	Tr	14	59
649	**Winkles**, *boiled, weighed with shell*	Samples cooked in sea or salt water	0.19	15.1	2.9	0.3	Tr	14	60

Fish and fish products continued

Composition of food per 100g

No.	Food	Total nitrogen g	Fatty acids			Cholest-erol mg	Starch g	Total sugars g	Dietary fibre	
			Satd g	Mono unsatd g	Poly unsatd g				Southgate method g	Englyst method g
Crustacea										
634	**Crab**, *boiled*	3.21	0.7	1.4	1.5	72	0	0	0	0
635	*boiled, weighed with shell*	0.64	0.1	0.3	0.3	11	0	0	0	0
636	*canned*	2.90	0.1	0.3	0.3	(72)	0	0	0	0
637	**Lobster**, *boiled*	3.54	N	N	N	150	0	0	0	0
638	*boiled, weighed with shell*	1.27	N	N	N	43	0	0	0	0
639	**Prawns**, *boiled*	3.62	0.4	0.5	0.4	81	0	0	0	0
640	*boiled, weighed with shell*	1.38	0.2	0.2	0.1	31	0	0	0	0
641	**Scampi**, *in breadcrumbs, frozen, fried*	1.95	1.7	6.0	8.1	110	(28.9)	Tr	1.1	N
642	**Shrimps**, *frozen, shell removed*	2.64	0.1	0.2	0.3	(200)	0	0	0	0
643	*canned, drained*	3.33	0.2	0.2	0.4	(200)	0	0	0	0
Molluscs										
644	**Cockles**, *boiled*	1.80	0.1	Tr	0.1	53	Tr	Tr	0	0
645	**Mussels**, *boiled*	2.75	0.4	0.3	0.7	58	Tr	Tr	0	0
646	*boiled, weighed with shell*	0.83	0.1	0.1	0.2	12	Tr	Tr	0	0
647	**Squid**, *frozen, raw*	2.10	N	N	N	N	0	0	0	0
648	**Whelks**, *boiled, weighed with shell*	0.44	0.1	0.1	0.1	19	Tr	Tr	0	0
649	**Winkles**, *boiled, weighed with shell*	0.47	Tr	0.1	0.1	20	Tr	Tr	0	0

Fish and fish products *continued*

Inorganic constituents per 100g

No.	Food	Na	K	Ca	Mg	P	Fe	Cu	Zn	Cl	Mn	Se	I
		mg										µg	
Crustacea													
634	**Crab**, *boiled*	370	270	29	48	350	1.3	4.80	5.5	570	0.17	(17)	N
635	*boiled, weighed with shell*	73	54	6	10	70	0.3	1.00	1.1	110	0.03	(3)	N
636	*canned*	550	100	120	32	140	2.8	0.42	5.0	830	N	N	N
637	**Lobster**, *boiled*	330	260	62	34	280	0.8	1.70	1.8	530	N	N	N
638	*boiled, weighed with shell*	120	93	22	12	100	0.3	0.65	0.6	190	N	N	N
639	**Prawns**, *boiled*	1590	260	150	42	350	1.1	(0.70)	(1.6)	2550	0.01	18	(28)
640	*boiled, weighed with shell*	610	99	55	16	130	0.4	(0.27)	(0.6)	970	Tr	7	(11)
641	**Scampi**, *in breadcrumbs, frozen, fried*	380	390	99	30	310	1.1	0.22	0.6	740	0.35	15	41
642	**Shrimps**, *frozen, shell removed*	375	75	128	47	150	2.6	0.15	1.1	520	0.15	49	N
643	*canned, drained*	980	100	110	49	150	5.1	0.23	2.4	1510	N	52	N
Molluscs													
644	**Cockles**, *boiled*	3520	43	130	51	200	26.0[a]	0.28	1.2	5220	0.69	45	160
645	**Mussels**, *boiled*	210	92	200	25	330	7.7	0.48	2.1	320	0.26	45	120
646	*boiled, weighed with shell*	63	28	59	8	99	2.3	0.16	0.6	95	0.05	9	23
647	**Squid**, *frozen, raw*	185	145	13	36	170	0.2	0.68	1.2	280	0.02	N	N
648	**Whelks**, *boiled, weighed with shell*	40	47	8	24	34	0.9	(1.10)	1.1	88	0.02	N	N
649	**Winkles**, *boiled, weighed with shell*	220	29	26	68	42	2.9	(0.25)	(1.1)	340	0.22	N	N

[a] The iron content of cockles can be as high as 40mg per 100g

Fish and fish products *continued*

No.	Food	Retinol µg	Carotene µg	Vitamin D µg	Vitamin E mg	Thiamin mg	Ribo-flavin mg	Niacin mg	Trypt 60 mg	Vitamin B6 mg	Vitamin B12 µg	Folate µg	Panto-thenate mg	Biotin µg	Vitamin C mg
Crustacea															
634	**Crab**, *boiled*	Tr	Tr	Tr	N	0.10	0.15	2.5	3.8	0.35	Tr	20	0.60	Tr	Tr
635	*boiled, weighed with shell*	Tr	Tr	Tr	N	0.02	0.03	0.5	0.8	0.07	Tr	4	0.12	Tr	Tr
636	*canned*	Tr	Tr	Tr	N	Tr	0.05	1.1	3.4	N	Tr	N	N	Tr	Tr
637	**Lobster**, *boiled*	Tr	Tr	Tr	1.50	0.08	0.05	1.5	4.1	N	1.0	17	1.63	5	Tr
638	*boiled, weighed with shell*	Tr	Tr	Tr	0.50	0.03	0.02	0.5	1.5	N	0	6	0.59	2	Tr
639	**Prawns**, *boiled*	Tr	Tr	Tr	N	(0.02)	(0.17)	(0.6)	4.2	(0.07)	N	N	(1.00)	(Tr)	Tr
640	*boiled, weighed with shell*	Tr	Tr	Tr	N	(0.01)	(0.06)	(0.2)	1.6	(0.03)	N	N	(0.40)	(Tr)	Tr
641	**Scampi**, *in breadcrumbs, frozen, fried*	Tr	Tr	Tr	N	0.08	0.05	1.3	2.3	0.09	1.1	N	0.26	1	Tr
642	**Shrimps**, *frozen, shell removed*	N	Tr	N	N	Tr	0.02	0.5	3.1	N	2.6	14	N	N	Tr
643	*canned, drained*	Tr	Tr	Tr	N	0.01	0.02	0.8	3.9	0.03	2.0	15	0.35	1	(1)
Molluscs															
644	**Cockles**, *boiled*	N	11	Tr	N	N	N	N	2.4	N	Tr	N	N	N	Tr
645	**Mussels**, *boiled*	N	Tr	Tr	N	0.02	0.38	1.3	3.7	0.06	22.1	N	0.40	9	Tr
646	*boiled, weighed with shell*	N	Tr	Tr	N	Tr	0.08	0.3	1.4	0.01	4.4	N	0.08	2	Tr
647	**Squid**, *frozen, raw*	N	0	N	N	0.05	0.02	2.1	2.8	N	2.5	2	N	N	0
648	**Whelks**, *boiled, weighed with shell*	N	Tr	Tr	0.10	0.01	0.03	0.2	0.6	0.01	3.1	N	N	N	Tr
649	**Winkles**, *boiled, weighed with shell*	N	Tr	Tr	N	0.05	0.07	0.3	0.6	0.02	6.9	N	0.07	1	Tr

Fish and fish products *continued*

Composition of food per 100g

Fish products and dishes

No.	Food	Description and main data sources	Edible proportion	Water g	Protein g	Fat g	Carbo-hydrate g	Energy value kcal	Energy value kJ
650	**Fish cakes,** *fried*	14 packets, 4 brands, white fish	1.00	63.3	9.1	10.5	15.1	188	785
651	**Fish fingers,** *fried in blended oil*	11 packets, 3 brands; coated in breadcrumbs	1.00	55.6	13.5	12.7	17.2	233	975
652	*fried in lard*	Calculated from fried in blended oil	1.00	55.6	13.5	12.7	17.2	233	975
653	*grilled*	Calculation from *fried in lard*	1.00	56.2	15.1	9.0	19.3	214	899
654	**Fish paste**	30 samples, sardine, crab, lobster, salmon	1.00	67.1	15.3	10.4	3.7	169	704
655	**Fish pie**	Recipe	1.00	75.7	8.0	3.0	12.3	105	443
656	**Kedgeree**	Recipe	1.00	65.6	14.2	7.9	10.5	166	701
657	**Roe,** cod, hard, *fried*	Parboiled, sliced and fried in crumbs	1.00	62.0	20.9	11.9	3.0	202	844
658	herring, soft, *fried*	Rolled in flour and fried	1.00	52.3	21.1[a]	15.8	4.7	244	1019
659	**Taramasalata**	10 assorted samples	1.00	35.9	3.2	46.4	4.1	446	1837

[a] (Total N – purine N) × 6.25

Fish and fish products *continued*

650 to 659

Composition of food per 100g

No.	Food	Total nitrogen g	Fatty acids			Cholesterol mg	Starch g	Total sugars g	Dietary fibre	
			Satd g	Mono unsatd g	Poly unsatd g				Southgate method g	Englyst method g
	Fish products and dishes									
650	**Fish cakes**, *fried*	1.45	1.0	3.5	4.8	17	(15.1)	(Tr)	0.6	N
651	**Fish fingers**, *fried in blended oil*	2.16	2.8	5.4	3.8	N	(17.2)	Tr	0.6	0.6
652	*fried in lard*	2.16	4.5	5.1	2.4	(37)	(17.2)	Tr	0.6	0.6
653	*grilled*	2.42	2.8	3.4	2.3	35	(19.3)	Tr	0.7	0.7
654	**Fish paste**	2.45	N	N	N	N	(3.7)	Tr	0.2	(0.2)
655	**Fish pie**	1.28	1.2	1.0	0.5	19	10.8	1.5	0.9	0.7
656	**Kedgeree**	2.29	2.3	3.0	1.8	116	10.4	0.1	0.3	Tr
657	**Roe**, cod, hard, *fried*	3.35	1.2	4.0	5.5	(500)	3.0	Tr	0.1	(0.1)
658	herring, soft, *fried*	3.85[a]	1.6	5.4	7.3	(700)	4.7	Tr	N	N
659	**Taramasalata**	0.51	3.2	22.6	12.8	37	4.1	Tr	N	N

[a] Includes 0.48g purine nitrogen per 100g

Inorganic constituents per 100g

No.	Food	mg													µg	
		Na	K	Ca	Mg	P	Fe	Cu	Zn	Cl	Mn		Se	I
	Fish products and dishes														
650	**Fish cakes**, *fried*	500	260	70	18	110	1.0	0.13	0.4	730	0.20		N	N
651	**Fish fingers**, *fried in blended*													
	oil	350	260	45	19	220	0.7	0.08	0.4	400	0.19		N	(110)
652	*fried in lard*	350	260	45	19	220	0.7	0.08	0.4	400	0.19		N	(110)
653	*grilled*	380	290	52	22	230	0.8	0.07	0.5	460	0.4		N	100
654	**Fish paste**	600	300	280	33	310	9.0	0.37	1.4	940	N		N	310
655	**Fish pie**	250	290	37	18	93	0.4	0.06	0.4	410	0.07		10	36
656	**Kedgeree**	870	180	46	18	180	0.8	0.08	0.7	1330	0.09		17	130
657	**Roe**, cod, hard, *fried*	130	260	17	11	500	1.6	N	N	190	N		N	N
658	*herring, soft, fried*	87	240	16	8	920	1.5	N	N	120	N		N	N
659	**Taramasalata**	650	60	21	6	50	0.4	5.80	0.4	1040	0.12		N	N

Fish and fish products *continued*

Vitamins per 100g

No.	Food	Retinol µg	Carotene µg	Vitamin D µg	Vitamin E mg	Thiamin mg	Ribo-flavin mg	Niacin mg	Trypt 60 mg	Vitamin B6 mg	Vitamin B12 µg	Folate µg	Panto-thenate mg	Biotin µg	Vitamin C mg
	Fish products and dishes														
650	**Fish cakes,** *fried*	Tr	Tr	Tr	N	0.06	0.06	1.1	1.7	N	N	N	N	N	Tr
651	**Fish fingers,** *fried in blended*														
	oil	Tr	Tr	Tr	N	0.08	0.07	1.4	2.5	0.21	2	16	(0.35)	(1)	Tr
652	*fried in lard*	Tr	Tr	Tr	N	0.08	0.07	1.4	2.5	0.21	2.0	16	(0.35)	(1)	Tr
653	*grilled*	Tr	Tr	Tr	N	0.10	0.06	1.3	2.8	0.25	1.0	16	0.35	1	Tr
654	**Fish paste**	34	Tr	N	0.87	0.02	0.20	4.1	2.9	N	N	N	N	N	Tr
655	**Fish pie**	28	15	0.2	0.40	0.12	0.06	0.8	1.6	0.31	0.7	9	0.35	1	2
656	**Kedgeree**	86	42	0.8	0.70	0.06	0.13	1.1	3.0	0.20	2.0	9	0.44	5	0
657	**Roe,** cod, hard, *fried*	41[a]	Tr	2.2	(6.90)	(1.30)	(0.90)	(1.3)	3.9	(0.28)	(11.0)	N	(2.60)	(15)	(26)
658	*herring, soft, fried*	N	Tr	N	N	(0.20)	(0.50)	(2.0)	3.9	N	(6.0)	N	(0.49)	N	(5)
659	**Taramasalata**	N	N	N	N	0.08	0.10	0.3	0.6	N	2.9	4	N	N	1

[a] 88 per cent is present as retinaldehyde

Section 2.7

Vegetables

The foods in this section of the Tables have been taken from the *'Vegetables, Herbs and Spices'* supplement with additional data for foods fried in a variety of fats. Because many of the vegetables and pulses eaten in this country are imported, a larger number of literature values from foreign sources have been used in this food group than many others in the Tables.

The Institute of Food Research (IFR) measured thiamin, riboflavin, niacin and vitamin B6 by HPLC (Kwiatkowska *et al.*, 1989) while the Laboratory of the Government Chemist (LGC) determined them microbiologically. HPLC often gave markedly higher values for thiamin and lower values for riboflavin than microbiological assays, even on very similar foods. The IFR also obtained higher values for fat because they included methanol in the extraction solvent. The laboratory conducting the analysis has therefore been shown where appropriate.

For most boiled vegetables, data is included for foods boiled in unsalted water. The amount of salt added to vegetables when boiled can vary considerably, where foods are included as boiled in salted water the water contained 0.5% salt. For fried foods the type of oil used for frying has been included in the name, this will determine the fatty acid profile of that particular food. Most values for cooked foods were obtained by analysis, but some were calculated from raw foods. For these, any nutrient losses were estimated using the factors shown in Section 3.4. The changes in weight of beans and some other vegetables when soaked and cooked are shown in Section 3.3.

Samples of the same or similar foods always vary somewhat in composition. Some nutrients differ in a consistent way between varieties of a vegetable and with season as shown for potatoes. There are also differences with the length of storage, the depth of peeling or the number of outer leaves removed, and with cooking conditions (such as the degree to which a vegetable is cut up, the amount of water and the length of cooking, although there is little or no difference between vegetables cooked with microwaves or by more conventional methods). Any differences arising from the method of cultivation, for example 'organic' methods, appear to be small and inconsistent. It is not practical to give specific nutrient values for each of these factors, and the tables therefore show average values for most products.

Vegetables

Composition of food per 100g

No.	Food	Description and main data sources	Edible proportion	Water g	Protein g	Fat g	Carbo-hydrate g	Energy value kcal	kJ
	Early potatoes								
660	**New potatoes**, *average, raw*	IFR; flesh only	0.89	81.7	1.7	0.3	16.1	70	298
661	*boiled in unsalted water*	IFR. Samples as raw; boiled 20 minutes	1.00	80.5	1.5	0.3	17.8	75	321
662	*in skins, boiled in unsalted water*	LGC; boiled 20 minutes	1.00	81.1	1.4	0.3	15.4	66	281
663	*canned, re-heated, drained*	LGC; 10 samples, 4 brands	0.65	81.3	1.5	0.1	15.1	63	271
	Main crop potatoes								
664	**Old potatoes**, *average, raw*	IFR; 4 varieties sampled over two years. Flesh only	0.80	79.0	2.1	0.2	17.2	75	318
665	*baked, flesh and skin*	LGC. Samples as raw; baked 90 minutes 200C	1.00	62.6	3.9	0.2	31.7	136	581
666	*-, flesh only*	IFR. Samples as raw; baked 90 minutes 200C	1.00	78.9	2.2	0.1	18.0	77	329
667	*-, flesh only, weighed with skin*	Calcd. from *flesh only*	0.67	52.9	1.5	0.1	12.1	52	223
668	*boiled in unsalted water*	IFR. Samples as raw; boiled 20 minutes	1.00	80.3	1.8	0.1	17.0	72	306
669	*mashed with butter*	Calculation from boiled (100g), butter (5g), milk (7g)	1.00	77.6	1.8	4.3	15.5	104	438
670	*mashed with margarine*	Calculation from boiled (100g), margarine (5g) and milk (7g)	1.00	77.6	1.8	4.3	15.5	104	438
671	*roast in blended oil*	Calculation from *roast in corn oil*	1.00	64.7	2.9	4.5	25.9	149	630
672	*roast in corn oil*	IFR. Samples as raw; roasted in shallow oil 90 minutes 200C	1.00	64.7	2.9	4.5	25.9	149	630
673	*roast in lard*	Calculation from *roast in corn oil*	1.00	64.7	2.9	4.5	25.9	149	630

Vegetables

Composition of food per 100g

No.	Food	Total nitrogen g	Fatty acids			Cholest-erol mg	Starch g	Total sugars g	Dietary fibre	
			Satd g	Mono unsatd g	Poly unsatd g				Southgate method g	Englyst method g
Early potatoes										
660	**New potatoes**, *average, raw*	0.28	0.1	Tr	0.1	0	14.8	1.3	1.3	1.0
661	*boiled in unsalted water*	0.24	0.1	Tr	0.1	0	16.7	1.1	1.2	1.1
662	*in skins, boiled in unsalted water*	0.23	0.1	Tr	0.1	0	14.4	1.0	(1.6)	1.5
663	*canned, re-heated, drained*	0.23	Tr	Tr	0.1	0	14.4	0.7	2.3	0.8
Main crop potatoes										
664	**Old potatoes**, *average, raw*	0.33	Tr	Tr	0.1	0	16.6	0.6	1.6	1.3
665	*baked, flesh and skin*	0.62	Tr	Tr	0.1	0	30.5	1.2	N	2.7
666	*-, flesh only*	0.35	Tr	Tr	0.1	0	17.3	0.7	1.7	1.4
667	*-, flesh only, weighed with skin*	0.23	Tr	Tr	0.1	0	11.6	0.5	1.1	0.9
668	*boiled in unsalted water*	0.29	Tr	Tr	0.1	0	16.3	0.7	1.4[a]	1.2
669	*mashed with butter*	0.29	2.8	1.0	0.2	12	14.5	1.0	1.3	1.1
670	*mashed with margarine*	0.29	1.4	1.6	1.1	0	14.5	1.0	1.3	1.1
671	*roast in blended oil*	0.46	0.4	2.2	1.6	0	25.3	0.6	2.4	1.8
672	*roast in corn oil*	0.46	0.6	1.1	2.6	0	25.3	0.6	2.4	1.8
673	*roast in lard*	0.46	1.8	2.0	0.4	4	25.3	0.6	2.4	1.8

[a] Analysis showed resistant starch present at 0.7g per 100g in boiled potatoes kept at room temperature for 2 hours

225

No.	Food	Na	K	Ca	Mg	P	Fe	Cu	Zn	Cl	Mn	Se	I
						mg						µg	
Early potatoes													
660	**New potatoes,** *average, raw*	11	320	6	14	34	0.3	0.09	0.20	57	(0.1)	(1)	(3)
661	*boiled in unsalted water*	9	250	5	12	28	0.3	0.06	0.1	43	(0.1)	(1)	(3)
662	*in skins, boiled in unsalted water*	10	430	13	18	54	1.6	0.06	0.3	(43)	0.2	(1)	(3)
663	*canned, re-heated, drained*	250	220	24	11	27	0.9	0.04	Tr	430	0.1	N	N
Main crop potatoes													
664	**Old potatoes,** *average, raw*	7	360	5	17	37	0.4	0.08	0.3	66	0.1	1	3
665	*baked, flesh and skin*	12	630	11	32	68	0.7	0.14	0.5	120	0.2	2	5
666	*-, flesh only*	7	360	7	18	40	0.4	0.08	0.3	72	0.1	1	3
667	*-, flesh only, weighed with skin*	5	240	5	12	27	0.3	0.05	0.2	48	0.1	1	2
668	*boiled in unsalted water*	7	280	5	14	31	0.4	0.07	0.3	45	0.1	1	3
669	*mashed with butter*	43	260	13	13	35	0.4	0.06	0.3	98	0.1	1	5
670	*mashed with margarine*	49	260	12	13	34	0.4	0.06	0.3	100	0.1	1	5
671	*roast in blended oil*	9	570	8	25	55	0.7	0.11	0.4	99	0.1	1	4
672	*roast in corn oil*	9	570	8	25	55	0.7	0.11	0.4	99	0.1	1	4
673	*roast in lard*	9	570	8	25	55	0.7	0.11	0.4	99	0.1	1	4

No.	Food	Retinol µg	Carotene µg	Vitamin D µg	Vitamin E mg	Thiamin mg	Riboflavin mg	Niacin mg	Trypt 60 mg	Vitamin B6 mg	Vitamin B12 µg	Folate µg	Pantothenate mg	Biotin µg	Vitamin C mg
Early potatoes															
660	**New potatoes,** average, raw	0	Tr	0	(0.06)	0.15	0.02	0.4	0.4	(0.44)	0	25	(0.37)	(0.3)	16
661	boiled in unsalted water	0	Tr	0	(0.06)	0.13	0.02	0.4	0.4	(0.33)	0	19	(0.38)	(0.3)	9
662	in skins, boiled in unsalted water	0	Tr	0	(0.06)	0.09	0.06	0.4	0.3	0.36	0	10	(0.38)	(0.3)	15
663	canned, re-heated, drained	0	Tr	0	(0.06)	(0.02)	(0.03)	(0.7)	0.3	(0.16)	0	(11)	N	Tr	5
Main crop potatoes															
664	**Old potatoes,** average, raw	0	Tr	0	0.06	0.21	0.02	0.6	0.5	0.44	0	35	0.37	0.3	11[a]
665	baked, flesh and skin	0	Tr	0	0.11	0.37	0.02	1.1	0.9	0.54	0	44	0.46	0.5	14
666	-, flesh only	0	Tr	0	0.06	0.21	0.01	0.6	0.5	0.31	0	25	0.26	0.3	8
667	-, flesh only, weighed with skin	0	Tr	0	0.04	0.14	0.01	0.4	0.3	0.21	0	17	0.17	0.2	5
668	boiled in unsalted water	0	Tr	0	0.06	0.18	0.01	0.5	0.4	0.33	0	26	0.38	0.3	6
669	mashed with butter	39	21	Tr	0.15	0.16	0.02	0.5	0.4	0.30	Tr	24	0.36	0.4	5
670	mashed with margarine	41	38	0.4	0.45	0.16	0.02	0.5	0.4	0.30	Tr	24	0.36	0.4	5
671	roast in blended oil	0	Tr	0	N	0.23	0.02	0.7	0.7	0.31	0	36	0.25	0.3	8
672	roast in corn oil	0	Tr	0	0.78	0.23	0.02	0.7	0.7	0.31	0	36	0.25	0.3	8
673	roast in lard	Tr	Tr	N	Tr	0.23	0.02	0.7	0.7	0.31	Tr	36	0.25	0.3	8

[a] Freshly dug potatoes contain 21mg vitamin C per 100g. This falls to 9mg per 100g after 3 months storage and to 7mg after 9 months

Chipped old potatoes

No.	Food	Description and main data sources	Edible proportion	Water g	Protein g	Fat g	Carbo-hydrate g	Energy value kcal	kJ
674	**Chips**, homemade, *fried in blended oil*	Calculation from *fried in corn oil*	1.00	56.5	3.9	6.7[a]	30.1	189	796
675	*fried in corn oil*	IFR. Samples as raw potatoes; deep fried 6 minutes 190C	1.00	56.5	3.9	6.7[a]	30.1	189	796
676	*fried in dripping*	Calculation from *fried in corn oil*	1.00	56.5	3.9	6.7[a]	30.1	189	796
677	retail, *fried in blended oil*	Calculation from *fried in vegetable oil*	1.00	52.3	3.2	12.4	30.5	239	1001
678	*fried in dripping*	Calculation from *fried in vegetable oil*	1.00	52.3	3.2	12.4	30.5	239	1001
679	*fried in vegetable oil*	5 samples from fish and chip shops	1.00	52.3	3.2	12.4	30.5	239	1001
680	French fries, retail	5 samples from burger outlets. Manufacturers' data	1.00	43.8	3.3	15.5	34.0	280	1174
681	straight cut, *frozen, fried in blended oil*	Calculation from *fried in corn oil*	1.00	40.3	4.1	13.5	36.0	273	1145
682	*frozen, fried in corn oil*	LGC; 10 samples, 10 brands. Deep fried 3-5 minutes	1.00	40.3	4.1	13.5	36.0	273	1145
683	-, *fried in dripping*	Calculation from *fried in corn oil*	1.00	40.3	4.1	13.5	36.0	273	1145
684	fine cut, *frozen, fried in blended oil*	Calculation from *fried in corn oil*	1.00	26.0	4.5	21.3	41.2	364	1524
685	*frozen, fried in corn oil*	LGC; 10 samples, 4 brands. Deep fried 1-4 minutes	1.00	26.0	4.5	21.3	41.2	364	1524
686	-, *fried in dripping*	Calculation from *fried in corn oil*	1.00	26.0	4.5	21.3	41.2	364	1524
687	**Oven chips**, *frozen, baked*	LGC; 10 samples, 7 brands. Oven baked 15-20 minutes	1.00	58.5	3.2	4.2	29.8	162	687

[a] The fat content of homemade chips will be variable and dependent on a number of factors related to their preparation

No.	Food	Total nitrogen g	Fatty acids			Cholest- erol mg	Starch g	Total sugars g	Dietary fibre	
			Satd g	Mono unsatd g	Poly unsatd g				Southgate method g	Englyst method g
	Chipped old potatoes									
674	**Chips**, homemade, *fried in blended oil*	0.63	0.6	3.3	2.4	0	29.5	0.6	3.0	2.2
675	*fried in corn oil*	0.63	0.9	1.7	3.9	0	29.5	0.6	3.0	2.2
676	*fried in dripping*	0.63	3.7	2.5	0.2	6	29.5	0.6	3.0	2.2
677	*retail, fried in blended oil*	0.51	1.1	6.2	4.5	0	28.8	1.7	(3.0)	(2.2)
678	*fried in dripping*	0.51	6.8	4.6	0.3	11	28.8	1.7	(3.0)	(2.2)
679	*fried in vegetable oil*	0.51	3.6	5.3	3.1	0	28.8	1.7	(3.0)	(2.2)
680	French fries, *retail*	0.54	5.8	6.9	2.1	N	32.7	1.3	(3.1)	(2.1)
681	straight cut, frozen, *fried in blended oil*	0.66	1.2	6.7	4.9	0	35.3	0.7	(3.5)	2.4
682	*frozen, fried in corn oil*	0.66	2.5	3.4	7.0	0	35.3	0.7	(3.5)	2.4
683	*-, fried in dripping*	0.66	7.5	5.0	0.3	12	35.3	0.7	(3.5)	2.4
684	fine cut, frozen, *fried in blended oil*	0.72	1.8	10.6	7.8	0	40.6	0.6	(4.0)	(2.4)
685	*frozen, fried in corn oil*	0.72	4.0	5.4	11.0	0	40.6	0.6	(4.0)	2.7
686	*-, fried in dripping*	0.72	11.8	7.9	0.5	19	40.6	0.6	(4.0)	2.7
687	**Oven chips**, *frozen, baked*	0.52	1.8	1.6	0.6	0	29.1	0.7	(2.8)	2.0

No.	Food	Na	K	Ca	Mg	P	Fe	Cu	Zn	Cl	Mn	Se	I
						mg						µg	

Chipped old potatoes

No.	Food	Na	K	Ca	Mg	P	Fe	Cu	Zn	Cl	Mn	Se	I
674	**Chips**, homemade, *fried in blended oil*	12	660	11	31	62	0.8	0.14	0.6	120	0.2	2	5
675	*fried in corn oil*	12	660	11	31	62	0.8	0.14	0.6	120	0.2	2	5
676	*fried in dripping*	12	660	11	31	63	0.8	0.14	0.6	120	0.2	2	5
677	*retail, fried in blended oil*	35	(660)	(11)	(31)	(62)	0.9	(0.14)	(0.6)	(120)	(0.2)	(2)	(5)
678	*fried in dripping*	35	(660)	(11)	(31)	(63)	0.9	(0.14)	(0.6)	(120)	(0.2)	(2)	(5)
679	*fried in vegetable oil*	35	(660)	(11)	(31)	(62)	0.9	(0.14)	(0.6)	(120)	(0.2)	(2)	(5)
680	**French fries**, retail	310[a]	650	14	(26)	(130)	1.0	(0.17)	(0.5)	480	(0.1)	N	N
681	*straight cut, frozen, fried in blended oil*	29	710	15	33	120	0.9	0.24	0.6	76	0.2	(2)	(7)
682	*frozen, fried in corn oil*	29	710	15	33	120	0.9	0.24	0.6	76	0.2	(2)	(7)
683	*-, fried in dripping*	30	710	15	33	120	0.9	0.24	0.6	76	0.2	(2)	(8)
684	*fine cut, frozen, fried in blended oil*	97	720	19	34	170	1.0	0.22	0.6	98	0.2	(3)	(8)
685	*frozen, fried in corn oil*	97	720	19	34	170	1.0	0.22	0.6	98	0.2	(3)	(8)
686	*-, fried in dripping*	98	720	19	34	170	1.0	0.22	0.6	98	0.2	(3)	(9)
687	**Oven chips**, *frozen, baked*	53	530	(12)	27	120	0.8	0.22	0.4	74	0.2	N	N

[a] Unsalted French fries contain approximately 35mg Na per 100g

Vegetables *continued*

Chipped old potatoes

No.	Food	Retinol µg	Carotene µg	Vitamin D µg	Vitamin E mg	Thiamin mg	Ribo- flavin mg	Niacin mg	Trypt 60 mg	Vitamin B6 mg	Vitamin B12 µg	Folate µg	Panto- thenate mg	Biotin µg	Vitamin C mg
674	**Chips**, homemade, *fried in blended oil*	0	Tr	0	N	0.24	0.02	0.7	0.9	0.32	0	43	0.25	0.4	9
675	*fried in corn oil*	0	Tr	0	4.90	0.24	0.02	0.7	0.9	0.32	0	43	0.25	0.4	9
676	*fried in dripping*	N	N	Tr	0.02	0.24	0.02	0.7	0.9	0.32	Tr	43	0.25	0.4	9
677	*retail, fried in blended oil*	Tr	Tr	0	N	0.08	0.01	(0.7)	0.8	(0.32)	0	N	(0.25)	(0.4)	(9)[a]
678	*fried in dripping*	N	N	Tr	0.04	0.08	0.01	(0.7)	0.8	(0.32)	Tr	N	(0.25)	(0.4)	(9)[a]
679	*fried in vegetable oil*	Tr	Tr	0	0.39	0.08	0.01	(0.7)	0.8	(0.32)	0	N	(0.25)	(0.4)	(9)[a]
680	*French fries, retail*	0	Tr	0	1.00	0.08	0.05	2.3	0.8	0.36	0	N	N	N	4
681	*straight cut, frozen, fried in blended oil*	0	Tr	0	N	0.16	0.08	2.1	1.0	0.46	0	30	N	N	16
682	*frozen, fried in corn oil*	0	Tr	0	3.27	0.16	0.08	2.1	1.0	0.46	0	30	N	N	16
683	*-, fried in dripping*	N	N	Tr	0.04	0.16	0.08	2.1	1.0	0.46	Tr	30	N	N	16
684	*fine cut, frozen, fried in blended oil*	0	Tr	0	N	0.18	0.09	2.4	1.1	0.52	0	34	N	N	12
685	*frozen, fried in corn oil*	0	Tr	0	(5.16)	0.18	0.09	2.4	1.1	0.52	0	34	N	N	12
686	*-, fried in dripping*	N	N	Tr	0.06	0.18	0.09	2.4	1.1	0.52	Tr	34	N	N	12
687	**Oven chips**, *frozen, baked*	0	Tr	0	0.44	0.11	0.04	2.2	0.8	0.37	0	21	N	N	12

[a] Storage of uncooked chips under some conditions may significantly reduce vitamin C levels which could approach zero

Vegetables *continued*

Composition of food per 100g

No.	Food	Description and main data sources	Edible proportion	Water g	Protein g	Fat g	Carbo-hydrate g	Energy value kcal	kJ
	Potato products								
688	**Instant potato powder,** *made up with water*	Calcd. from ingredients; made up as packet directions	1.00	83.3	1.5	0.1	13.5	57	245
689	*made up with whole milk*	Calcd. from ingredients; made up as packet directions	1.00	80.0	2.4	1.2	14.8	76	322
690	**Potato croquettes,** *fried in blended oil*	LGC; 10 samples, 5 brands. Shallow fried 5-7 minutes	1.00	58.2	3.7	13.1	21.6	214	893
691	**Potato waffles,** *frozen, cooked*	IFR. 10 samples (Birds Eye); grilled, shallow and deep fried in corn oil, oven baked	1.00	52.7	3.2	8.2	30.3	200	842
	Beans and lentils								
692	**Aduki beans,** *dried, raw*	LGC; 6 samples, whole beans	1.00	12.7	19.9	0.5	50.1[a]	272	1158
693	*dried, boiled in unsalted water*	LGC analysis and calculation from dried	1.00	59.4	9.3	0.2	22.5[a]	123	525
694	**Baked beans,** canned in tomato sauce, *re-heated*	LGC; 10 cans, 7 brands	1.00	71.5	5.2	0.6	15.3	84	355
695	*reduced sugar, reduced salt*	LGC; 5 cans, 2 own brands	1.00	73.6	5.4	0.6	12.5	73	311
696	**Beansprouts,** mung, *raw*	IFR; as purchased	1.00	90.4	2.9	0.5	4.0	31	131
697	*stir-fried in blended oil*	LGC. 6 samples; stir-fried 2 minutes. And calcd. from raw	1.00	88.4	1.9	6.1	2.5	72	298
698	**Black gram,** urad gram, *dried, raw*	Whole beans. Literature sources	1.00	11.5	24.9	1.4	40.8[a]	275	1169
699	*dried, boiled in unsalted water*	As raw; soaked and boiled	1.00	71.3	7.8	0.4	13.6[a]	89	379

[a] Including oligosaccharides

Vegetables continued

Composition of food per 100g

No.	Food	Total nitrogen g	Fatty acids Satd g	Mono unsatd g	Poly unsatd g	Cholesterol mg	Starch g	Total sugars g	Dietary fibre Southgate method g	Englyst method g
	Potato products									
688	**Instant potato powder**, *made up with water*	0.24	Tr	Tr	0.1	0	12.7	0.7	2.7	1.0
689	*made up with whole milk*	0.37	0.7	0.3	0.1	4	12.7	2.0	2.7	1.0
690	**Potato croquettes**, *fried in blended oil*	0.59	1.7	3.2	7.6	0	21.1	0.5	N	1.3
691	**Potato waffles**, *frozen, cooked*	0.51	1.0	2.0	4.7	0	29.8	0.6	N	2.3
	Beans and lentils									
692	**Aduki beans**, *dried, raw*	3.18	N	N	N	0	44.7	1.0[a]	N	11.1
693	*dried, boiled in unsalted water*	1.48	N	N	N	0	20.8	0.5[a]	N	5.5
694	**Baked beans**, *canned in tomato sauce, re-heated*	0.83	0.1	0.1	0.3	0	9.4	5.9	6.9	3.7
695	*reduced sugar, reduced salt*	0.85	0.1	0.1	0.3	0	9.7	2.8	7.1	3.8
696	**Beansprouts**, *mung, raw*	0.47	0.1	0.1	0.2	0	1.8	2.2	(5.6)	1.5
697	*stir-fried in blended oil*	0.30	0.5	3.0	2.2	0	1.1	1.4	(3.4)	0.9
698	**Black gram**, urad gram, *dried, raw*	3.98	0.2	0.2	0.7	0	37.6	1.3[a]	17.9	N
699	*dried, boiled in unsalted water*	1.25	0.1	0.1	0.2	0	13.0	0.3[a]	5.8	N

[a] Not including oligosaccharides

233

No.	Food	Na	K	Ca	Mg	P	Fe	Cu	Zn	Cl	Mn	Se	I
						mg						µg	
Potato products													
688	**Instant potato powder**, *made up with water*	200	260	13	12	41	0.4	0.04	0.2	290	0.1	N	N
689	*made up with whole milk*	210	290	44	15	66	0.4	0.04	0.3	310	0.1	N	N
690	**Potato croquettes**, *fried in blended oil*	420	360	44	19	49	0.9	0.08	0.3	650	0.2	N	N
691	**Potato waffles**, *frozen, cooked*	430	480	32	21	120	0.5	Tr	0.3	630	0.1	N	N
Beans and lentils													
692	**Aduki beans**, *dried, raw*	5	1220	84	130	380	4.2	1.09	5.0	N	1.7	2	N
693	*dried, boiled in unsalted water*	2	570	39	60	180	1.9	0.51	2.3	N	0.8	1	N
694	**Baked beans**, *canned in tomato sauce, re-heated*	530	310	53	31	100	1.4	0.03	0.5	820	0.3	2	3
695	*reduced sugar, reduced salt*	330	270	45	29	90	1.2	0.10	0.5	480	0.3	2	3
696	**Beansprouts**, *mung, raw*	5	74	20	18	48	1.7	0.08	0.3	15	0.3	N	N
697	*stir-fried in blended oil*	3	45	12	11	29	1.0	0.05	0.2	9	0.2	N	N
698	**Black gram**, *urad gram, dried, raw*	40	800	150	160	370	6.3	0.72	2.8	N	1.2	N	N
699	*dried, boiled in unsalted water*	13	260	49	52	120	2.0	0.23	0.9	N	0.4	N	N

No.	Food	Retinol μg	Carotene μg	Vitamin D μg	Vitamin E mg	Thiamin mg	Ribo-flavin mg	Niacin mg	Trypt 60 mg	Vitamin B6 mg	Vitamin B12 μg	Folate μg	Panto-thenate mg	Biotin μg	Vitamin C mg
	Potato products														
688	**Instant potato powder**, *made up with water*	0	3	0	0.05	0.01	0.03	1.2	0.4	0.15	0	2	N	N	23
689	*made up with whole milk*	14	8	Tr	0.06	0.02	0.07	1.2	0.5	0.17	0.1	4	N	N	23
690	**Potato croquettes**, *fried in blended oil*	0	N	0	N	0.08	0.08	1.4	1.0	0.22	0	2	N	N	2
691	**Potato waffles**, *frozen, cooked*	0	Tr	0	N	N	N	N	0.7	N	0	N	N	N	36
	Beans and lentils														
692	**Aduki beans**, *dried, raw*	0	12	0	N	0.45	0.22	2.6	3.2	N	0	N	N	N	Tr
693	*dried, boiled in unsalted water*	0	6	0	N	0.14	0.08	0.9	1.5	N	0	N	N	N	Tr
694	**Baked beans**, canned in tomato *sauce, re-heated*	0	74	0	0.37	0.09	0.06	0.5	0.8	0.14	0	22	0.18	2.5	Tr
695	*reduced sugar, reduced salt*	0	77	0	0.39	0.09	0.06	0.5	0.9	0.14	0	23	0.19	2.6	Tr
696	**Beansprouts**, mung, *raw*	0	40	0	N	0.11	0.04	0.5	0.5	0.10	0	61	0.38	N	7
697	*stir-fried in blended oil*	0	24	0	N	0.06	0.02	0.5	0.3	0.07	0	37	0.23	N	7
698	**Black gram**, urad gram, *dried, raw*	0	38	0	N	0.42	0.37	2.0	4.6	N	0	132	N	N	Tr
699	*dried, boiled in unsalted water*	0	12	0	N	(0.11)	(0.09)	(0.5)	1.5	N	0	(33)	N	N	Tr

Composition of food per 100g

Beans and lentils

No.	Food	Description and main data sources	Edible proportion	Water g	Protein g	Fat g	Carbo-hydrate g	Energy value kcal	kJ
700	**Blackeye beans**, *dried, raw*	Whole beans. Analysis and literature sources	1.00	10.7	23.5	1.6	54.1[a]	311	1324
701	*dried, boiled in unsalted water*	As raw; soaked and boiled	1.00	66.2	8.8	0.7	19.9[a]	116	494
702	**Broad beans**, *frozen, boiled in unsalted water*	LGC; 10 samples, 7 brands. Boiled 3-10 minutes	1.00	73.8	7.9	0.6	11.7[a]	81	344
703	**Butter beans**, canned, re-heated, drained	LGC; 10 cans, 5 brands	0.57	74.0	5.9	0.5	13.0[a]	77	327
704	**Chick peas**, whole, *dried, raw*	Analytical and literature sources. Kabuli variety	1.00	10.0	21.3	5.4	49.6[a]	320	1355
705	*dried, boiled in unsalted water*	As raw. Soaked and boiled	1.00	65.8	8.4	2.1	18.2[a]	121	512
706	*canned, re-heated, drained*	LGC. Whole peas; 10 samples, 5 brands	0.60	67.5	7.2	2.9	16.1[a]	115	487
707	**Green beans/French beans**, *raw*	IFR; pods and beans, ends trimmed	0.83	90.7	1.9	0.5	3.2	24	99
708	*frozen, boiled in unsalted water*	LGC; 10 samples, 8 brands. Boiled 3-8 minutes	1.00	90.0	1.7	0.1	4.7	25	108
709	**Hummus**	LGC. Chick pea spread; 10 samples, retail and homemade	1.00	61.4	7.6	12.6	11.6[a]	187	781
710	**Lentils**, green and brown, whole, *dried, raw*	LGC; 10 samples, 6 brands. Continental type	1.00	10.8	24.3	1.9	48.8[a]	297	1264
711	*dried, boiled in salted water*	LGC; as raw. Boiled 10 minutes, simmered 25 minutes	1.00	66.7	8.8	0.7	16.9[a]	105	446
712	red, split, *dried, raw*	LGC; as purchased	1.00	11.1	23.8	1.3	56.3[a]	318	1353
713	*dried, boiled in unsalted water*	LGC. As purchased; boiled 20 minutes	1.00	72.1	7.6	0.4	17.5[a]	100	424

[a] Including oligosaccharides

No.	Food	Total nitrogen g	Fatty acids			Cholest-erol mg	Starch g	Total sugars g	Dietary fibre	
			Satd g	Mono unsatd g	Poly unsatd g				Southgate method g	Englyst method g
	Beans and lentils									
700	**Blackeye beans**, *dried, raw*	3.76	0.5	0.1	0.7	0	47.5	2.9[a]	N	8.2
701	*dried, boiled in unsalted water*	1.41	0.2	0.1	0.3	0	18.0	1.1[a]	N	3.5
702	**Broad beans**, *frozen, boiled in unsalted water*	1.27	0.1	0.1	0.3	0	10.0	1.3[a]	N	6.5
703	**Butter beans**, *canned, re-heated, drained*	0.95	0.1	Tr	0.2	0	10.9	1.1[a]	N	4.6
704	**Chick peas**, *whole, dried, raw*	3.42	0.5	1.1	2.7	0	43.8	2.6[a]	13.5	10.7
705	*dried, boiled in unsalted water*	1.35	0.2	0.4	1.0	0	16.6	1.0[a]	5.1	4.3
706	*canned, re-heated, drained*	1.15	0.3	0.7	1.3	0	15.1	0.4[a]	N	4.1
707	**Green beans/French beans,** *raw*	0.31	(0.1)	Tr	(0.3)	0	0.9	2.3	3.0	2.2
708	*frozen, boiled in unsalted water*	0.28	Tr	Tr	Tr	0	2.6	2.1	N	4.1
709	**Hummus**	1.22	N	N	N	0	9.3	1.9[a]	3.2	2.4
710	**Lentils,** *green and brown, whole, dried, raw*	3.90	0.2	0.3	0.8	0	44.5	1.2[a]	N	8.9
711	*dried, boiled in salted water*	1.41	0.1	0.1	0.3	0	15.9	0.4[a]	N	3.8
712	*red, split, dried, raw*	3.80	0.2	0.2	0.5	0	50.8	2.4[a]	10.5	4.9
713	*dried, boiled in unsalted water*	1.22	Tr	0.1	0.2	0	16.2	0.8[a]	3.3	1.9

[a] Not including oligosaccharides

No.	Food	Na	K	Ca	Mg	P	Fe	Cu	Zn	Cl	Mn	Se	I
						mg						µg	

Beans and lentils

No.	Food	Na	K	Ca	Mg	P	Fe	Cu	Zn	Cl	Mn	Se	I
700	**Blackeye beans**, *dried, raw*	16	1170	81	140	410	7.6	0.75	3.2	N	1.3	7	N
701	*dried, boiled in unsalted water*	5	320	21	45	140	1.9	0.22	1.1	N	0.5	3	N
702	**Broad beans**, *frozen, boiled in unsalted water*	8	280	56	36	150	1.6	0.32	1.0	15	0.3	N	(6)
703	**Butter beans**, *canned, re-heated, drained*	420	290	15	27	68	1.5	0.14	0.6	660	0.3	N	N
704	**Chick peas**, whole, *dried, raw*	39	1000	160	130	310	5.5	0.95	3.0	60	2.4	2	N
705	*dried, boiled in unsalted water*	5	270	46	37	83	2.1	0.28	1.2	7	0.7	1	N
706	*canned, re-heated, drained*	220	110	43	24	81	1.5	0.05	0.8	280	0.8	1	N
707	**Green beans/French beans**, *raw*	Tr	230	36	17	38	1.2	0.01	0.2	9	N	N	N
708	*frozen, boiled in unsalted water*	8	160	56	17	33	0.6	0.05	0.2	21	0.2	N	N
709	**Hummus**	670	190	41	62	160	1.9	0.30	1.4	670	0.5	N	N
710	**Lentils**, green and brown, whole, *dried, raw*	12	940	71	110	350	11.1	1.02	3.9	87	1.4	105	N
711	*dried, boiled in salted water*	3	310	22	34	130	3.5	0.33	1.4	26	0.5	40	N
712	red, split, *dried, raw*	36	710	51	83	320	7.6	0.58	3.1	64	N	(6)	N
713	*dried, boiled in unsalted water*	12	220	16	26	100	2.4	0.19	1.0	20	N	(2)	N

No.	Food	Retinol µg	Carotene µg	Vitamin D µg	Vitamin E mg	Thiamin mg	Ribo-flavin mg	Niacin mg	Trypt 60 mg	Vitamin B6 mg	Vitamin B12 µg	Folate µg	Panto-thenate mg	Biotin µg	Vitamin C mg
	Beans and lentils														
700	**Blackeye beans**, *dried, raw*	0	35	0	N	0.87	0.19	2.1	5.0	0.36	0	630	1.50	18.4	1
701	*dried, boiled in unsalted water*	0	13	0	N	0.19	0.05	0.5	1.9	0.10	0	210	0.30	7.0	Tr
702	**Broad beans**, *frozen, boiled in unsalted water*	0	225	0	0.61	(0.03)	(0.06)	(3.0)	1.3	(0.08)	0	(32)	(3.80)	(2.1)	8
703	**Butter beans**, *canned, re-heated, drained*	0	Tr	0	0.33	0.05	0.03	0.2	0.9	0.05	0	12	N	N	Tr
704	**Chick peas**, *whole, dried, raw*	0	60	0	2.88	0.39	0.24	1.9	2.9	0.53	0	180	1.59	N	Tr
705	*dried, boiled in unsalted water*	0	23	0	1.10	0.10	0.07	0.7	1.1	0.14	0	54	0.29	N	Tr
706	*canned, re-heated, drained*	0	21	0	1.55	0.05	0.03	0.2	1.0	0.04	0	11	N	N	Tr
707	**Green beans/French beans**, *raw*	0	(330)	0	0.20	0.05	0.07	0.9	0.5	0.05	0	(80)	0.09	1.0	12
708	*frozen, boiled in unsalted water*	0	180	0	0.12	0.05	0.09	0.4	0.4	0.06	0	56	N	N	7
709	**Hummus**	0	N	0	N	0.16	0.05	1.1	1.0	N	0	N	N	N	1
710	**Lentils**, green and brown, whole, *dried, raw*	0	N	0	N	0.41	0.27	2.2	3.3	0.93	0	110	N	N	Tr
711	*dried, boiled in salted water*	0	N	0	N	0.14	0.08	0.6	1.2	0.28	0	30	N	N	Tr
712	*red, split, dried, raw*	0	(60)	0	N	0.50	0.20	2.0	3.2	0.60	0	35	1.36	N	Tr
713	*dried, boiled in unsalted water*	0	(20)	0	N	0.11	0.04	0.4	1.0	0.11	0	5	0.31	N	Tr

Vegetables *continued*

714 to 724

Composition of food per 100g

Beans and lentils

No.	Food	Description and main data sources	Edible proportion	Water g	Protein g	Fat g	Carbo-hydrate g	Energy value kcal	kJ
714	**Mung beans**, whole, *dried, raw*	Literature sources	1.00	11.0	23.9	1.1	46.3[a]	279	1188
715	*dried, boiled in unsalted water*	As raw, soaked and boiled	1.00	69.3	7.6	0.4	15.3[a]	91	389
716	**Red kidney beans**, *dried, raw*	Whole beans. Analytical and literature sources	1.00	11.2	22.1	1.4	44.1[a]	266	1133
717	*dried, boiled in unsalted water*	As raw, soaked and boiled	1.00	66.0	8.4	0.5	17.4[a]	103	440
718	*canned, re-heated, drained*	LGC; 10 cans, 6 brands	0.64	67.5	6.9	0.6	17.8[a]	100	424
719	**Runner beans**, *raw*	IFR; ends and sides trimmed	0.86	91.2	1.6	0.4	3.2	22	93
720	*boiled in unsalted water*	IFR. Sliced and boiled 20 minutes	1.00	92.8	1.2	0.5	2.3	18	76
721	**Soya beans**, *dried, raw*	Whole beans. Analysis and literature sources	1.00	8.5	35.9	18.6	15.8[a]	370	1551
722	*dried, boiled in unsalted water*	As raw	1.00	64.3	14.0	7.3	5.1[a]	141	590
723	**Tofu**, soya bean, *steamed*	LGC. Soya bean curd; 7 assorted samples	1.00	85.0	8.1	4.2	0.7[a]	73	304
724	*steamed, fried*	Calcd. from *steamed* and ref. Haytowitz and Matthews (1986)	1.00	51.0	23.5	17.7	2.0[a]	261	1086

[a] Including oligosaccharides

Vegetables *continued*

Composition of food per 100g

No.	Food	Total nitrogen g	Fatty acids Satd g	Fatty acids Mono unsatd g	Fatty acids Poly unsatd g	Cholest-erol mg	Starch g	Total sugars g	Dietary fibre Southgate method g	Dietary fibre Englyst method g
	Beans and lentils									
714	**Mung beans**, whole, *dried, raw*	3.80	0.3	0.1	0.5	0	40.9	1.5[a]	13.9	10.0
715	*dried, boiled in unsalted water*	1.21	0.1	Tr	0.2	0	14.1	0.5[a]	4.8	3.0
716	**Red kidney beans**, *dried, raw*	3.54	0.2	0.1	0.8	0	38.0	2.5[a]	(23.4)	15.7
717	*dried, boiled in unsalted water*	1.35	0.1	Tr	0.3	0	14.5	1.0[a]	(9.0)	6.7
718	*canned, re-heated, drained*	1.11	0.1	0.1	0.3	0	12.8	3.6[a]	(8.5)	6.2
719	**Runner beans**, *raw*	0.26	0.1	Tr	0.2	0	0.4	2.8	2.6	2.0
720	*boiled in unsalted water*	0.19	0.1	Tr	0.3	0	0.3	2.0	3.1	1.9
721	**Soya beans**, *dried, raw*	5.74	2.3	3.5	9.1	0	4.8	5.5[a]	N	15.7
722	*dried, boiled in unsalted water*	2.24	0.9	1.4	3.5	0	1.9	2.1[a]	N	6.1
723	**Tofu**, soya bean, *steamed*	1.29	0.5	0.8	2.0	0	0.3	0.3[a]	0.3	N
724	*steamed, fried*	3.76	N	N	N	0	0.9	0.9[a]	0.9	N

[a] Not including oligosaccharides

Inorganic constituents per 100g

No.	Food	mg											μg		
		Na	K	Ca	Mg	P	Fe	Cu	Zn	Cl	Mn	Se	I
Beans and lentils															
714	**Mung beans**, whole, *dried, raw*	12	1250	89	150	360	6.0	0.47	2.7	12	0.8	16	N
715	*dried, boiled in unsalted water*	2	270	24	43	81	1.4	0.19	0.9	4	0.3	5	N
716	**Red kidney beans**, *dried, raw*	18	1370	100	150	410	6.4	0.68	3.0	(2)	1.2	16	N
717	*dried, boiled in unsalted water*	2	420	37	45	130	2.5	0.23	1.0	(1)	0.5	6	N
718	*canned, re-heated, drained*	390	280	71	30	130	2.0	Tr	0.7	640	0.3	6	N
719	**Runner beans**, *raw*	Tr	220	33	19	34	1.2	0.02	0.2	21	0.2	N	2
720	*boiled in unsalted water*	1	130	22	14	21	1.0	0.01	0.2	5	0.2	N	Tr
721	**Soya beans**, *dried, raw*	5	1730	240	250	660	9.7	1.55	4.3	7	2.6	14	6
722	*dried, boiled in unsalted water*	1	510	83	63	250	3.0	0.32	0.9	3	0.7	5	2
723	**Tofu**, soya bean, *steamed*	4	63	510[a]	23[a]	95	1.2	0.20	0.7	16	0.4	N	N
724	*steamed, fried*	12	180	1480	67	270	3.5	0.58	2.0	46	1.2	N	N

[a] If nigari is used as a coagulent Ca and Mg are 150 and 59 mg per 100g respectively

Beans and lentils

No.	Food	Retinol µg	Carotene µg	Vitamin D µg	Vitamin E mg	Thiamin mg	Ribo-flavin mg	Niacin mg	Trypt 60 mg	Vitamin B6 mg	Vitamin B12 µg	Folate µg	Panto-thenate mg	Biotin µg	Vitamin C mg
714	Mung beans, whole, *dried, raw*	0	24	0	N	0.36	0.26	2.1	3.8	0.38	0	140	1.91	N	Tr
715	*dried, boiled in unsalted water*	0	12	0	N	0.09	0.07	0.5	1.2	0.07	0	35	0.41	N	Tr
716	Red kidney beans, *dried, raw*	0	11	0	0.52	0.65	0.19	2.1	3.5	0.40	0	130	0.78	N	4
717	*dried, boiled in unsalted water*	0	4	0	0.20	0.17	0.05	0.6	1.3	0.12	0	42	0.22	N	1
718	*canned, re-heated, drained*	0	4	0	0.19	0.21	0.06	0.6	1.1	0.11	0	8	0.15	N	Tr
719	Runner beans, *raw*	0	145	0	0.23	0.06	0.03	Tr	0.4	0.08	0	60	0.05	0.7	18
720	*boiled in unsalted water*	0	120	0	0.23	0.05	0.02	Tr	0.3	0.04	0	42	0.04	0.5	10
721	Soya beans, *dried, raw*	0	12	0	2.90	0.61	0.27	2.2	5.7	0.38	0	370	0.79	N	Tr
722	*dried, boiled in unsalted water*	0	6	0	1.13	0.12	0.09	0.5	2.2	0.23	0	54	0.18	65.0	Tr
723	Tofu, soya bean, *steamed*	0	2	0	0.95	0.06	0.02	0.1	1.3	0.07	0	15	0.05	25.0	0
724	*steamed, fried*	0	2	0	N	0.09	0.02	0.1	3.8	0.10	0	27	0.14	N	0

Peas

No.	Food	Description and main data sources	Edible proportion	Water g	Protein g	Fat g	Carbohydrate g	Energy value kcal	kJ
725	**Mange-tout peas**, *raw*	LGC. Whole pods, ends trimmed; 10 samples	0.92	88.7	3.6	0.2	4.2	32	136
726	*boiled in salted water*	LGC. As raw; boiled 3 minutes. And calcd. from raw	1.00	89.2	3.2	0.1	3.3	26	111
727	*stir-fried in blended oil*	LGC. As raw; stir-fried 5 minutes. And calcd. from raw	1.00	83.6	3.8	4.8	3.5	71	298
728	**Mushy peas**, *canned, re-heated*	LGC; 10 samples, 3 brands	1.00	76.5	5.8	0.7	13.8[a]	81	345
729	**Peas**, *raw*	IFR; whole peas, no pods	0.37	74.6	6.9	1.5	11.3[a]	83	344
730	*boiled in unsalted water*	IFR. As raw; boiled 20 minutes	1.00	75.6	6.7	1.6	10.0[a]	79	329
731	*frozen, boiled in salted water*	Based on *frozen, boiled in unsalted water*	1.00	78.3	6.0	0.9	9.7[a]	69	291
732	*-, boiled in unsalted water*	LGC; 10 samples, 8 brands. Boiled 2-5 minutes	1.00	78.3	6.0	0.9	9.7[a]	69	291
733	*canned, re-heated, drained*	LGC; 10 samples, 9 brands	0.67	77.9	5.3	0.9	13.5[a]	80	339
734	**Petit pois**, *frozen, boiled in salted water*	LGC; 10 samples, 4 brands. Boiled 2-5 minutes	1.00	81.1	5.0	0.9	5.5[a]	49	206
735	*frozen, boiled in unsalted water*	Based on *boiled in salted water*	1.00	81.1	5.0	0.9	5.5[a]	49	206
736	**Processed peas**, *canned, re-heated, drained*	LGC; 10 samples, 7 brands	0.65	69.6	6.9	0.7	17.5[a]	99	423

[a] Including oligosaccharides

No.	Food	Total nitrogen g	Fatty acids			Cholest- erol mg	Starch g	Total sugars g	Dietary fibre	
			Satd g	Mono unsatd g	Poly unsatd g				Southgate method g	Englyst method g
Peas										
725	**Mange-tout peas**, *raw*	0.58	Tr	Tr	0.1	0	0.8	3.4	4.2	2.3
726	*boiled in salted water*	0.51	Tr	Tr	Tr	0	0.5	2.8	4.0	2.2
727	*stir-fried in blended oil*	0.61	0.4	2.4	1.8	0	0.2	3.3	4.4	2.4
728	**Mushy peas**, *canned, re-heated*	0.92	0.1	0.1	0.3	0	10.7	1.7[a]	N	1.8
729	**Peas**, *raw*	1.10	0.3	0.2	0.7	0	7.0	2.3[a]	4.7	4.7
730	*boiled in unsalted water*	1.07	0.3	0.2	0.8	0	7.6	1.2[a]	4.7	4.5
731	*frozen, boiled in salted water*	0.95	0.2	0.1	0.5	0	4.7	2.7[a]	(7.3)	5.1
732	*-, boiled in unsalted water*	0.95	0.2	0.1	0.5	0	4.7	2.7[a]	(7.3)	5.1
733	*canned, re-heated, drained*	0.85	0.2	0.1	0.4	0	6.3	3.9[a]	5.7	5.1
734	**Petit pois**, *frozen, boiled in salted water*	0.80	0.2	0.1	0.5	0	Tr	3.0[a]	(6.4)	4.5
735	*frozen, boiled in unsalted water*	0.80	0.2	0.1	0.5	0	Tr	3.0[a]	(6.4)	4.5
736	**Processed peas**, *canned, re-heated, drained*	1.10	0.1	0.1	0.3	0	14.7	1.5[a]	7.1	4.8

[a] Not including oligosaccharides

Vegetables *continued*

725 to 736

Inorganic constituents per 100g

No.	Food	Na	K	Ca	Mg	P	Fe	Cu	Zn	Cl	Mn	Se	I
						mg						μg	
Peas													
725	**Mange-tout peas**, *raw*	2	200	44	28	62	0.8	0.06	0.5	28	0.3	Tr	N
726	*boiled in salted water*	42	170	35	22	55	0.8	0.06	0.4	72	0.3	Tr	N
727	*stir-fried in blended oil*	2	210	46	29	65	0.8	0.06	0.5	29	0.3	Tr	N
728	**Mushy peas**, *canned, re-heated*	340	170	14	22	100	1.3	0.11	0.7	490	0.2	N	N
729	**Peas**, *raw*	1	330	21	34	130	2.8	0.05	1.1	39	0.4	(1)	2
730	*boiled in unsalted water*	Tr	230	19	29	130	1.5	0.03	1.0	8	0.4	(1)	2
731	*frozen, boiled in salted water*	(94)	150	35	21	99	1.6	(0.03)	0.7	(120)	0.3	(1)	(2)
732	*-, boiled in unsalted water*	2	150	35	21	99	1.6	(0.03)	0.7	9	0.3	(1)	(2)
733	*canned, re-heated, drained*	250	130	30	20	81	1.9	0.02	0.6	360	0.2	N	N
734	**Petit pois**, *frozen, boiled in salted water*	69	130	42	24	95	1.6	0.11	0.9	110	0.3	(1)	(2)
735	*frozen, boiled in unsalted water*	(2)	130	42	24	95	1.6	0.11	0.9	(9)	0.3	(1)	(2)
736	**Processed peas**, *canned, re-heated, drained*	380	150	33	25	89	1.8	0.09	0.7	520	0.3	N	13

Peas

No.	Food	Retinol µg	Carotene µg	Vitamin D µg	Vitamin E mg	Thiamin mg	Ribo-flavin mg	Niacin mg	Trypt 60 mg	Vitamin B6 mg	Vitamin B12 µg	Folate µg	Panto-thenate mg	Biotin µg	Vitamin C mg
725	**Mange-tout peas**, *raw*	0	695	0	0.39	0.22	0.15	0.6	0.6	0.18	0	10	0.72	5.3	54
726	*boiled in salted water*	0	665	0	0.37	0.14	0.16	0.4	0.5	0.14	0	6	0.67	3.7	28
727	*stir-fried in blended oil*	0	725	0	N	0.17	0.14	0.6	0.6	0.17	0	9	0.68	5.0	51
728	**Mushy peas**, *canned, re-heated*	0	N	0	(0.30)	N	N	N	0.9	N	0	N	N	Tr	Tr
729	**Peas**, *raw*	0	300	0	0.21	0.74	0.02	2.5	1.1	0.12	0	62	(0.15)	0.5	24
730	*boiled in unsalted water*	0	250	0	0.21	0.70	0.03	1.8	1.1	0.09	0	27	0.15	0.4	16
731	*frozen, boiled in salted water*	0	405	0	0.18	0.26	0.09	1.6	0.9	0.09	0	47	0.14	0.4	12
732	*-, boiled in unsalted water*	0	405	0	0.18	0.26	0.09	1.6	0.9	0.09	0	47	0.14	0.4	12
733	*canned, re-heated, drained*	0	450	0	0.22	0.09	0.07	1.2	0.9	0.06	0	20	(0.04)	Tr	1
734	**Petit pois**, *frozen, boiled in salted water*	0	(405)	0	(0.18)	0.13	0.12	1.5	0.9	0.09	0	50	(0.14)	(0.4)	8
735	*frozen, boiled in unsalted water*	0	(405)	0	(0.18)	0.13	0.12	1.5	0.9	0.09	0	50	(0.14)	(0.4)	8
736	**Processed peas**, *canned, re-heated, drained*	0	60	0	0.30	0.10	0.04	0.4	1.1	0.10	0	11	(0.04)	Tr	Tr

Vegetables *continued*

737 to 753

Composition of food per 100g

No.	Food	Description and main data sources	Edible proportion	Water g	Protein g	Fat g	Carbo-hydrate g	Energy value kcal	Energy value kJ
	Vegetables, general								
737	**Asparagus**, *raw*	IFR; tough base of stems removed	0.75	91.4	2.9	0.6	2.0	25	103
738	*boiled in salted water*	IFR. Soft tips only; boiled 15 minutes	0.48	91.5	3.4	0.8	1.4	26	110
739	**Aubergine**, *raw*	IFR; ends trimmed	0.96[a]	92.9	0.9	0.4	2.2	15	64
740	*fried in corn oil*	IFR. Sliced; shallow fried 10 minutes	1.00	59.5	1.2	31.9	2.8	302	1246
741	**Beetroot**, *raw*	IFR; top and root trimmed, peeled	0.80	87.1	1.7	0.1	7.6	36	154
742	*boiled in salted water*	IFR. As raw; boiled 45 minutes	0.80	82.4	2.3	0.1	9.5	46	195
743	*pickled, drained*	LGC; 10 samples, 5 brands. Whole and sliced	0.65	88.6	1.2	0.2	5.6	28[b]	117[b]
744	**Broccoli**, *green, raw*	IFR; tough stems removed	0.61	88.2	4.4	0.9	1.8[c]	33	138
745	*boiled in unsalted water*	IFR. As raw, cut into florets; boiled 15 minutes	1.00	91.1	3.1	0.8	1.1[c]	24	100
746	**Brussels sprouts**, *raw*	IFR; base trimmed, outer leaves removed	0.69	84.3	3.5	1.4	4.1[c]	42	177
747	*boiled in unsalted water*	IFR. As raw; boiled 15 minutes	1.00	86.9	2.9	1.3	3.5[c]	35	153
748	*frozen, boiled in unsalted water*	LGC. 10 samples, 8 brands; boiled 5-10 minutes	1.00	86.8	3.5	(1.3)	2.5[c]	35	148
749	**Cabbage**, *raw, average*	Average of January King, Savoy, summer and white	0.77	90.1	1.7	0.4	4.1	26	109
750	*boiled in unsalted water, average*	As raw	1.00	93.1	1.0	0.4	2.2	16	67
751	*January King, raw*	IFR; outer leaves and stem removed	0.66	89.7	1.8	0.4	3.9	25	107
752	*boiled in salted water*	IFR. As raw; shredded and boiled 19 minutes	1.00	92.5	0.8	0.6	2.5	18	75
753	*white, raw*	IFR; outer leaves and stem removed	0.91	90.7	1.4	0.2	5.0	27	113

[a] If peeled = 0.77
[b] Acetic acid from vinegar will contribute to the energy value
[c] Including oligosaccharides

No.	Food	Total nitrogen g	Fatty acids			Cholesterol mg	Starch g	Total sugars g	Dietary fibre	
			Satd g	Mono unsatd g	Poly unsatd g				Southgate method g	Englyst method g
	Vegetables, general									
737	**Asparagus**, raw	0.47	0.1	0.1	0.2	0	0.1	1.9	1.7	1.7
738	boiled in salted water	0.55	0.1	0.2	0.3	0	Tr	1.4	1.4	1.4
739	**Aubergine**, raw	0.14	0.1	Tr	0.2	0	0.2	2.0	2.3	2.0
740	fried in corn oil	0.19	4.1	7.9	18.5	0	0.2	2.6	2.9	2.3
741	**Beetroot**, raw	0.27	Tr	Tr	0.1	0	0.6	7.0	2.8	1.9
742	boiled in salted water	0.37	Tr	Tr	0.1	0	0.7	8.8	2.3	1.9
743	pickled, drained	0.19	Tr	Tr	0.1	0	Tr	5.6	2.5	1.7
744	**Broccoli**, green, raw	0.71	0.2	0.1	0.5	0	0.1	1.5[a]	N	2.6
745	boiled in unsalted water	0.50	0.2	0.1	0.4	0	Tr	0.9[a]	N	2.3
746	**Brussels sprouts**, raw	0.56	0.3	0.1	0.7	0	0.8	3.1[a]	3.8	4.1
747	boiled in unsalted water	0.46	0.3	0.1	0.7	0	0.3	3.0[a]	2.6	3.1
748	frozen, boiled in unsalted water	0.56	0.3	0.1	0.7	0	0.4	2.4[a]	N	4.3
749	**Cabbage**, raw, average	0.28	0.1	Tr	0.3	0	0.1	4.0	(2.9)	2.4
750	boiled in unsalted water, average	0.16	0.1	Tr	0.3	0	0.1	2.0	(2.3)	1.8
751	January King, raw	0.29	0.1	Tr	0.3	0	0.1	3.8	(3.1)	2.3
752	boiled in salted water	0.13	0.1	0.1	0.4	0	0.2	2.3	(2.5)	2.1
753	white, raw	0.23	Tr	Tr	0.1	0	0.1	4.9	2.4	2.1

[a] Not including oligosaccharides

Inorganic constituents per 100g

No.	Food	Na	K	Ca	Mg	P	Fe	Cu	Zn	Cl	Mn	Se	I
						mg						µg	
												Se	I

Vegetables, general

No.	Food	Na	K	Ca	Mg	P	Fe	Cu	Zn	Cl	Mn	Se	I
737	**Asparagus**, *raw*	1	260	27	13	72	0.7	0.08	0.7	60	0.2	(1)	Tr
738	*boiled in salted water*	60	220	25	13	50	0.6	0.08	0.7	110	0.2	(1)	Tr
739	**Aubergine**, *raw*	2	210	10	11	16	0.3	0.01	0.2	14	0.1	(1)	1
740	*fried in corn oil*	2	170	8	8	25	0.5	0.03	0.1	16	0.2	(1)	(1)
741	**Beetroot**, *raw*	66	380	20	11	51	1.0	0.02	0.4	59	0.7	Tr	N
742	*boiled in salted water*	110	510	29	16	87	0.8	0.03	0.5	N	0.9	Tr	N
743	*pickled, drained*	120	190	19	13	17	0.5	0.04	0.3	210	0.2	Tr	N
744	**Broccoli**, *green, raw*	8	370	56	22	87	1.7	0.02	0.6	100	0.2	Tr	2
745	*boiled in unsalted water*	(13)	170	40	13	57	1.0	0.02	0.4	(23)	0.2	N	2
746	**Brussels sprouts**, *raw*	6	450	26	8	77	0.7	0.02	0.5	38	0.2	N	1
747	*boiled in unsalted water*	2	310	20	13	61	0.5	0.03	0.3	16	0.2	N	1
748	*frozen, boiled in unsalted water*	8	340	29	17	66	0.6	0.04	0.4	15	0.2	N	(1)
749	**Cabbage**, *raw, average*	5	270	52	8	41	0.7	0.02	0.3	37	0.2	(1)	2
750	*boiled in unsalted water, average*	8	120	33	4	25	0.3	0.01	0.1	9	0.2	(2)	2
751	*January King, raw*	3	270	68	6	46	0.6	0.02	0.4	45	0.2	(2)	2
752	*boiled in salted water*	100	130	42	3	31	0.3	0.01	0.1	170	0.2	(2)	2
753	*white, raw*	7	240	49	6	29	0.5	0.01	0.2	40	0.2	Tr	2

No.	Food	Retinol µg	Carotene µg	Vitamin D µg	Vitamin E mg	Thiamin mg	Riboflavin mg	Niacin mg	Trypt 60 mg	Vitamin B6 mg	Vitamin B12 µg	Folate µg	Pantothenate mg	Biotin µg	Vitamin C mg
	Vegetables, general														
737	**Asparagus**, *raw*	0	315	0	1.16	0.16	0.06	1.0	0.5	0.09	0	175	0.17	(0.4)	12
738	*boiled in salted water*	0	530	0	(1.16)	0.12	0.06	0.8	0.6	0.07	0	155	0.16	0.4	10
739	**Aubergine**, *raw*	0	70	0	0.03	0.02	0.01	0.1	0.2	0.08	0	18	0.08	N	4
740	*fried in corn oil*	0	125	0	5.50	0	Tr	Tr	0.2	0.07	0	(5)	(0.07)	N	1
741	**Beetroot**, *raw*	0	20	0	Tr	0.01	0.01	0.1	0.3	0.03	0	150	0.12	Tr	5
742	*boiled in salted water*	0	27	0	Tr	0.01	0.01	0.1	0.3	0.04	0	110	0.10	Tr	5
743	*pickled, drained*	0	Tr	0	Tr	0.02	0.03	0.1	0.2	0.04	0	2	(0.10)	Tr	N
744	**Broccoli**, green, *raw*	0	575	0	(1.30)	0.10	0.06	0.9	0.8	0.14	0	90	N	N	87
745	*boiled in unsalted water*	0	475	0	(1.10)	0.05	0.05	0.7	0.6	0.11	0	64	N	N	44
746	**Brussels sprouts**, *raw*	0	215	0	1.00	0.15	0.11	0.2	0.7	0.37	0	135	1.00	0.4	115
747	*boiled in unsalted water*	0	320	0	0.90	0.07	0.09	Tr	0.5	0.19	0	110	0.28	0.3	60
748	*frozen, boiled in unsalted water*	0	320	0	(0.90)	0.09	0.08	Tr	0.7	0.28	0	59	(0.28)	(0.3)	69
749	**Cabbage**, *raw, average*	0	385a	0	0.20b	0.15	0.02	0.5	0.3	0.17	0	75	0.21	0.1	49
750	*boiled in unsalted water, average*	0	210a	0	0.20b	0.08	0.01	0.3	0.2	0.08	0	29	0.15	Tr	20
751	*January King, raw*	0	340a	0	0.20b	0.22	0.02	0.3	0.3	0.22	0	78	0.21	0.1	49
752	*boiled in salted water*	0	100a	0	0.20b	0.12	0.02	0.3	0.1	0.06	0	43	0.15	Tr	19
753	*white, raw*	0	40	0	0.20	0.12	0.01	0.3	0.2	0.18	0	34	0.21	0.1	35

[a] Average figures. The amount of carotene in leafy vegetables depends on the amount of chlorophyll, and the outer green leaves may contain 50 times as much as inner white ones

[b] The value for inner leaves. Outer leaves contain 7.0mg α-tocopherol per 100g

Vegetables *continued*

Composition of food per 100g

Vegetables, general

No.	Food	Description and main data sources	Edible proportion	Water g	Protein g	Fat g	Carbo-hydrate g	Energy value kcal	Energy value kJ
754	**Carrots**, old, *raw*	IFR; ends trimmed, peeled	0.70	89.8	0.6	0.3	7.9[a]	35	146
755	*boiled in unsalted water*	IFR. As raw; sliced and boiled 12.5 minutes	1.00	90.5	0.6	0.4	4.9[a]	24	100
756	young, *raw*	IFR; ends trimmed, scrubbed	0.87	88.8	0.7	0.5	6.0[a]	30	125
757	*boiled in unsalted water*	IFR. As raw; sliced and boiled 15 minutes	1.00	90.7	0.6	0.4	4.4[a]	22	93
758	*canned, re-heated, drained*	LGC; 10 cans, 5 brands	0.61	91.9	0.5	0.3	4.2[a]	20	87
759	**Cauliflower**, *raw*	IFR; florets only	0.45	88.4	3.6	0.9	3.0[a]	34	142
760	*boiled in unsalted water*	IFR. As raw; boiled 13 minutes	1.00	90.6	2.9	0.9	2.1[a]	28	117
761	**Celery**, *raw*	IFR; stem only	0.91	95.1	0.5	0.2	0.9	7	32
762	*boiled in salted water*	IFR. Stem only; boiled 20 minutes	1.00	95.2	0.5	0.3	0.8	8	34
763	**Chicory**, *raw*	IFR. Stem and inner leaves; pale variety	0.80	94.3	0.5	0.6	2.8	11[b]	45[b]
764	**Courgette**, *raw*	IFR; ends trimmed	0.88	93.7	1.8	0.4	1.8	18	74
765	*boiled in unsalted water*	Analysis and calculation from raw	1.00	93.0	2.0	0.4	2.0	19	81
766	*fried in corn oil*	IFR. As raw; sliced and shallow fried 5 minutes	1.00	86.8	2.6	4.8	2.6	63	265
767	**Cucumber**, *raw*	IFR; ends trimmed, not peeled	0.97[c]	96.4	0.7	0.1	1.5	10	40
768	**Curly kale**, *raw*	IFR; main ribs and stalks removed	0.85	88.4	3.4	1.6	1.4	33	140
769	*boiled in salted water*	IFR. As raw; shredded and boiled 7 minutes	1.00	90.9	2.4	1.1	1.0	24	100
770	**Fennel**, Florence, *raw*	IFR; inner leaves and bulb only	0.80	94.2	0.9	0.2	1.8	12	50
771	*boiled in salted water*	IFR. As raw; boiled 14 minutes	1.00	94.4	0.9	0.2	1.5	11	47

[a] Including oligosaccharides

[b] Contains inulin; 32 per cent total carbohydrate taken to be available for energy purposes

[c] If peeled = 0.77

Vegetables *continued*

Composition of food per 100g 754 to 771

No.	Food	Total nitrogen g	Fatty acids			Cholesterol mg	Starch g	Total sugars g	Dietary fibre	
			Satd g	Mono unsatd g	Poly unsatd g				Southgate method g	Englyst method g
Vegetables, general										
754	**Carrots**, old, *raw*	0.10	0.1	Tr	0.2	0	0.3	7.4[a]	2.6	2.4
755	*boiled in unsalted water*	0.10	0.1	Tr	0.2	0	0.2	4.6[a]	2.8	2.5
756	*young, raw*	0.11	0.1	Tr	0.3	0	0.2	5.6[a]	(2.6)	2.4
757	*boiled in unsalted water*	0.09	0.1	Tr	0.2	0	0.2	4.2[a]	2.7	2.3
758	*canned, re-heated, drained*	0.09	(0.1)	Tr	(0.2)	0	0.4	3.7[a]	N	1.9
759	**Cauliflower**, *raw*	0.58	0.2	0.1	0.5	0	0.4	2.5[a]	1.9	1.8
760	*boiled in unsalted water*	0.47	0.2	0.1	0.5	0	0.2	1.8[a]	1.6	1.6
761	**Celery**, *raw*	0.08	Tr	Tr	0.1	0	Tr	0.9	1.6	1.1
762	*boiled in salted water*	0.08	0.1	0.1	0.1	0	Tr	0.8	2.0	1.2
763	**Chicory**, *raw*	0.09	0.2	Tr	0.3	0	0.2	0.7	N	0.9
764	**Courgette**, *raw*	0.29	0.1	Tr	0.2	0	0.1	1.7	N	0.9
765	*boiled in unsalted water*	0.32	0.1	Tr	0.2	0	0.1	1.9	N	1.2
766	*fried in corn oil*	0.41	0.6	1.2	2.8	0	0.1	2.5	N	1.2
767	**Cucumber**, *raw*	0.11	Tr	Tr	Tr	0	0.1	1.4[b]	(0.7)	0.6
768	**Curly kale**, *raw*	0.55	0.2	0.1	0.9	0	0.1	1.3	(3.3)	3.1
769	*boiled in salted water*	0.39	0.2	0.1	0.6	0	0.1	0.9	(2.6)	2.8
770	**Fennel**, Florence, *raw*	0.15	Tr	Tr	Tr	0	0.1	1.7	N	2.4
771	*boiled in salted water*	0.14	Tr	Tr	Tr	0	0.1	1.4	N	2.3

[a] Not including oligosaccharides
[b] Peeled cucumbers contain approximately 2.0g total sugars per 100g as equal quantities of glucose and fructose

Vegetables continued

Inorganic constituents per 100g

No.	Food	Na	K	Ca	Mg	P	Fe	Cu	Zn	Cl	Mn	Se	I
						mg						µg	
												Se	I
	Vegetables, general												
754	**Carrots**, old, *raw*	25	170	25	3	15	0.3	0.02	0.1	33	0.1	1	2
755	*boiled in unsalted water*	50	120	24	3	17	0.4	0.01	0.1	31	0.1	1	2
756	young, *raw*	40	240	34	9	25	0.4	0.02	0.2	39	(0.1)	(1)	(2)
757	*boiled in unsalted water*	23	160	30	6	15	0.4	0.02	0.2	28	(0.1)	(1)	(2)
758	canned, re-heated, drained	370	110	25	5	14	0.6	0.04	0.1	490	0.1	(1)	N
759	**Cauliflower**, *raw*	9	380	21	17	64	0.7	0.03	0.6	28	0.3	Tr	Tr
760	*boiled in unsalted water*	4	120	17	12	52	0.4	0.02	0.4	14	0.2	Tr	Tr
761	**Celery**, *raw*	60	320	41	5	21	0.4	0.01	0.1	130	0.1	(3)	N
762	*boiled in salted water*	160	230	45	4	20	0.3	0.01	0.1	250	0.1	(3)	N
763	**Chicory**, *raw*	1	170	21	6	27	0.4	0.05	0.2	25	0.3	N	N
764	**Courgette**, *raw*	1	360	25	22	45	0.8	0.02	0.3	45	0.1	(1)	N
765	*boiled in unsalted water*	1	210	19	17	36	0.6	0.01	0.2	26	0.1	(1)	N
766	*fried in corn oil*	1	490	38	32	61	1.4	0.05	0.5	65	0.1	(1)	N
767	**Cucumber**, *raw*	3	140	18	8	49	0.3	0.01	0.1	17	0.1	Tr	3
768	**Curly kale**, *raw*	43	450	130	34	61	1.7	0.03	0.4	68	0.8	(2)	N
769	*boiled in salted water*	100	160	150	8	39	2.0	0.02	0.2	N	0.4	(2)	N
770	**Fennel**, Florence, *raw*	11	440	24	8	26	0.3	0.02	0.5	27	N	N	N
771	*boiled in salted water*	96	300	20	7	21	0.2	0.01	0.4	120	N	N	N

No.	Food	Retinol μg	Carotene μg	Vitamin D μg	Vitamin E mg	Thiamin mg	Ribo-flavin mg	Niacin mg	Trypt 60 mg	Vitamin B6 mg	Vitamin B12 μg	Folate μg	Panto-thenate mg	Biotin μg	Vitamin C mg
	Vegetables, general														
754	**Carrots**, old, *raw*	0	8115[a]	0	0.56	0.10	0.01	0.2	0.1	0.14	0	12	0.25	0.6	6
755	*boiled in unsalted water*	0	7560	0	0.56	0.09	Tr	Tr	0.1	0.10	0	16	0.18	0.4	2
756	*young, raw*	0	5330	0	(0.56)	0.04	0.02	0.2	0.1	0.07	0	28	(0.25)	(0.6)	4
757	*boiled in unsalted water*	0	4425	0	(0.56)	0.05	0.01	0.1	0.1	0.05	0	17	0.18	0.4	2
758	*canned, re-heated, drained*	0	3370	0	0.64	0.01	0.02	0.2	0.1	0.07	0	8	0.10	0.4	1
759	**Cauliflower**, *raw*	0	50	0	0.22	0.17	0.05	0.6	0.9	0.28	0	66	0.60	1.5	43
760	*boiled in unsalted water*	0	60	0	0.11	0.07	0.04	0.4	0.7	0.15	0	51	0.42	1.0	27
761	**Celery**, *raw*	0	50	0	0.20	0.06	0.01	0.3	0.1	0.03	0	16	0.40	0.1	8
762	*boiled in salted water*	0	50	0	0.20	0.06	0.01	Tr	0.1	0.03	0	10	0.28	Tr	4
763	**Chicory**, *raw*	0	120	0	N	0.14	Tr	0.1	0.2	0.01	0	14	N	N	5
764	**Courgette**, *raw*	0	610	0	N	0.12	0.02	0.3	0.3	0.15	0	52	0.08	N	21
765	*boiled in unsalted water*	0	(440)	0	N	0.08	0.02	0.2	0.3	0.09	0	31	0.11	N	11
766	*fried in corn oil*	0	500	0	0.83	0.10	0.01	0.4	0.4	0.09	0	42	N	N	15
767	**Cucumber**, *raw*	0	60[b]	0	0.07	0.03	0.01	0.2	0.1	0.04	0	9	0.30	0.9	2
768	**Curly kale**, *raw*	0	3145	0	(1.70)	0.08	0.09	1.0	0.7	0.26	0	120	0.09	0.5	110
769	*boiled in salted water*	0	3375	0	(1.33)	0.02	0.06	0.8	0.5	0.13	0	86	0.05	0.4	71
770	**Fennel**, Florence, *raw*	0	140	0	N	0.06	0.01	0.6	N	0.06	0	42	N	N	5
771	*boiled in salted water*	0	60	0	N	0.05	0.01	0.4	N	0.08	0	26	N	N	2

[a] Levels ranged from 4300 to 11000 μg carotene per 100g

[b] Carotene can be as high as 260μg per 100g. In peeled cucumbers the carotene ranges from 0 to 35μg per 100g

Composition of food per 100g

No.	Food	Description and main data sources	Edible proportion	Water g	Protein g	Fat g	Carbo-hydrate g	Energy value kcal	Energy value kJ
	Vegetables, general								
772	**Garlic**, raw	IFR; peeled cloves	0.79	64.3	7.9	0.6	16.3	98	411
773	**Gherkins**, pickled, *drained*	LGC; 10 samples, 5 brands	0.67	92.8	0.9	0.1	2.6	14[a]	61[a]
774	**Gourd**, karela, *raw*	LGC; 5 samples. Ends trimmed	0.93	93.3	1.6	0.2	0.8	11	47
775	**Leeks**, raw	IFR; trimmed and outer leaves removed	0.57[b]	90.8	1.6	0.5	2.9[c]	22	93
776	*boiled in unsalted water*	IFR. As raw; chopped and boiled 22 minutes	1.00	92.2	1.2	0.7	2.6[c]	21	87
777	**Lettuce**, average, *raw*	Average of 4 varieties	0.74	95.1	0.8	0.5	1.7	14	59
778	butterhead, *raw*	IFR; outer leaves removed	0.76	94.4	0.9	0.6	1.2	12	52
779	iceberg, *raw*	IFR; outer leaves removed	0.83	95.6	0.7	0.3	1.9	13	53
780	**Marrow**, raw	IFR; flesh only, seeds removed	0.54	95.6	0.5	0.2	2.2	12	51
781	*boiled in unsalted water*	IFR. As raw; cut and boiled 19 minutes	1.00	95.9	0.4	0.2	1.6	9	38
782	**Mixed vegetables**, *frozen, boiled in salted water*	LGC; 10 samples. Assorted varieties. Simmered 3-7 minutes	1.00	85.8	3.3	0.5	6.6	42	180
783	**Mushrooms**, common, *raw*	IFR; stalks trimmed where necessary	0.97[d]	92.6	1.8[e]	0.5	0.4	13	55
784	*boiled in salted water*	LGC; 10 samples, button and sliced; boiled 5-10 minutes. And calcd. from raw	1.00	92.7	1.8[e]	0.3	0.4	11	48
785	*fried in blended oil*	Calculation from *fried in corn oil*	1.00	74.8	2.4[e]	16.2	0.3	157	645
786	*fried in butter*	Calculation from *fried in corn oil*	1.00	74.8	2.4[e]	16.2	0.3	157	645
787	*fried in corn oil*	IFR. As raw; sliced and fried 8 minutes	1.00	74.8	2.4[e]	16.2	0.3	157	645

[a] Acetic acid from vinegar will contribute to the energy value
[b] Bulb only = 0.36
[c] Including oligosaccharides
[d] If peeled = 0.75
[e] (Total N – non-protein N) × 6.25

No.	Food	Total nitrogen g	Fatty acids			Cholesterol mg	Starch g	Total sugars g	Dietary fibre	
			Satd g	Mono unsatd g	Poly unsatd g				Southgate method g	Englyst method g
	Vegetables, general									
772	**Garlic**, *raw*	1.27	0.1	Tr	0.3	0	14.7	1.6	N	4.1
773	**Gherkins**, pickled, *drained*	0.14	Tr	Tr	Tr	0	0.2	2.4	1.2	(1.2)
774	**Gourd**, karela, *raw*	0.26	N	N	N	0	0.8	Tr	3.6	2.6
775	**Leeks**, *raw*	0.26	0.1	Tr	0.3	0	0.3	2.2[a]	2.8	2.2
776	*boiled in unsalted water*	0.20	0.1	Tr	0.4	0	0.2	2.0[a]	2.4	1.7
777	**Lettuce**, *average, raw*	0.13	0.1	Tr	0.3	0	Tr	1.7	1.3	0.9
778	*butterhead, raw*	0.14	0.1	Tr	0.4	0	Tr	1.0	(1.3)	1.2
779	*Iceberg, raw*	0.11	Tr	Tr	0.2	0	Tr	1.9	(1.3)	0.6
780	**Marrow**, *raw*	0.08	Tr	Tr	Tr	0	0.1	2.1	1.1	0.5
781	*boiled in unsalted water*	0.07	Tr	Tr	Tr	0	0.2	1.4	1.0	0.6
782	**Mixed vegetables**, *frozen, boiled in salted water*	0.53	N	N	N	0	3.0	3.6	N	N
783	**Mushrooms**, common, *raw*	0.64[b]	0.1	Tr	0.3	0	0.2	0.2	2.3	1.1
784	*boiled in salted water*	0.64[b]	0.1	Tr	0.2	0	0.2	0.2	2.3	1.1
785	*fried in blended oil*	0.95[b]	1.4	8.1	5.9	0	0.2	0.1	3.0	1.5
786	*fried in butter*	0.95[b]	10.7	3.9	0.5	37	0.2	0.1	3.0	1.5
787	*fried in corn oil*	0.95[b]	2.1	4.0	9.4	0	0.2	0.1	3.0	1.5

[a] Not including oligosaccharides

[b] 60 per cent of this nitrogen is non-protein nitrogen

Inorganic constituents per 100g

No.	Food	Na	K	Ca	Mg	P	Fe	Cu	Zn	Cl	Mn	Se	I
						mg						μg	
Vegetables, general													
772	**Garlic**, *raw*	4	620	19	25	170	1.9	0.06	1.0	73	0.5	2	3
773	**Gherkins**, pickled, *drained*	690	110	20	11	22	0.7	0.10	0.3	1060	0.1	N	N
774	**Gourd**, karela, *raw*	1	330	19	31	48	1.4	0.27	0.4	21	0.3	N	N
775	**Leeks**, *raw*	2	260	24	3	44	1.1	0.02	0.2	59	0.2	(1)	N
776	*boiled in unsalted water*	6	150	20	2	32	0.7	0.02	0.2	43	0.2	(1)	N
777	**Lettuce**, *average, raw*	3	220	28	6	28	0.7	0.01	0.2	47	0.3	(1)	2
778	*butterhead, raw*	5	360	53	8	43	1.5	0.04	0.4	67	0.3	(1)	2
779	*iceberg, raw*	2	160	19	5	18	0.4	0.01	0.1	42	0.3	(1)	2
780	**Marrow**, *raw*	1	140	18	10	17	0.2	0.02	0.2	30	N	N	N
781	*boiled in unsalted water*	1	110	14	7	18	0.1	0.01	0.2	14	N	N	N
782	**Mixed vegetables**, *frozen, boiled in salted water*	96	130	35	16	57	0.8	0.02	0.4	140	0.2	N	N
783	**Mushrooms**, common, *raw*	5	320	6	9	80	0.6	0.72	0.4	69	0.1	9	3
784	*boiled in salted water*	71	250	5	7	64	0.5	0.40	0.3	N	0.1	7	2
785	*fried in blended oil*	4	340	8	19	100	1.0	0.40	0.5	89	0.1	12	4
786	*fried in butter*	150	340	11	19	110	1.0	0.40	0.5	320	0.1	12	11
787	*fried in corn oil*	4	340	8	19	100	1.0	0.40	0.5	89	0.1	12	4

Vegetables, general

No.	Food	Retinol µg	Carotene µg	Vitamin D µg	Vitamin E mg	Thiamin mg	Ribo-flavin mg	Niacin mg	Trypt 60 mg	Vitamin B6 mg	Vitamin B12 µg	Folate µg	Panto-thenate mg	Biotin µg	Vitamin C mg
772	**Garlic**, *raw*	0	Tr	0	0.01	0.13	0.03	0.3	1.9	0.38	0	5	N	N	17
773	**Gherkins**, pickled, *drained*	0	2	0	N	Tr	0.02	0.1	0.1	N	0	6	N	N	1
774	**Gourd**, karela, *raw*	0	345	0	N	0.09	0.05	0.4	0.3	N	0	45	N	N	185
775	**Leeks**, *raw*	0	735	0	0.92	0.29	0.05	0.4	0.2	0.48	0	56	0.12	1.4	17
776	*boiled in unsalted water*	0	575	0	0.78	0.02	0.02	0.4	0.2	0.05	0	40	0.10	1.0	7
777	**Lettuce**, *average, raw*	0	355[a]	0	0.57	0.12	0.02	0.4	0.1	0.04	0	55	(0.18)	0.7	5
778	butterhead, *raw*	0	910[a]	0	0.57	0.15	0.03	0.5	0.1	0.08	0	57	0.18	0.7	7
779	iceberg, *raw*	0	50[a]	0	0.57	0.11	0.01	0.3	0.1	0.03	0	53	(0.18)	0.7	3
780	**Marrow**, *raw*	0	110	0	Tr	0.08	Tr	0.2	0.1	0.03	0	13	0.10	0.4	11
781	*boiled in unsalted water*	0	110	0	Tr	0.08	Tr	0.2	0.1	0.01	0	15	0.07	0.4	3
782	**Mixed vegetables**, *frozen, boiled in salted water*	0	2520	0	N	0.12	0.09	0.8	0.5	0.11	0	52	N	N	13
783	**Mushrooms**, common, *raw*	0	0	0	0.12	0.09	0.31	3.2	0.3	0.18	0	44	2.00	12	1
784	*boiled in salted water*	0	0	0	0.12	0.07	0.35	2.3	0.3	0.06	0	8	1.40	8.4	1
785	*fried in blended oil*	0	Tr	0	N	0.09	0.34	2.3	0.4	0.19	0	11	1.40	8.0	1
786	*fried in butter*	160	85	0.1	0.40	0.09	0.34	2.3	0.4	0.19	Tr	11	1.40	8.0	1
787	*fried in corn oil*	0	Tr	0	2.84	0.09	0.34	2.3	0.4	0.19	0	11	1.40	8.0	1

[a] Average figures. The outer green leaves may contain 50 times as much carotene as the inner white ones

Vegetables continued

Composition of food per 100g

No.	Food	Description and main data sources	Edible proportion	Water g	Protein g	Fat g	Carbo-hydrate g	Energy value kcal	kJ
	Vegetables, general								
788	**Mustard and cress**, *raw*	IFR; leaves and cut stems	1.00[a]	95.3	1.6	0.6	0.4	13	56
789	**Okra**, *raw*	IFR and literature sources. Ends trimmed	0.74	86.6	2.8	1.0	3.0	31	130
790	*boiled in unsalted water*	Calcd. from raw	1.00	87.9	2.5	0.9	2.7	28	119
791	*stir-fried in corn oil*	IFR. As raw; sliced and fried 5 minutes	1.00	54.5	4.3	26.1	4.4	269	1122
792	**Onions**, *raw*	IFR; flesh only	0.91	89.0	1.2	0.2	7.9[b]	36	150
793	*boiled in unsalted water*	Calcd. from raw	1.00	92.6	0.6	0.1	3.7[b]	17	73
794	*fried in blended oil*	Calculation from *fried in corn oil*	1.00	65.7	2.3	11.2[c]	14.1[b]	164	684
795	*fried in corn oil*	IFR. As raw: sliced into rings and fried 15 minutes	1.00	65.7	2.3	11.2[c]	14.1[b]	164	684
796	*fried in lard*	Calculation from *fried in corn oil*	1.00	65.7	2.3	11.2[c]	14.1[b]	164	684
797	*pickled, drained*	LGC; 10 samples, 7 brands	0.58	90.6	0.9	0.2	4.9[b]	24[d]	101[d]
798	*cocktail/silverskin, drained*	LGC; 10 samples, 8 brands	0.59	91.8	0.6	0.1	3.1[b]	15[d]	63[d]
799	**Parsnip**, *raw*	LGC; ends trimmed and peeled	0.72	79.3	1.8	1.1	12.5[b]	64	271
800	*boiled in unsalted water*	LGC. As raw; sliced and boiled 12 minutes	1.00	78.7	1.6	1.2[e]	12.9[b]	66	278
801	**Peppers**, capsicum, chilli, green, *raw*	Refs. Cashel et al. (1989), Gopalan et al. (1980)	0.90	85.7	2.9	0.6	0.7	20	83
802	capsicum, green, *raw*	IFR; stalk and seeds removed	0.84	93.3	0.8	0.3	2.6[b]	15	65
803	*boiled in salted water*	IFR. As raw; sliced and boiled 15 minutes	1.00	92.6	1.0	0.5	2.6[b]	18	76
804	capsicum, red, *raw*	IFR; stalk and seeds removed	0.83	90.4	1.0	0.4	6.4[b]	32	134
805	*boiled in salted water*	Calculation from raw	1.00	89.5	1.1	0.4	7.0[b]	34	145

[a] If purchased on soil block = 0.27

[c] The fat content of fried onions can vary considerably

[e] Roast parsnips contain 6.5g fat per 100g

[b] Including oligosaccharides

[d] Acetic acid from vinegar will contribute to the energy value

Vegetables *continued*

788 to 805

Composition of food per 100g

No.	Food	Total nitrogen g	Fatty acids			Cholest-erol mg	Starch g	Total sugars g	Dietary fibre	
			Satd g	Mono unsatd g	Poly unsatd g				Southgate method g	Englyst method g
	Vegetables, general									
788	**Mustard and cress,** *raw*	0.26	Tr	0.2	0.2	0	Tr	0.4	3.3	1.1
789	**Okra,** *raw*	0.40	0.3	0.1	0.3	0	0.5	2.5	4.5	4.0
790	*boiled in unsalted water*	0.40	0.3	0.1	0.3	0	0.5	2.3	4.1	3.6
791	*stir-fried in corn oil*	0.69	3.3	6.5	15.1	0	0.8	3.6	7.0	6.3
792	**Onions,** *raw*	0.20	Tr	Tr	0.1	0	Tr	5.6[a]	1.5	1.4
793	*boiled in unsalted water*	0.09	Tr	Tr	Tr	0	Tr	2.6[a]	0.7	0.7
794	*fried in blended oil*	0.37	1.0	5.6	4.1	0	0.1	10.0[a]	3.2	3.1
795	*fried in corn oil*	0.37	1.4	2.8	6.5	0	0.1	10.0[a]	3.2	3.1
796	*fried in lard*	0.37	4.6	4.9	1.1	10	0.1	10.0[a]	3.2	3.1
797	*pickled, drained*	0.14	Tr	Tr	0.1	0	Tr	3.5[a]	1.3	1.2
798	*cocktail/silverskin, drained*	0.10	Tr	Tr	Tr	0	Tr	2.2[a]	0.6	N
799	**Parsnip,** *raw*	0.29	0.2	0.5	0.2	0	6.2	5.7[a]	4.3	4.6
800	*boiled in unsalted water*	0.26	0.2	0.5	0.2	0	6.4	5.9[a]	4.4	4.7
801	**Peppers,** capsicum, chilli, green, *raw*	0.46	N	N	N	0	Tr	0.7	N	N
802	capsicum, green, *raw*	0.13	0.1	Tr	0.2	0	0.1	2.4[a]	(1.9)	1.6
803	*boiled in salted water*	0.16	0.1	Tr	0.3	0	0.2	2.3[a]	(2.1)	1.8
804	capsicum, red, *raw*	0.17	0.1	Tr	0.2	0	0.1	6.1[a]	(1.9)	1.6
805	*boiled in salted water*	0.19	0.1	Tr	0.2	0	0.1	6.7[a]	2.1	1.7

[a] Not including oligosaccharides

Vegetables continued

Inorganic constituents per 100g

No.	Food	Na	K	Ca	Mg	P	Fe	Cu	Zn	Cl	Mn	Se	I
						mg						µg	
Vegetables, general													
788	**Mustard and cress**, *raw*	19	110	50	22	33	1.0	0.01	0.3	39	N	N	N
789	**Okra**, *raw*	8	330	160	71	59	1.1	0.13	0.6	41	N	(1)	N
790	*boiled in unsalted water*	5	310	120	57	54	0.6	0.09	0.5	N	N	(1)	N
791	*stir-fried in corn oil*	13	480	220	110	89	1.5	0.19	1.0	64	N	(2)	N
792	**Onions**, *raw*	3	160	25	4	30	0.3	0.05	0.2	25	0.1	(1)	3
793	*boiled in unsalted water*	2	90	19	3	16	0.3	0.05	0.2	6	Tr	Tr	1
794	*fried in blended oil*	4	370	47	8	44	0.8	0.04	0.3	53	0.2	(2)	6
795	*fried in corn oil*	4	370	47	8	44	0.8	0.04	0.3	53	0.2	(2)	6
796	*fried in lard*	4	370	47	8	44	0.8	0.04	0.3	53	0.2	(2)	6
797	*pickled, drained*	450	93	22	5	23	0.2	0.04	0.1	730	0.1	(1)	(3)
798	*cocktail/silverskin, drained*	620	60	29	5	16	0.5	0.04	0.1	990	0.1	N	N
799	**Parsnip**, *raw*	10	450	41	23	74	0.6	0.05	0.3	49	0.5	2	N
800	*boiled in unsalted water*	4	350	50	23	76	0.6	0.04	0.3	33	0.3	N	N
801	**Peppers**, capsicum, chilli, green, *raw*	7	220	30	24	80	1.2	N	0.4	15	N	N	N
802	*capsicum, green, raw*	4	120	8	10	19	0.4	0.02	0.1	19	0.1	Tr	1
803	*boiled in salted water*	70	140	9	10	23	0.4	0.03	0.2	100	0.1	Tr	1
804	*capsicum, red, raw*	4	160	8	14	22	0.3	0.01	0.1	24	0.1	Tr	(1)
805	*boiled in salted water*	70	180	9	14	26	0.3	0.01	0.2	100	0.1	Tr	(1)

Vegetables continued

Vegetables, general

No.	Food	Retinol µg	Carotene µg	Vitamin D µg	Vitamin E mg	Thiamin mg	Riboflavin mg	Niacin mg	Trypt 60 mg	Vitamin B6 mg	Vitamin B12 µg	Folate µg	Pantothenate mg	Biotin µg	Vitamin C mg
788	**Mustard and cress**, raw	0	(1280)	0	0.70	0.04	0.04	1.0	0.3	0.15	0	60	N	N	33
789	**Okra**, raw	0	515	0	N	0.20	0.06	1.0	0.4	0.21	0	88	0.25	N	21
790	boiled in unsalted water	0	465	0	N	0.13	0.05	0.9	0.3	0.19	0	46	0.21	N	16
791	stir-fried in corn oil	0	560	0	4.50	0.17	0.06	0.9	0.6	0.20	0	83	0.23	N	21
792	**Onions**, raw	0	10	0	0.31	0.13	Tr	0.7	0.3	0.20	0	17	0.11	0.9	5
793	boiled in unsalted water	0	5	0	0.15	0.09	Tr	0.3	0.1	0.12	0	9	0.07	0.6	3
794	fried in blended oil	0	40	0	N	0.08	0.01	Tr	0.5	0.10	0	38	0.12	1.3	3
795	fried in corn oil	0	40	0	1.93	0.08	0.01	Tr	0.5	0.10	0	38	0.12	1.3	3
796	fried in lard	Tr	40	N	Tr	0.08	0.01	Tr	0.5	0.10	0	38	0.12	1.3	3
797	pickled, drained	0	(10)	0	(0.31)	0.02	Tr	0.1	0.2	0.10	0	14	N	N	Tr
798	cocktail/silverskin, drained	0	Tr	0	N	N	Tr	N	0.1	N	0	N	N	N	Tr
799	**Parsnip**, raw	0	30	0	1.00	0.23	0.01	1.0	0.5	0.11	0	87	0.50	0.1	17
800	boiled in unsalted water	0	30	0	1.00	0.07	0.01	0.7	0.4	0.09	0	48	0.35	Tr	10
801	**Peppers**, capsicum, chilli, green, raw	0	175	0	N	0.07	0.08	1.1	0.5	N	0	29	N	N	120
802	capsicum, green, raw	0	265	0	0.80	0.01	0.01	0.1	0.1	0.30	0	36	0.08	N	120
803	boiled in salted water	0	240	0	0.80	0.01	0.02	Tr	0.2	0.26	0	19	0.06	N	69
804	capsicum, red, raw	0	3840	0	0.80	0.01	0.03	1.3	0.2	0.36	0	21	0.08	N	140
805	boiled in salted water	0	3780	0	0.90	0.01	0.03	0.9	0.2	0.31	0	11	0.06	N	81

Vegetables continued

Composition of food per 100g

No.	Food	Description and main data sources	Edible proportion	Water g	Protein g	Fat g	Carbo-hydrate g	Energy value kcal	kJ
	Vegetables, general								
806	**Plantain**, *raw*	Literature sources. Green flesh	0.65	67.5	1.1	0.3	29.4	117	500
807	*boiled in unsalted water*	10 samples. Flesh only; boiled 30 minutes. And literature sources	1.00	68.5	0.8	0.2	28.5	112	477
808	*ripe, fried in vegetable oil*	8 samples	1.00	34.7	1.5	9.2	47.5	267	1126
809	**Pumpkin**, *raw*	IFR; flesh only, peeled thickly, seeds removed	0.67	95.0	0.7	0.2	2.2	13	55
810	*boiled in salted water*	IFR. As raw; boiled 15 minutes	1.00	94.9	0.6	0.3	2.1	13	56
811	**Quorn**, myco-protein	Manufacturer's data (Marlow Foods)	1.00	75.0	11.8[a]	3.5	2.0[b]	86	360
812	**Radish**, red, *raw*	IFR; ends trimmed, flesh and skin	0.81	95.4	0.7	0.2	1.9	12	49
813	**Spinach**, *raw*	IFR; ribs and stems removed	0.81	89.7	2.8	0.8	1.6	25	103
814	*boiled in unsalted water*	IFR. As raw; shredded	1.00	91.8	2.2	0.8	0.8	19	79
815	*frozen, boiled in unsalted water*	LGC; 10 samples, 8 brands. Boiled 2-10 minutes	1.00	91.6	3.1	(0.8)	0.5	21	90
816	**Spring greens**, *raw*	IFR; main ribs and stems removed	0.84	86.2	3.0	1.0	3.1	33	136
817	*boiled in unsalted water*	IFR. As raw; boiled 12 minutes	1.00	92.2	1.9	0.7	1.6	20	82
818	**Spring onions**, *bulbs and tops, raw*	IFR; peeled bulb and leaves	0.69	92.2	2.0	0.5	3.0	23	98
819	**Swede**, *raw*	IFR; flesh only, peeled thinly	0.73	91.2	0.7	0.3	5.0	24	101
820	*boiled in unsalted water*	IFR. As raw; diced and boiled 22 minutes	1.00	95.8	0.3	0.1	2.3	11	46
821	**Sweet potato**, *raw*	IFR; flesh only, yellow variety	0.84	73.7	1.2	0.3	21.3	87	372
822	*boiled in salted water*	IFR. As raw; boiled 27 minutes	1.00	74.7	1.1	0.3	20.5	84	358

[a] N x 6.22   [b] Including oligosaccharides

Vegetables continued

| No. | Food | Total nitrogen g | Fatty acids | | | Cholesterol mg | Starch g | Total sugars g | Dietary fibre | |
			Satd g	Mono unsatd g	Poly unsatd g				Southgate method g	Englyst method g
	Vegetables, general									
806	**Plantain**, *raw*	0.18	0.1	Tr	0.1	0	23.7	5.7	2.3	1.3
807	*boiled in unsalted water*	0.13	0.1	Tr	0.1	0	23.0	5.5	2.2	1.2
808	*ripe, fried in vegetable oil*	0.24	1.0	3.3	4.5	0	36.0	11.5	4.0	2.3
809	**Pumpkin**, *raw*	0.12	0.1	Tr	Tr	0	0.3	1.7	0.5	1.0
810	*boiled in salted water*	0.10	0.1	Tr	Tr	0	0.1	1.8	0.5	1.1
811	**Quorn**, myco-protein	1.90[a]	0.6	0.7	1.3	0	Tr	1.1[b]	N	4.8
812	**Radish**, red, *raw*	0.11	0.1	Tr	0.1	0	Tr	1.9	0.9	0.9
813	**Spinach**, *raw*	0.45	0.1	0.1	0.5	0	0.1	1.5	(3.9)	2.1
814	*boiled in unsalted water*	0.35	0.1	0.1	0.5	0	Tr	0.8	3.1	2.1
815	*frozen, boiled in unsalted water*	0.50	0.1	0.1	0.5	0	0.2	0.3	(3.1)	(2.1)
816	**Spring greens**, *raw*	0.48	0.1	0.1	0.6	0	0.4	2.7	(6.1)	3.4
817	*boiled in unsalted water*	0.30	0.1	0.1	0.4	0	0.2	1.4	3.4	2.6
818	**Spring onions**, *bulbs and tops, raw*	0.32	0.1	0.1	0.2	0	0.2	2.8	N	1.5
819	**Swede**, *raw*	0.11	Tr	Tr	0.2	0	0.1	4.9	2.4	1.9
820	*boiled in unsalted water*	0.05	Tr	Tr	0.1	0	0.1	2.2	1.2	0.7
821	**Sweet potato**, *raw*	0.19	0.1	Tr	0.1	0	15.6	5.7	2.3	2.4
822	*boiled in salted water*	0.18	0.1	Tr	0.1	0	8.9	11.6	2.1	2.3

[a] Additional non-protein nitrogen from chitin is present in variable amounts

[b] Not including oligosaccharides

265

Vegetables, general

No.	Food	Na	K	Ca	Mg	P	Fe	Cu	Zn	Cl	Mn	Se	I
						mg						µg	
806	Plantain, raw	4	500	9	37	36	0.5	0.08	0.1	(80)	N	(2)	N
807	boiled in unsalted water	4	400	5	33	31	0.5	0.08	0.2	50	N	(2)	N
808	ripe, fried in vegetable oil	3	610	6	54	66	0.8	0.20	0.4	110	N	(3)	N
809	Pumpkin, raw	Tr	130	29	10	19	0.4	0.02	0.2	37	(0.1)	N	N
810	boiled in salted water	76	84	23	7	15	0.1	0.02	0.2	N	(0.1)	N	N
811	Quorn, myco-protein	240	N	N	N	N	N	N	N	N	N	N	N
812	Radish, red, raw	11	240	19	5	20	0.6	0.01	0.2	37	0.1	(2)	(1)
813	Spinach, raw	140	500	170	54	45	2.1	0.04	0.7	98	0.6	(1)	2
814	boiled in unsalted water	120	230	160	34	28	1.6	0.01	0.5	56	0.5	(1)	2
815	frozen, boiled in unsalted water	16	340	150	31	48	1.7	0.09	0.6	31	0.2	(1)	(2)
816	Spring greens, raw	20	370	210	19	91	3.0	0.02	0.4	78	N	N	N
817	boiled in unsalted water	10	160	75	8	29	1.4	0.02	0.3	16	N	N	N
818	Spring onions, bulbs and tops, raw	7	260	39	12	29	1.9	0.06	0.4	31	0.2	N	N
819	Swede, raw	15	170	53	9	40	0.1	0.01	0.3	31	(0.1)	(1)	N
820	boiled in unsalted water	14	86	26	4	11	0.1	Tr	0.1	9	(0.1)	(1)	N
821	Sweet potato, raw	40	370	24	18	50	0.7	0.14	0.3	65	0.4	(1)	2
822	boiled in salted water	32	300	23	45	50	0.7	0.14	0.3	52	0.4	(1)	2

Vegetables *continued*

No.	Food	Retinol µg	Carotene µg	Vitamin D µg	Vitamin E mg	Thiamin mg	Ribo-flavin mg	Niacin mg	Trypt 60 mg	Vitamin B6 mg	Vitamin B12 µg	Folate µg	Panto-thenate mg	Biotin µg	Vitamin C mg
	Vegetables, general														
806	**Plantain**, *raw*	0	(360)	0	(0.20)	0.10	0.05	0.7	0.2	0.30	0	22	0.26	N	15
807	*boiled in unsalted water*	0	(350)	0	(0.20)	0.03	0.04	0.5	0.1	0.24	0	22	0.25	N	9
808	*ripe, fried in vegetable oil*	0	N	0	N	0.11	0.02	0.6	0.2	(1.0)	0	37	0.73	N	12
809	**Pumpkin**, *raw*	0	450	0	1.06	0.16	Tr	0.1	0.1	0.02	0	10	0.40	(0.4)	14
810	*boiled in salted water*	0	955	0	(1.06)	0.14	Tr	0.1	0.1	0.03	0	10	0.30	(0.4)	7
811	**Quorn**, *myco-protein*	0	0	0	0	36.60	0.15	0.3	N	Tr	0.3	7	0.14	9.0	0
812	**Radish, red**, *raw*	0	Tr	0	0	0.03	Tr	0.4	0.1	0.07	0	38	0.18	N	17
813	**Spinach**, *raw*	0	3535	0	1.71	0.07	0.09	1.2	0.7	0.17	0	150	(0.27)	(0.1)	26
814	*boiled in unsalted water*	0	3840	0	(1.71)	0.06	0.05	0.9	0.6	0.09	0	(90)	0.21	0.1	8
815	*frozen, boiled in unsalted water*	0	(3840)	0	(1.71)	(0.06)	(0.05)	(0.9)	0.8	(0.09)	0	(90)	(0.21)	(0.1)	6
816	**Spring greens**, *raw*	0	2630	0	N	0.07	0.11	1.5	0.5	0.23	0	92	0.39	(0.4)	180
817	*boiled in unsalted water*	0	2270	0	N	0.05	0.06	1.2	0.3	0.18	0	66	0.30	(0.4)	77
818	**Spring onions, bulbs and tops,** *raw*	0	620	0	N	0.05	0.03	0.5	0.5	0.13	0	54	0.07	N	26
819	**Swede**, *raw*	0	350	0	Tr	0.15	Tr	1.2	0.1	0.21	0	31	0.11	0.1	31
820	*boiled in unsalted water*	0	165	0	Tr	0.13	0.01	1.0	0.1	0.04	0	18	0.07	Tr	15
821	**Sweet potato**, *raw*	0	3930[a]	0	4.56	0.17	Tr	0.5	0.3	0.09	0	17	0.59	N	23
822	*boiled in salted water*	0	3960[b]	0	4.39	0.07	0.01	0.5	0.3	0.05	0	8	0.53	N	17

[a] Value for orange fleshed varieties. Carotene can range from 1820 to 16000µg per 100g. White fleshed varieties contain approximately 69µg per 100g

[b] White fleshed varieties contain approximately 66µg per 100g

Vegetables *continued*

Vegetables, general

No.	Food	Description and main data sources	Edible proportion	Water g	Protein g	Fat g	Carbo-hydrate g	Energy value kcal	kJ
823	**Sweetcorn**, baby, canned, *drained* Ref. Wu Leung *et al.* (1972)		0.53	92.5	2.9	0.4	2.0	23	96
824	kernels, canned, *re-heated, drained*	LGC; 10 samples, 5 brands	0.82	72.3	2.9	1.2	26.6[a]	122	519
825	on-the-cob, *whole, boiled in unsalted water*	IFR; boiled 19 minutes	0.59	41.2	2.5	1.4	11.6[a]	66	280
826	**Tomato purée**	LGC; 10 samples, 8 brands	1.00	71.9	4.5	0.2	12.9	68	290
827	**Tomatoes**, *raw*	IFR; flesh, skin and seeds	1.00	93.1	0.7	0.3	3.1	17	73
828	*fried in blended oil*	Calculation from *fried in corn oil*	1.00	84.4	0.7	7.7	5.0	91	377
829	*fried in corn oil*	IFR. As raw; sliced and fried 10 minutes	1.00	84.4	0.7	7.7	5.0	91	377
830	*fried in lard*	Calculation from *fried in corn oil*	1.00	84.4	0.7	7.7	5.0	91	377
831	*grilled*	Calcd. from raw using water loss of 13%	1.00[b]	80.1	2.0	0.9	8.9	49	210
832	canned, *whole contents*	LGC; 10 samples, 10 brands. Tomatoes and juice	1.00	94.0	1.0	0.1	3.0	16	69
833	**Turnip**, *raw*	IFR; flesh only, peeled thinly	0.75	91.2	0.9	0.3	4.7	23	98
834	*boiled in unsalted water*	IFR. As raw; diced and boiled 19 minutes	1.00	93.1	0.6	0.2	2.0	12	51
835	**Watercress**, *raw*	IFR; large stalks removed	0.62	92.5	3.0	1.0	0.4	22	94
836	**Yam**, *raw*	IFR; flesh only	0.81	67.2	1.5	0.3	28.2	114	488
837	*boiled in unsalted water*	IFR. As raw; boiled 25 minutes	1.00	64.4	1.7	0.3	33.0	133	568

[a] Including oligosaccharides

[b] Drained = 0.60

Vegetables *continued*

Composition of food per 100g

No.	Food	Total nitrogen g	Fatty acids			Cholest-erol mg	Starch g	Total sugars g	Dietary fibre	
			Satd g	Mono unsatd g	Poly unsatd g				Southgate method g	Englyst method g
	Vegetables, general									
823	**Sweetcorn**, baby, canned, *drained*	0.46	N	N	N	0	0.6	1.4	N	1.5
824	kernels, canned, *re-heated, drained*	0.47	0.2	0.3	0.5	0	16.6	9.6[a]	3.9	1.4
825	on-the-cob, whole, *boiled in unsalted water*	0.40	0.2	0.3	0.5	0	10.0	1.4[a]	2.5	1.3
826	**Tomato purée**	0.71	Tr	Tr	0.1	0	0.3	12.6	N	2.8
827	**Tomatoes**, *raw*	0.11	0.1	0.1	0.2	0	Tr	3.1	1.3	1.0
828	*fried in blended oil*	0.13	0.7	3.8	2.8	0	0.1	4.9	2.8	1.3
829	*fried in corn oil*	0.13	1.0	1.9	4.5	0	0.1	4.9	2.8	1.3
830	*fried in lard*	0.13	4.2	2.8	0.2	7	0.1	4.9	2.8	1.3
831	*grilled*	0.32	0.3	0.3	0.6	0	Tr	8.9	3.7	2.9
832	*canned, whole contents*	0.16	Tr	Tr	Tr	0	0.2	2.8	0.8	0.7
833	**Turnip**, *raw*	0.14	Tr	Tr	0.2	0	0.2	4.5	2.5	2.4
834	*boiled in unsalted water*	0.10	Tr	Tr	0.1	0	0.1	1.9	2.0	1.9
835	**Watercress**, *raw*	0.48	0.3	0.1	0.4	0	Tr	0.4	3.0	1.5
836	**Yam**, *raw*	0.25	0.1	Tr	0.1	0	27.5	0.7	3.7	1.3
837	*boiled in unsalted water*	0.27	0.1	Tr	0.1	0	32.3	0.7	3.5	1.4

[a] Not including oligosaccharides

Vegetables, general

No.	Food	mg										µg	
		Na	K	Ca	Mg	P	Fe	Cu	Zn	Cl	Mn	Se	I
823	**Sweetcorn**, baby, canned, *drained*	1140	180	8	N	N	1.2	N	N	1760	N	N	N
824	kernels, canned, *re-heated, drained*	270	220	4	23	79	0.5	Tr	0.5	390	0.1	Tr	N
825	on-the-cob, whole, *boiled in unsalted water*	1	140	2	20	48	0.3	0.02	0.2	8	0.1	Tr	N
826	**Tomato purée**	240[a]	1150	48	48	94	1.6	0.53	0.7	490	0.3	N	N
827	**Tomatoes**, *raw*	9	250	7	7	24	0.5	0.01	0.1	55	0.1	Tr	2
828	*fried in blended oil*	10	300	12	13	24	0.5	0.02	0.1	57	0.1	Tr	2
829	*fried in corn oil*	10	300	12	13	24	0.5	0.02	0.1	57	0.1	Tr	2
830	*fried in lard*	10	300	12	13	24	0.5	0.02	0.1	57	0.1	Tr	2
831	*grilled*	26	720	20	20	69	1.4	0.03	0.3	160	0.3	Tr	6
832	*canned, whole contents*	39	250	12	11	19	0.4	0.07	0.1	93	0.1	Tr	(3)
833	**Turnip**, *raw*	15	280	48	8	41	0.2	0.01	0.1	39	0.1	(1)	N
834	*boiled in unsalted water*	28	200	45	6	31	0.2	0.01	0.1	31	0.1	(1)	N
835	**Watercress**, *raw*	49	230	170	15	52	2.2	0.01	0.7	170	0.6	N	N
836	**Yam**, *raw*	2	380	15	15	27	0.7	0.01	0.3	10	0.1	N	N
837	*boiled in unsalted water*	17	260	12	12	21	0.4	0.03	0.4	40	Tr	N	N

[a] The sodium content of unsalted tomato purée is approximately 20mg per 100g

Vegetables *continued*

No.	Food	Retinol µg	Carotene µg	Vitamin D µg	Vitamin E mg	Thiamin mg	Ribo-flavin mg	Niacin mg	Trypt 60 mg	Vitamin B6 mg	Vitamin B12 µg	Folate µg	Panto-thenate mg	Biotin µg	Vitamin C mg
	Vegetables, general														
823	**Sweetcorn**, baby, canned, drained	0	(140)	0	N	0.02	0.04	0.1	0.3	N	0	N	N	N	14
824	kernels, canned, re-heated, drained	0	110	0	0.46	0.04	0.06	1.5	0.5	0.13	0	8	0.22	N	1
825	on-the-cob, whole, boiled in unsalted water	0	71	0	0.52	0.11	0.03	1.2	0.3	0.09	0	20	0.37	N	4
826	**Tomato purée**	0	1300	0	5.37	0.22	0.12	3.5	0.6	0.44	0	54	1.00	6.1	38
827	**Tomatoes**, raw	0	640	0	1.22	0.09	0.01	1.0	0.1	0.14	0	17	0.25	1.5	17
828	fried in blended oil	0	765	0	N	0.09	0.01	0.5	0.1	0.10	0	17	(0.25)	(1.5)	16
829	fried in corn oil	0	765	0	N	0.09	0.01	0.5	0.1	0.10	0	17	(0.25)	(1.5)	16
830	fried in lard	Tr	765	N	N	0.09	0.01	0.5	0.1	0.10	Tr	17	(0.25)	(1.5)	16
831	grilled	0	1840	0	4.04	0.25	0.03	2.9	0.3	0.38	0	44	0.72	4.3	44
832	canned, whole contents	0	220	0	1.22	0.05	0.02	0.7	0.1	0.11	0	11	0.20	1.5	12
833	**Turnip**, raw	0	20	0	Tr	0.05	0.01	0.4	0.2	0.08	0	14	0.20	0.1	17
834	boiled in unsalted water	0	20	0	Tr	0.05	0.02	0.2	0.1	0.04	0	8	0.14	Tr	10
835	**Watercress**, raw	0	2520	0	1.46	0.16	0.06	0.3	0.5	0.23	0	N	0.10	0.4	62
836	**Yam**, raw	0	Tr[a]	0	N	0.16	0.01	0.2	0.3	0.16	0	8	0.31	N	4
837	boiled in unsalted water	0	Tr	0	N	0.14	0.01	0.2	0.4	0.12	0	6	0.31	N	4

a Yellow fleshed varieties contain 400 to 1440µg carotene per 100g

Herbs and spices

The foods in this section of the Tables have been taken from the *'Vegetables, Herbs and Spices'* supplement. The majority of values are derived from the literature.

Many of the values for carbohydrate are not analysed but are calculated 'by difference'. For spices this carbohydrate is likely to include much woody material and aromatic oils, resulting in an overestimate of both the carbohydrate and energy value. For some spices energy values and carbohydrate have therefore been given as 'unknown', i.e. as 'N'.

Taxonomic names for foods in this part of the Tables can be found in Section 3.6.

Herbs and spices

No.	Food	Description and main data sources	Edible proportion	Water g	Protein g	Fat g	Carbo-hydrate g	Energy value kcal	kJ
838	**Chilli powder**[a]	Ref. Marsh et al. (1977)	1.00	7.8	12.3	16.8	N	N	N
839	**Cinnamon**, ground	Ref. Marsh et al. (1977)	1.00	9.5	3.9	3.2	N	N	N
840	**Curry powder**[b]	2 samples	1.00	8.5	9.5	10.8	26.1	233	979
841	**Garam masala**	Ref. Wharton et al. (1983)	1.00	10.1	15.6	15.1	45.2	379	1592
842	**Mint**, fresh	Literature sources	1.00	86.4	3.8	0.7	5.3	43	181
843	**Mustard powder**	2 brands	1.00	(8.0)	28.9	28.7	20.7	452	1884
844	**Nutmeg**, ground	Ref. Marsh et al. (1977)	1.00	6.2	5.8	36.3	N	N	N
845	**Paprika**	Ref. Marsh et al. (1977)	1.00	9.5	14.8	13.0	34.9	289	1209
846	**Parsley**, fresh	IFR; tough stalks removed	0.80	83.1	3.0	1.3	2.7	34	141
847	**Pepper**, black	Ref. Marsh et al. (1977)	1.00	10.5	10.9	3.3	N	N	N
848	white	Ref. Marsh et al. (1977)	1.00	11.4	10.4	2.1	N	N	N
849	**Rosemary**, dried	Ref. Marsh et al. (1977)	1.00	9.3	4.9	15.2	46.4	331	1387
850	**Sage**, dried, ground	Ref. Marsh et al. (1977)	1.00	8.0	10.6	12.7	42.7	315	1317
851	**Thyme**, dried, ground	Ref. Marsh et al. (1977)	1.00	7.8	9.1	7.4	45.3	276	1156

[a] Mix of chilli pepper 83%, cumin 9%, oregano 4%, salt 2.5% and garlic powder 1.5%

[b] Composition will vary according to variety

No.	Food	Total nitrogen g	Fatty acids			Cholest-erol mg	Starch g	Total sugars g	Dietary fibre	
			Satd g	Mono unsatd g	Poly unsatd g				Southgate method g	Englyst method g
838	**Chilli powder**	1.96	N	N	N	0	N	N	N	N
839	**Cinnamon,** *ground*	0.62	0.7	0.5	0.5	0	N	N	N	N
840	**Curry powder**	1.52	N	N	N	0	N	N	N	23.0
841	**Garam masala**	2.50	N	N	N	0	N	N	N	N
842	**Mint,** *fresh*	0.61	N	N	N	0	N	N	N	N
843	**Mustard powder**	4.62	1.5	19.8	5.4	0	N	N	N	N
844	**Nutmeg,** *ground*	1.10	25.9	3.2	0.3	0	N	N	N	N
845	**Paprika**	2.36	1.9	1.4	7.1	0	N	N	N	N
846	**Parsley,** *fresh*	0.47	N	N	N	0	0.4	2.3	8.2	5.0
847	**Pepper,** black	2.05	N	N	N	0	Tr	N	N	N
848	white	1.95	N	N	N	0	Tr	N	N	N
849	**Rosemary,** dried	0.78	N	N	N	0	N	N	N	N
850	**Sage,** dried, *ground*	1.70	7.0	1.9	1.8	0	N	N	N	N
851	**Thyme,** dried, *ground*	1.46	2.7	0.5	1.2	0	N	N	N	N

No.	Food	Na	K	Ca	Mg	P	Fe	Cu	Zn	Cl	Mn	Se	I
						mg						µg	
838	**Chilli powder**	1010	1920	280	170	300	14.3	0.43	2.7	1510	2.2	N	N
839	**Cinnamon**, *ground*	26	500	1230	56	61	38.1[a]	0.46	2.0	N	5.7	(15)	N
840	**Curry powder**	450	1830	640	280	270	58.3	1.04	4.1	470	4.7	N	N
841	**Garam masala**	97	1450	760	330	390	32.6	1.62	3.8	N	6.0	N	N
842	**Mint**, *fresh*	15	260	210	N	75	9.5	N	N	34	1.4	N	N
843	**Mustard powder**	5	940	330	260	180	9.5	0.20	(6.5)	62	1.7	N	N
844	**Nutmeg**, *ground*	16	350	180	180	210	3.0	1.03	2.2	N	2.9	N	N
845	**Paprika**	34	2340	180	190	350	23.6	0.61	4.1	N	0.8	N	N
846	**Parsley**, *fresh*	33	760	200	23	64	7.7	0.03	0.7	160	0.2	(1)	N
847	**Pepper**, *black*	44	1260	430	190	170	11.2	1.13	1.4	60	6.5	(3)	N
848	*white*	5	73	270	90	180	14.3	1.13	1.1	60	4.5	(3)	N
849	**Rosemary**, *dried*	50	950	1280	220	70	29.3	0.55	3.2	N	0.5	N	N
850	**Sage**, *dried, ground*	11	1070	1650	430	91	28.1	0.76	4.7	N	25.0	N	N
851	**Thyme**, *dried, ground*	55	810	1890	220	200	123.6	0.86	6.2	N	7.6	N	N

[a] Whole unground cinnamon contains 4mg Fe per 100g

Herbs and spices

No.	Food	Retinol µg	Carotene µg	Vitamin D µg	Vitamin E mg	Thiamin mg	Ribo-flavin mg	Niacin mg	Trypt 60 mg	Vitamin B6 mg	Vitamin B12 µg	Folate µg	Panto-thenate mg	Biotin µg	Vitamin C mg
838	**Chilli powder**	0	(21000)	0	N	0.35	0.79	7.9	2.6	N	0	0	N	N	0
839	**Cinnamon**, *ground*	0	(155)	0	N	0.08	0.14	1.3	N	N	0	0	N	N	0
840	**Curry powder**	0	(100)	0	N	0.25	0.28	3.5	N	N	0	0	N	N	0
841	**Garam masala**	0	(340)	0	N	0.35	0.33	2.5	N	N	0	0	N	N	0
842	**Mint**, *fresh*	0	(740)	0	5.00	0.12	0.33	1.1	N	N	0	110	N	N	31
843	**Mustard powder**	0	N	0	N	N	N	N	8.5	N	0	0	N	N	0
844	**Nutmeg**, *ground*	0	(60)	0	N	0.35	0.06	1.3	N	N	0	0	N	N	0
845	**Paprika**	0	36250	0	N	0.65	1.74	15.3	3.1	N	0	0	N	N	0
846	**Parsley**, *fresh*	0	4040	0	1.70	0.23	0.06	1.0	0.5	0.09	0	170	0.30	0.4	190
847	**Pepper**, black	0	(115)	0	N	0.11	0.24	1.1	N	N	0	0	N	N	0
848	white	0	Tr	0	N	0.02	0.13	0.2	N	N	0	0	N	N	0
849	**Rosemary**, dried	0	(1880)	0	N	N	N	1.0	N	N	0	0	N	N	0
850	**Sage**, dried, ground	0	(3540)	0	N	N	0.34	5.7	N	N	0	0	N	N	0
851	**Thyme**, dried, ground	0	(2280)	0	N	N	0.40	4.9	3.1	N	0	0	N	N	0

Fruit

The data in this section of the Tables have been revised and now cover more varieties of fresh and processed fruit. Because many of the fruit eaten in this country are imported, a larger number of literature values from foreign sources have been used in this food group than many others in the Tables.

In general the word 'raw' has not been included in the food name unless there is a processed or cooked version of the same food. The description of 'whole' refers to fruit with both skin and pips, but excluding any inedible stone.

The nutrient content of fruit samples can vary widely, often being greater within the same fruit type than between different varieties of fruit.

During the process of stewing fruit, sucrose becomes inverted to glucose and fructose, the extent depending on the length of cooking time and level of acidity. A factor of 10% inversion of sucrose has been applied to all stewed fruit. The nutrient values for stewed fruits have been derived from both analysis and calculation. The proportions of sugar used for cooking and the method of calculation of the data have been included in the description of the food and are those for average consumption of sugar. However, for fruit cooked with a different proportion of sugar, the values for fruit 'stewed without sugar' can be used, with the appropriate quantity of sugar added. Corrections have been made for both vitamin losses (see Section 3.4 for the factors used) and evaporative losses of 10% during stewing.

Values for canned fruit include either syrup or juice, unless it is stated that the contents have been drained. It has been found by analysis that sugar diffuses between the syrup or juice and the fruit until it reaches an equilibrium, so that there are no significant differences between the levels of sugars in the fruit and the syrup or juice. One study found that the only significant differences between the fruit and its canning liquid were that the fruit contained higher levels of carotenoids and fibre.

Taxonomic names for foods included in this part of the Tables can be found in Section 3.6.

Composition of food per 100g

No.	Food	Description and main data sources	Edible proportion	Water g	Protein g	Fat g	Carbo-hydrate g	Energy value kcal	kJ
852	**Apples**, cooking, raw, peeled	Bramley variety; flesh only	1.00	87.7	0.3	0.1	8.9	35	151
853	raw, peeled, weighed with skin and core	Calculated from No. 852	0.73	63.1	0.2	0.1	6.4	26	109
854	stewed with sugar	Samples as raw. 1000g fruit, 100g water, 120g sugar	1.00	77.7	0.3	0.1	19.1	74	314
855	stewed without sugar	Samples as raw. 1000g fruit, 100g water and calculation from No. 854	1.00	87.5	0.3	0.1	8.1	33	138
856	eating, average, raw	15 varieties; flesh and skin	1.00	84.5	0.4	0.1	11.8	47	199
857	average, raw, weighed with core	Calculated from No. 856	0.89	75.2	0.4	0.1	10.5	42	179
858	-, raw, peeled	Literature sources and calculation from No. 856; flesh only	1.00	85.4	0.4	0.1	11.2	45	190
859	-, raw, peeled, weighed with skin and core	Calculated from No. 858	0.76	64.9	0.3	0.1	8.5	34	145
860	**Apricots**, raw	18 samples; flesh and skin	1.00	87.2	0.9	0.1	7.2	31	134
861	raw, weighed with stones	Calculated from No. 860	0.92	80.2	0.8	0.1	6.6	29	123
862	ready-to-eat	10 samples, no stones; semi-dried	1.00	29.7	4.0	0.6	36.5	158	674
863	canned in syrup	10 samples, 9 brands	1.00	80.0	0.4	0.1	16.1	63	268
864	canned in juice	10 samples, 5 brands	1.00	87.5	0.5	0.1	8.4	34	147
865	**Avocado**, average	Average of Fuerte and Hass varieties	1.00	72.5[a]	1.9	19.5[b]	1.9[c]	190	784
866	average, weighed with skin and stone	Calculated from No. 865	0.71	51.5	1.3	13.8	1.3[c]	134	553

[a] Water can range from 50 to 80g per 100g
[b] Fat can range from 10 to 40g per 100g
[c] Including mannoheptulose

Fruit

Composition of food per 100g

No.	Food	Total nitrogen g	Fatty acids Satd g	Mono unsatd g	Poly unsatd g	Cholesterol mg	Starch g	Total sugars g	Dietary fibre Southgate method g	Englyst method g
852	**Apples,** cooking, raw, peeled	0.05	Tr	Tr	0.1	0	Tr	8.9	2.2	1.6
853	raw, peeled, weighed with skin and core	0.04	Tr	Tr	0.1	0	Tr	6.4	1.6	1.1
854	stewed with sugar	0.05	Tr	Tr	0.1	0	Tr	19.1	1.8	1.2
855	stewed without sugar	0.04	Tr	Tr	0.1	0	Tr	8.1	2.0	1.5
856	eating, average, raw	0.06	Tr	Tr	0.1	0	Tr	11.8[a]	(2.0)	1.8
857	average, raw, weighed with core	0.06	Tr	Tr	Tr	0	Tr	10.5	(1.8)	1.6
858	-, raw, peeled	0.06	Tr	Tr	0.1	0	Tr	11.2	1.8[b]	1.6
859	-, raw, peeled, weighed with skin and core	0.05	Tr	Tr	Tr	0	Tr	8.5	1.4	1.2
860	**Apricots,** raw	0.14	Tr	Tr	Tr	0	0	7.2	1.9	1.7
861	raw, weighed with stones	0.13	Tr	Tr	Tr	0	0	6.6	1.7	1.6
862	ready-to-eat	0.63	N	N	N	0	Tr	36.5	18.1	6.3
863	canned in syrup	0.07	Tr	Tr	Tr	0	0	16.1	1.2	0.9
864	canned in juice	0.08	Tr	Tr	Tr	0	0	8.4	(1.2)	0.9
865	**Avocado,** average	0.30	4.1	12.1	2.2	0	Tr	0.5[c]	N	3.4
866	average, weighed with skin and stone	0.21	2.9	8.6	1.6	0	Tr	0.4[c]	N	2.4

[a] Levels ranged from 9.5 to 13.0g total sugars per 100g

[b] Apple peel contains 3.3g Southgate fibre per 100g

[c] Not including mannoheptulose

281

Fruit

Inorganic constituents per 100g

No.	Food	mg										µg	
		Na	K	Ca	Mg	P	Fe	Cu	Zn	Cl	Mn	Se	I
852	**Apples**, cooking, raw, peeled	2	88	4	3	7	0.1	0.02	Tr	2	Tr	Tr	Tr
853	raw, peeled, weighed with skin and core	1	63	3	2	5	0.1	0.01	Tr	1	Tr	Tr	Tr
854	stewed with sugar	4	140	4	3	7	0.1	0.02	Tr	2	Tr	Tr	Tr
855	stewed without sugar	4	150	4	3	8	0.1	0.02	Tr	2	Tr	Tr	Tr
856	eating, average, raw	3	120	4	5	11	0.1	0.02	0.1	Tr	0.1	Tr	Tr
857	average, raw, weighed with core	3	110	4	4	10	0.1	0.02	0.1	Tr	0.1	Tr	Tr
858	-, raw, peeled	3	100	3	3	8	0.1	0.02	0.1	Tr	0.1	Tr	Tr
859	-, raw, peeled, weighed with skin and core	2	76	2	2	6	0.1	0.01	0.1	Tr	0.1	Tr	Tr
860	**Apricots**, raw	2	270	15	11	20	0.5	0.06	0.1	3	0.1	(1)	N
861	raw, weighed with stones	2	250	14	10	18	0.5	0.05	0.1	3	0.1	(1)	N
862	ready-to-eat	14	1380	73	43	82	3.4	0.35	0.5	29	0.3	(5)	N
863	canned in syrup	10	150	19	5	8	0.2	Tr	0.1	2	Tr	Tr	7
864	canned in juice	5	170	21	7	12	0.4	0.03	0.1	2	Tr	Tr	7
865	**Avocado**, average	6	450	11	25	39	0.4	0.19	0.4	6	0.2	Tr	2
866	average, weighed with skin and stone	4	320	8	18	28	0.3	0.13	0.3	4	0.1	Tr	1

No.	Food	Retinol μg	Carotene μg	Vitamin D μg	Vitamin E mg	Thiamin mg	Ribo-flavin mg	Niacin mg	Trypt 60 mg	Vitamin B6 mg	Vitamin B12 μg	Folate μg	Panto-thenate mg	Biotin μg	Vitamin C mg
852	**Apples,** cooking, raw, peeled	0	(17)	0	0.27	0.04	0.02	0.1	0.1	0.06	0	5	Tr	1.2	14[a]
853	raw, peeled, weighed with skin and core	0	(12)	0	0.19	0.03	0.01	0.1	0.1	0.04	0	4	Tr	0.9	10
854	stewed with sugar	0	(14)	0	0.22	0.01	0.01	0.1	0.1	0.05	0	Tr	Tr	0.8	10[b]
855	stewed without sugar	0	(15)	0	0.25	0.01	0.01	0.1	Tr	0.05	0	Tr	Tr	0.9	11
856	eating, average, raw	0	18	0	0.59	0.03	0.02	0.1	0.1	0.06	0	1	Tr	1.2	6[c]
857	average, raw, weighed with core	0	16	0	0.53	0.03	0.02	0.1	0.1	0.05	0	1	Tr	1.1	5
858	-, raw, peeled	0	17	0	0.27	0.03	0.02	0.1	0.1	0.06	0	1	Tr	1.1	4
859	-, raw, peeled, weighed with skin and core	0	13	0	0.21	0.02	0.01	0.1	0.1	0.05	0	1	Tr	0.8	3
860	**Apricots,** raw	0	405[d]	0	N	0.04	0.05	0.5	0.1	0.08	0	5	0.24	N	6
861	raw, weighed with stones	0	375	0	N	0.04	0.05	0.5	0.1	0.07	0	5	0.22	N	5
862	ready-to-eat	0	545	0	N	Tr	0.16	2.3	0.5	0.14	0	11	0.58	N	1
863	canned in syrup	0	155	0	N	0.01	0.01	0.3	0.1	(0.06)	0	2	(0.06)	(0.4)	5
864	canned in juice	0	210	0	N	0.02	0.01	0.3	0.1	0.06	0	2	0.06	0.4	14
865	**Avocado,** average	0	16	0	3.20	0.10	0.18	1.1	0.3	0.36	0	11	1.10	3.6	6
866	average, weighed with skin and stone	0	11	0	2.27	0.07	0.13	0.8	0.2	0.26	0	8	0.78	2.6	4

[a] Unpeeled cooking apples contain 20mg vitamin C per 100g

[b] Frozen apple slices, stewed with sugar, contain 12mg vitamin C per 100g

[c] Levels ranged from 3 to 20mg vitamin C per 100g

[d] Levels ranged from 200 to 3370μg carotene per 100g

No.	Food	Description and main data sources	Edible proportion	Water g	Protein g	Fat g	Carbo-hydrate g	Energy value kcal	kJ
867	**Bananas**	10 samples; flesh only	1.00	75.1	1.2	0.3	23.2	95	403
868	*weighed with skin*	Calculated from No. 867	0.66	49.6	0.8	0.2	15.3	62	266
869	**Blackberries,** *raw*	Cultivated and wild berries; whole fruit	1.00	85.0	0.9	0.2	5.1	25	104
870	*stewed with sugar*	Calculated from 700g fruit, 210g water, 84g sugar	1.00	78.9	0.7	0.2	13.8	56	239
871	*stewed without sugar*	Calculated from 700g fruit, 210g water	1.00	87.2	0.8	0.2	4.4	21	88
872	**Blackcurrants,** *raw*	Whole fruit, stalks removed	0.98	77.4	0.9	Tr	6.6	28	121
873	*stewed with sugar*	Calculated from 700g fruit, 210g water, 84g sugar	1.00	72.9	0.7	Tr	15.0	58	252
874	*canned in juice*	4 samples, 2 brands	1.00	84.0	0.8	Tr	7.6	31	135
875	*canned in syrup*	3 samples of the same brand (Hartley's)	1.00	75.0	0.7	Tr	18.4	72	306
876	**Cherries,** *raw*	10 samples of black and red cherries; flesh and skin	1.00	82.8	0.9	0.1	11.5	48	203
877	*raw, weighed with stones*	Calculated from No. 876	0.83	68.7	0.7	0.1	9.5	39	168
878	*canned in syrup*	10 samples, red and black	1.00	77.8	0.5	Tr	18.5	71	305
879	*glacé*	10 samples, 8 brands; red and multicoloured	1.00	23.6	0.4	Tr	66.4	251	1069
880	**Cherry pie filling**	10 samples, 7 brands	1.00	75.8	0.4	Tr	21.5	82	351
881	**Clementines**	10 samples; flesh only	1.00	87.5	0.9	0.1	8.7	37	158
882	*weighed with peel and pips*	Calculated from No. 881	0.75	65.6	0.7	0.1	6.5	28	120
883	**Currants**	10 samples, 9 brands	1.00	15.7	2.3	0.4	67.8	267	1139

No.	Food	Total nitrogen g	Fatty acids Satd g	Mono unsatd g	Poly unsatd g	Cholesterol mg	Starch g	Total sugars g	Dietary fibre Southgate method g	Englyst method g
867	**Bananas**	0.19	0.1	Tr	0.1	0	2.3[a]	20.9[a]	3.1[b]	1.1
868	*weighed with skin*	0.13	0.1	Tr	0.1	0	1.5	13.8	2.0	0.7
869	**Blackberries**, *raw*	0.14	Tr	0.1	0.1	0	0	5.1	6.6	3.1
870	*stewed with sugar*	0.11	Tr	0.1	0.1	0	0	13.8	5.2	2.4
871	*stewed without sugar*	0.12	Tr	0.1	0.1	0	0	4.4	5.6	2.6
872	**Blackcurrants**, *raw*	0.15	Tr	Tr	Tr	0	0	6.6	7.8	3.6
873	*stewed with sugar*	0.12	Tr	Tr	Tr	0	0	15.0	6.1	2.8
874	*canned in juice*	0.13	N	N	N	0	0	7.6	(4.2)	3.1
875	*canned in syrup*	0.11	N	N	N	0	0	18.4	(3.6)	2.6
876	**Cherries**, *raw*	0.14	Tr	Tr	Tr	0	0	11.5	1.5	0.9
877	*raw, weighed with stones*	0.12	Tr	Tr	Tr	0	0	9.5	1.2	0.7
878	*canned in syrup*	0.09	Tr	Tr	Tr	0	0	18.5	(0.7)	0.6
879	*glacé*	0.07	Tr	Tr	Tr	0	0	66.4	(1.5)	0.9
880	**Cherry pie filling**	0.07	Tr	Tr	Tr	0	3.9	17.6	N	0.4
881	**Clementines**	0.14	Tr	Tr	Tr	0	0	8.7	(1.7)	1.2
882	*weighed with peel and pips*	0.11	Tr	Tr	Tr	0	0	6.5	(1.3)	0.9
883	**Currants**	0.37	N	N	N	0	0	67.8	5.9	1.9

[a] These are proportions for yellow ripe bananas. The starch content falls and the sugar content rises on ripening
[b] Bananas contain significant amounts of resistant starch

Fruit *continued*

867 to 883

Inorganic constituents per 100g

No.	Food	Na	K	Ca	Mg	P	Fe	Cu	Zn	Cl	Mn	Se	I
						mg						µg	
867	**Bananas**	1	400	6	34	28	0.3	0.10	0.2	79	0.4	(1)	8
868	*weighed with skin*	1	270	4	22	18	0.2	0.66	0.1	52	0.3	(1)	5
869	**Blackberries**, *raw*	2	160	41	23	31	0.7	0.11	0.2	22	1.4	Tr	N
870	*stewed with sugar*	1	130	32	17	24	0.5	0.09	0.2	17	1.1	Tr	N
871	*stewed without sugar*	1	140	35	19	26	0.6	0.09	0.2	18	1.2	Tr	N
872	**Blackcurrants**, *raw*	3	370	60	17	43	1.3	0.14	0.3	15	0.3	N	N
873	*stewed with sugar*	2	290	47	13	33	1.0	0.11	0.3	11	0.2	N	N
874	*canned in juice*	Tr	190	26	13	29	5.2	0.04	0.1	20	0.2	N	N
875	*canned in syrup*	Tr	130	25	10	27	4.7	0.02	0.1	17	0.2	N	N
876	**Cherries**, *raw*	1	210	13	10	21	0.2	0.07	0.1	Tr	0.1	(1)	Tr
877	*raw, weighed with stones*	1	170	11	8	17	0.2	0.06	0.1	Tr	0.1	(1)	Tr
878	*canned in syrup*	8	120	15	7	13	2.9	Tr	Tr	N	0.1	(1)	N[a]
879	*glacé*	27	24	56	5	9	0.9	0.08	0.1	N	Tr	Tr	N[a]
880	**Cherry pie filling**	30	75	28	5	17	2.6	Tr	Tr	N	0.1	Tr	N
881	**Clementines**	4	130	31	10	18	0.1	0.01	0.1	(2)	Tr	N	N
882	*weighed with peel and pips*	3	97	23	7	13	0.1	0.01	0.1	(1)	Tr	N	N
883	**Currants**	14	720	93	30	71	1.3	0.81	0.3	16	0.7	N	N

[a] Iodine from erythrosine is present but largely unavailable

Fruit *continued*

No.	Food	Retinol µg	Carotene µg	Vitamin D µg	Vitamin E mg	Thiamin mg	Riboflavin mg	Niacin mg	Trypt 60 mg	Vitamin B6 mg	Vitamin B12 µg	Folate µg	Pantothenate mg	Biotin µg	Vitamin C mg
867	**Bananas**	0	21	0	0.27	0.04	0.06	0.7	0.2	0.29	0	14	0.36	2.6	11
868	*weighed with skin*	0	14	0	0.18	0.03	0.04	0.5	0.1	0.19	0	9	0.24	1.7	7
869	**Blackberries,** *raw*	0	80	0	2.37	0.02	0.05	0.5	0.1	0.05	0	34	0.25	0.4	15
870	*stewed with sugar*	0	62	0	1.85	0.01	0.03	0.3	0.1	0.03	0	5	0.15	0.2	9
871	*stewed without sugar*	0	68	0	2.03	0.01	0.03	0.3	0.1	0.03	0	5	0.16	0.3	10
872	**Blackcurrants,** *raw*	0	100	0	1.00	0.03	0.06	0.3	0.1	0.08	0	N	0.40	2.4	200[a]
873	*stewed with sugar*	0	78	0	0.78	0.02	0.04	0.2	0.1	0.05	0	N	0.23	1.4	115
874	*canned in juice*	0	(29)	0	0.54	0.01	0.03	0.2	0.1	0.11	0	4	(0.14)	(0.9)	37
875	*canned in syrup*	0	(29)	0	(0.54)	(0.01)	(0.03)	(0.2)	0.1	(0.11)	0	(4)	(0.14)	(0.9)	57
876	**Cherries,** *raw*	0	25	0	0.13	0.03	0.03	0.2	0.1	0.05	0	5	0.26	0.4	11
877	*raw, weighed with stones*	0	21	0	0.11	0.02	0.02	0.2	0.1	0.04	0	4	0.22	0.3	9
878	*canned in syrup*	0	17	0	(0.06)	0.02	0.01	0.1	Tr	(0.22)	0	5	(0.08)	(0.1)	1
879	*glacé*	0	7	0	Tr	Tr	Tr	Tr	Tr	Tr	0	Tr	Tr	Tr	Tr
880	**Cherry pie filling**	0	18	0	N	0.02	0.01	0.2	0.1	N	0	2	N	N	1
881	**Clementines**	0	75	0	N	0.09	0.04	0.3	0.1	(0.07)	0	33	(0.20)	N	54
882	*weighed with peel and pips*	0	57	0	N	0.07	0.03	0.2	0.1	(0.05)	0	25	(0.15)	N	41
883	**Currants**	0	6	0	N	0.16	0.05	0.9	0.2	0.23	0	4	0.07	4.8	Tr

[a] Levels ranged from 150 to 230mg vitamin C per 100g

No.	Food	Description and main data sources	Edible proportion	Water g	Protein g	Fat g	Carbohydrate g	Energy value kcal	kJ
884	Damsons, *raw, weighed with stones*	Calculated from raw damsons, *weighed without stones*	0.90	69.7	0.5	Tr	8.6	34	146
885	*stewed with sugar*	Calculated from 1050g fruit, 210g water, 126g sugar	1.00	70.6	0.4	Tr	19.3	74	316
886	Dates, *raw, weighed with stones*	Calculated from whole raw dates	0.86	52.2	1.3	0.1	26.9	107	456
887	*dried, weighed with stones*	Calculated from whole dried dates	0.84	12.3	2.8	0.2	57.1	227	969
888	Dried mixed fruit[a]	Calculated from recipe proportions	1.00	15.5	2.3	0.4	68.1	268	1144
889	Figs, *dried*	Analysis and literature sources; whole fruit	1.00	16.8	3.6	1.6	52.9	227	967
890	*ready-to-eat*	6 samples; semi-dried	1.00	23.6	3.3	1.5	48.6	209	889
891	Fruit cocktail, canned in juice	10 samples, 6 brands	1.00	86.9	0.4	Tr	7.2	29	122
892	canned in syrup[b]	Analysis and calculation from recipe proportions	1.00	81.8	0.4	Tr	14.8	57	244
893	Fruit pie filling	10 samples, 7 brands. Assorted flavours	1.00	79.5	0.4	Tr	20.1	77	328
894	Fruit salad, *homemade*	Calculated from equal proportions of bananas, oranges, apples, pears and grapes	1.00	82.3	0.7	0.1	13.8	55	237
895	Gooseberries, *cooking, raw*	Tops and tails removed	0.91	90.1	1.1	0.4	3.0	19	81
896	*stewed with sugar*	1000g fruit, 150g water, 120g sugar	1.00	82.1	0.7	0.3	12.9	54	229
897	*stewed without sugar*	500g fruit, 100g water and calculation from No. 896	1.00	90.6	0.9	0.3	2.5	16	66
898	*dessert, canned in syrup*	4 samples, 2 brands	1.00	78.9	0.4	0.2	18.5	73	310

[a] Calculated as sultanas 49%, currants 24%, raisins 18% and peel 9%
[b] Calculated as pears 42%, peaches 41%, pineapple 8%, grapes 5% and cherries 4%

Fruit *continued*

Composition of food per 100g

No.	Food	Total nitrogen g	Satd g	Mono unsatd g	Poly unsatd g	Cholest- erol mg	Starch g	Total sugars g	Southgate method g	Englyst method g
				Fatty acids					Dietary fibre	
884	**Damsons**, *raw, weighed with stones*	0.07	Tr	Tr	Tr	0	0	8.6	3.3	(1.6)
885	*stewed with sugar*	0.07	Tr	Tr	Tr	0	0	19.3	3.0	(1.5)
886	**Dates**, *raw, weighed with stones*	0.21	Tr	Tr	Tr	0	0	26.9	3.1	1.5
887	*dried, weighed with stones*	0.45	0.1	0.1	Tr	0	0	57.1	6.5	3.4
888	**Dried mixed fruit**	0.37	N	N	N	0	0	68.1	5.6	2.2
889	**Figs**, *dried*	0.57	N	N	N	0	0	52.9	12.4	7.5
890	*ready-to-eat*	0.52	N	N	N	0	0	48.6	11.4	6.9
891	**Fruit cocktail**, *canned in juice*	0.07	Tr	Tr	Tr	0	0	7.2	1.0	1.0
892	*canned in syrup*	0.06	Tr	Tr	Tr	0	0	14.8	1.0	1.0
893	**Fruit pie filling**	0.06	N	N	N	0	5.5	14.6	1.6	1.0
894	**Fruit salad**, *homemade*	0.11	Tr	Tr	Tr	0	0.5	13.3	(1.5)	1.5
895	**Gooseberries**, *cooking, raw*	0.18	N	N	N	0	0	3.0	2.9	2.4
896	*stewed with sugar*	0.11	N	N	N	0	0	12.9	2.3	1.9
897	*stewed without sugar*	0.15	N	N	N	0	0	2.5	2.4	2.0
898	*dessert, canned in syrup*	0.06	Tr	0.1	0.1	0	0	18.5	1.7	1.7

No.	Food	Na	K	Ca	Mg	P	Fe	Cu	Zn	Cl	Mn	Se	I
						mg						μg	
884	**Damsons**, *raw, weighed with stones*	2	260	22	10	14	0.4	0.07	(0.1)	Tr	N	Tr	N
885	*stewed with sugar*	1	240	19	9	13	0.3	0.07	(0.1)	Tr	N	Tr	N
886	**Dates**, *raw, weighed with stones*	6	350	21	21	24	0.3	0.10	0.2	180	0.2	(1)	N
887	*dried, weighed with stones*	8	590	38	34	50	1.1	0.22	0.3	310	0.3	(3)	N
888	**Dried mixed fruit**	48	880	73	29	73	2.2	0.47	0.4	13	0.4	N	N
889	**Figs**, *dried*	62	970	250	80	89	4.2	0.30	0.7	170	0.5	Tr	N
890	*ready-to-eat*	57	890	230	73	82	3.9	0.27	0.6	160	0.5	Tr	N[a]
891	**Fruit cocktail**, *canned in juice*	3	95	9	7	14	0.4	0.04	0.1	2	0.1	Tr	N[a]
892	*canned in syrup*	3	95	5	5	9	0.3	0.02	0.1	3	0.1	Tr	N
893	**Fruit pie filling**	43	84	30	5	15	1.0	0.02	Tr	45	0.1	Tr	N
894	**Fruit salad**, *homemade*	2	210	16	12	18	0.2	0.07	0.1	16	0.1	1	2
895	**Gooseberries**, *cooking, raw*	2	210	28	7	34	0.3	0.06	0.1	7	0.1	Tr	Tr
896	*stewed with sugar*	7	140	19	6	22	0.3	0.07	0.1	5	0.3	Tr	Tr
897	*stewed without sugar*	2	170	23	6	28	0.3	0.05	0.1	6	0.1	Tr	Tr
898	*dessert, canned in syrup*	2	66	12	4	12	0.2	Tr	0.3	3	0.1	Tr	Tr

[a] Iodine from erythrosine is present but largely unavailable

Fruit *continued*

No.	Food	Retinol µg	Carotene µg	Vitamin D µg	Vitamin E mg	Thiamin mg	Riboflavin mg	Niacin mg	Trypt 60 mg	Vitamin B6 mg	Vitamin B12 µg	Folate µg	Pantothenate mg	Biotin µg	Vitamin C mg
884	**Damsons**, *raw, weighed with stones*	0	(265)	0	0.60	0.09	0.03	0.3	0.1	(0.05)	0	(3)	0.24	0.1	(5)
885	*stewed with sugar*	0	(240)	0	0.57	0.06	0.02	0.2	0.1	(0.03)	0	Tr	0.17	0.1	(3)
886	**Dates**, *raw, weighed with stones*	0	(15)	0	N	0.05	0.06	0.6	0.6	0.10	0	21	0.18	N	12
887	*dried, weighed with stones*	0	(34)	0	N	0.06	0.08	1.5	1.3	0.16	0	11	0.65	N	Tr
888	**Dried mixed fruit**	0	9	0	N	0.10	0.05	0.7	0.2	0.22	0	15	0.09	3.9	Tr
889	**Figs**, *dried*	0	(64)	0	N	0.08	0.10	0.8	0.5	0.26	0	9	0.51	N	1
890	*ready-to-eat*	0	(59)	0	N	0.07	0.09	0.7	0.4	0.24	0	8	0.47	N	1
891	**Fruit cocktail**, canned in juice	0	54	0	N	0.01	0.01	0.3	0.1	0.04	0	6	0.05	0.3	14
892	*canned in syrup*	0	(54)	0	N	0.02	0.01	0.4	0.1	0.03	0	5	0.05	0.1	4
893	**Fruit pie filling**	0	17	0	N	0.01	0.01	0.2	0.1	N	0	3	N	N	7
894	**Fruit salad**, *homemade*	0	20	0	0.32	0.05	0.03	0.3	0.1	0.11	0	9	0.17	1.1	16
895	**Gooseberries**, cooking, raw	0	110	0	0.37	0.03	0.03	0.3	0.2	0.02	0	(8)	0.29	0.5	14
896	*stewed with sugar*	0	41	0	0.29	0.01	0.02	0.2	0.1	0.01	0	6	0.17	0.3	11
897	*stewed without sugar*	0	43	0	0.31	0.01	0.02	0.2	0.1	0.01	0	6	0.18	0.3	11
898	*dessert, canned in syrup*	0	(18)	0	(0.20)	(0.01)	(0.01)	(0.1)	0.1	(0.01)	0	(2)	(0.01)	Tr	27

Composition of food per 100g

No.	Food	Description and main data sources	Edible proportion	Water g	Protein g	Fat g	Carbo-hydrate g	Energy value kcal	kJ
899	**Grapefruit,** *raw*	10 samples; flesh only	1.00	89.0	0.8	0.1	6.8	30	126
900	*raw, weighed with peel and pips*	Calculated from No.899	0.68	60.5	0.5	0.1	4.6	20	86
901	canned in juice	10 samples, 8 brands	1.00	88.6	0.6	Tr	7.3	30	120
902	canned in syrup	10 samples	1.00	81.8	0.5	Tr	15.5	60	257
903	**Grapes,** *average*[a]	10 samples, white, black and seedless	1.00	81.8	0.4	0.1	15.4	60	257
904	*weighed with pips*	Calculated from No. 903	0.95	77.7	0.4	0.1	14.6	57	244
905	**Guava,** *raw*	Literature sources	1.00	84.7	0.8	0.5	5.0	26	112
906	*raw, weighed with skin and pips*	Calculated from No. 905	0.90	76.2	0.7	0.5	4.5	24	102
907	canned in syrup	10 samples	1.00	77.6	0.4	Tr	15.7	60	258
908	**Kiwi fruit**	Analysis and literature sources, flesh and seeds	1.00	84.0	1.1	0.5	10.6	49	207
909	*weighed with skin*	Calculated from No. 908	0.86	72.2	1.0	0.4	9.1	42	177
910	**Lemons,** *whole, without pips*	Analysis and literature sources; includes peel but no pips	0.99	86.3	1.0	0.3	3.2	19	79
911	**Lychees,** *raw*	Analysis and literature sources; flesh only	1.00	81.1	0.9	0.1	14.3	58	248
912	*raw, weighed with skin and stone*	Calculated from No. 911	0.62	50.3	0.5	0.1	8.9	36	155
913	canned in syrup	Analysis and literature sources	1.00	79.3	0.4	Tr	17.7	68	290
914	**Mandarin oranges,** *canned in* juice	10 samples, 4 brands	1.00	89.6	0.7	Tr	7.7	32	135
915	canned in syrup	10 samples, 10 brands	1.00	84.8	0.5	Tr	13.4	52	223

[a] Few significant differences reported between varieties

Fruit continued

Composition of food per 100g

No.	Food	Total nitrogen g	Fatty acids Satd g	Mono unsatd g	Poly unsatd g	Cholest-erol mg	Starch g	Total sugars g	Dietary fibre Southgate method g	Englyst method g
899	**Grapefruit**, raw	0.13	Tr	Tr	Tr	0	0	6.8	(1.6)	1.3
900	raw, weighed with peel and pips	0.09	Tr	Tr	Tr	0	0	4.6	(1.1)	0.9
901	canned in juice	0.09	Tr	Tr	Tr	0	0	7.3	(0.8)	0.4
902	canned in syrup	0.08	Tr	Tr	Tr	0	0	15.5	(0.9)	0.6
903	**Grapes**, average	0.06	Tr	Tr	Tr	0	0	15.4	0.8	0.7
904	weighed with pips	0.06	Tr	Tr	Tr	0	0	14.6	0.8	0.7
905	**Guava**, raw	0.13	N	N	N	0	0.1	4.9	4.7	3.7
906	raw, weighed with skin and pips	0.12	N	N	N	0	0.1	4.4	4.2	3.3
907	canned in syrup	0.06	Tr	Tr	Tr	0	Tr	15.7	3.2	3.0
908	**Kiwi fruit**	0.18	N	N	N	0	0.3	10.3	N	1.9
909	weighed with skin	0.15	N	N	N	0	0.3	8.9	N	1.6
910	**Lemons**, whole, without pips	0.16	0.1	Tr	0.1	0	0	3.2	4.7	N
911	**Lychees**, raw	0.14	Tr	Tr	Tr	0	0	14.3	1.5	0.7
912	raw, weighed with skin and stone	0.09	Tr	Tr	Tr	0	0	8.9	0.9	0.4
913	canned in syrup	0.06	Tr	Tr	Tr	0	0	17.7	0.7	0.5
914	**Mandarin oranges**, canned in juice	0.11	Tr	Tr	Tr	0	0	7.7	(0.3)	0.3
915	canned in syrup	0.08	Tr	Tr	Tr	0	0	13.4	0.3	0.2

Fruit *continued*

Inorganic constituents per 100g

No.	Food	Na	K	Ca	Mg	P	Fe	Cu	Zn	Cl	Mn	Se	I
						mg						µg	
899	**Grapefruit**, *raw*	3	200	23	9	20	0.1	0.02	Tr	3	Tr	(1)	N
900	*raw, weighed with peel and pips*	2	140	16	6	14	0.1	0.01	Tr	2	Tr	(1)	N
901	*canned in juice*	10	72	22	8	16	0.3	0.01	Tr	(5)	Tr	Tr	N
902	*canned in syrup*	10	79	17	7	13	0.7	(0.01)	0.4	(5)	Tr	Tr	N
903	**Grapes**, *average*	2	210	13	7	18	0.3	0.12	0.1	Tr	0.1	(1)	1
904	*weighed with pips*	2	200	12	7	17	0.3	0.11	0.1	Tr	0.1	(1)	1
905	**Guava**, *raw*	5	230	13	12	25	0.4	0.10	0.2	4	0.1	N	N
906	*raw, weighed with skin and pips*	5	210	12	11	23	0.4	0.09	0.2	4	0.1	N	N
907	*canned in syrup*	7	120	8	6	11	0.5	0.10	0.4	10	N	N	N
908	**Kiwi fruit**	4	290	25	15	32	0.4	0.13	0.1	39	0.1	N	N
909	*weighed with skin*	3	250	21	13	27	0.3	0.11	0.1	33	0.1	N	N
910	**Lemons**, *whole, without pips*	5	150	85	12	18	0.5	0.26	0.1	5	N	(1)	N
911	**Lychees**, *raw*	1	160	6	9	30	0.5	0.15	0.3	3	0.1	N	N
912	*raw, weighed with skin and stone*	1	99	4	6	19	0.3	0.09	0.2	2	Tr	N	N
913	*canned in syrup*	2	75	4	6	12	0.7	0.11	0.2	(5)	N	N	N
914	**Mandarin oranges**, canned in *juice*	6	85	17	9	13	0.5	Tr	0.1	2	Tr	Tr	Tr
915	*canned in syrup*	6	49	17	7	8	0.2	Tr	Tr	2	Tr	Tr	Tr

Fruit continued

No.	Food	Retinol μg	Carotene μg	Vitamin D μg	Vitamin E mg	Thiamin mg	Ribo-flavin mg	Niacin mg	Trypt 60 mg	Vitamin B6 mg	Vitamin B12 μg	Folate μg	Panto-thenate mg	Biotin μg	Vitamin C mg
899	**Grapefruit**, *raw*	0	17[a]	0	(0.19)	0.05	0.02	0.3	0.1	0.03	0	26	0.28	(1.0)	36
900	*raw, weighed with peel and pips*	0	11	0	(0.14)	0.03	0.01	0.2	0.1	0.02	0	18	0.19	(0.7)	24
901	*canned in juice*	0	Tr	0	(0.10)	0.04	0.01	0.3	0.1	(0.02)	0	6	(0.12)	(1.0)	33
902	*canned in syrup*	0	Tr	0	(0.11)	0.04	0.01	0.2	0.1	0.02	0	4	0.12	1.0	30
903	**Grapes**, *average*	0	17	0	Tr	0.05	0.01	0.2	Tr	0.10	0	2	0.05	0.3	3
904	*weighed with pips*	0	16	0	Tr	0.05	0.01	0.2	Tr	0.09	0	2	0.05	0.3	3
905	**Guava**, *raw*	0	435[b]	0	N	0.04	0.04	1.0	0.1	0.14	0	N	0.15	N	230[c]
906	*raw, weighed with skin and pips*	0	390	0	N	0.04	0.04	0.9	0.1	0.13	0	N	0.13	N	210
907	*canned in syrup*	0	(145)	0	N	(0.02)	(0.02)	(0.6)	0.1	(0.09)	0	N	(0.09)	N	180
908	**Kiwi fruit**	0	37	0	N	0.01	0.03	0.3	0.3	0.15	0	N	N	N	59
909	*weighed with skin*	0	32	0	N	0.01	0.03	0.3	0.3	0.13	0	N	N	N	51
910	**Lemons**, *whole, without pips*	0	18	0	N	0.05	0.04	0.2	0.1	0.11	0	N	0.23	0.5	58
911	**Lychees**, *raw*	0	0	0	N	0.04	0.06	0.5	0.1	N	0	N	N	N	45
912	*raw, weighed with skin and stone*	0	0	0	N	0.02	0.04	0.3	Tr	N	0	N	N	N	28
913	*canned in syrup*	0	0	0	N	Tr	0.04	Tr	0.1	N	0	N	N	N	8
914	**Mandarin oranges**, *canned in juice*	0	95	0	Tr	0.08	0.01	0.2	0.1	(0.03)	0	12	(0.15)	(0.8)	20
915	*canned in syrup*	0	105	0	Tr	0.06	0.01	0.2	Tr	0.03	0	12	(0.15)	(0.8)	15

[a] Pink varieties contain approximately 280μg carotene per 100g

[b] Peel included on analysis

[c] Levels ranged from 9 to 410mg vitamin C per 100g

No.	Food	Description and main data sources	Edible proportion	Water g	Protein g	Fat g	Carbo-hydrate g	Energy value kcal	Energy value kJ
916	**Mangoes**, ripe, *raw*	Literature sources; flesh only	1.00	82.4	0.7	0.2	14.1	57	245
917	*raw, weighed with skin and stone*	Calculated from No. 916	0.68	56.0	0.5	0.1	9.6	39	166
918	canned in syrup	10 samples	1.00	74.8	0.3	Tr	20.3	77	330
919	**Melon**, Canteloupe-type	10 samples, Canteloupe, Charantais and Rock; flesh only	1.00	92.1	0.6	0.1	4.2	19	81
920	*weighed with skin*	Calculated from No. 919; no pips	0.66	60.8	0.4	0.1	2.8	13	55
921	Galia	11 samples; flesh only	1.00	91.7	0.5	0.1	5.6	24	102
922	*weighed with skin*	Calculated from No. 921; no pips	0.68	63.4	0.3	0.1	3.8	16	70
923	Honeydew	10 samples; flesh only	1.00	92.2	0.6	0.1	6.6	28	119
924	*weighed with skin*	Calculated from No. 923; no pips	0.65	59.9	0.4	0.1	4.3	19	77
925	**watermelon**	Literature sources; flesh only	1.00	92.3	0.5	0.3	7.1	31	133
926	**Mixed peel**	10 samples, 9 brands	1.00	20.9	0.3	0.9	59.1	231	984
927	**Nectarines**	10 samples; flesh and skin	1.00	88.9	1.4	0.1	9.0	40	171
928	*weighed with stones*	Calculated from No. 927	0.89	79.1	1.2	0.1	8.0	36	152

Fruit *continued*

Composition of food per 100g

No.	Food	Total nitrogen g	Fatty acids Satd g	Mono unsatd g	Poly unsatd g	Cholest-erol mg	Starch g	Total sugars g	Dietary fibre Southgate method g	Englyst method g
916	**Mangoes**, *ripe, raw*	0.11	0.1	Tr	Tr	0	0.3	13.8	(2.9)	2.6
917	*raw, weighed with skin and stone*	0.07	Tr	Tr	Tr	0	0.2	9.4	(2.0)	1.8
918	canned in syrup	0.05	Tr	Tr	Tr	0	0.1	20.2	0.9	0.7
919	**Melon**, Canteloupe-type	0.10	Tr	Tr	Tr	0	0	4.2	0.9	1.0
920	*weighed with skin*	0.07	Tr	Tr	Tr	0	0	2.8	0.6	0.7
921	Galia	0.08	Tr	Tr	Tr	0	0	5.6	(0.9)	0.4
922	*weighed with skin*	0.05	Tr	Tr	Tr	0	0	3.8	(0.6)	0.3
923	Honeydew	0.10	Tr	Tr	Tr	0	0	6.6	0.8	0.6
924	*weighed with skin*	0.07	Tr	Tr	Tr	0	0	4.3	0.5	0.4
925	watermelon	0.07	0.1	0.1	0.1	0	0	7.1	0.3	0.1
926	**Mixed peel**	0.05	N	N	N	0	0	59.1	N	4.8
927	**Nectarines**	0.22	Tr	Tr	Tr	0	0	9.0	2.2	1.2
928	*weighed with stones*	0.20	Tr	Tr	Tr	0	0	8.0	2.0	1.1

Fruit *continued*

Inorganic constituents per 100g

No.	Food	Na	K	Ca	Mg	P	Fe	Cu	Zn	Cl	Mn	Se	I
						mg						µg	
916	**Mangoes**, ripe, raw	2	180	12	13	16	0.7	0.12	0.1	N	0.3	N	N
917	*raw, weighed with skin and stone*	1	120	8	9	11	0.5	0.08	0.1	N	0.2	N	N
918	*canned in syrup*	3	100	10	7	10	0.4	0.09	0.3	(5)	N	N	N
919	**Melon**, Canteloupe-type	8	210	20	11	13	0.3	Tr	Tr	44	Tr	Tr	(4)
920	*weighed with skin*	5	140	13	7	9	0.2	Tr	Tr	29	Tr	Tr	(3)
921	Galia	31	150	13	12	10	0.2	Tr	0.1	75	Tr	Tr	N
922	*weighed with skin*	21	100	9	8	7	0.1	Tr	0.1	51	Tr	Tr	N
923	Honeydew	32	210	9	10	16	0.1	Tr	Tr	45	Tr	Tr	N
924	*weighed with skin*	21	140	6	7	10	0.1	Tr	Tr	29	Tr	Tr	N
925	*watermelon*	2	100	7	8	9	0.3	0.03	0.2	N	Tr	N	Tr
926	**Mixed peel**	280	21	130	12	6	1.3	0.15	0.2	N	0.1	N	N
927	**Nectarines**	1	170	7	10	22	0.4	0.06	0.1	5	0.1	(1)	3
928	*weighed with stones*	1	150	6	9	20	0.4	0.05	0.1	4	0.1	(1)	3

Fruit *continued*

No.	Food	Retinol μg	Carotene μg	Vitamin D μg	Vitamin E mg	Thiamin mg	Ribo-flavin mg	Niacin mg	Trypt/60 mg	Vitamin B6 mg	Vitamin B12 μg	Folate μg	Panto-thenate mg	Biotin μg	Vitamin C mg
916	**Mangoes**, *ripe, raw*	0	1800[a]	0	1.05	0.04	0.05	0.5	1.3	0.13	0	N	0.16	N	37
917	*raw, weighed with skin and stone*	0	1225	0	0.71	0.03	0.03	0.3	0.1	0.09	0	N	0.11	N	25
918	*canned in syrup*	0	1470	0	0.64	(0.02)	(0.03)	(0.2)	0.1	(0.07)	0	N	(0.04)	N	10
919	**Melon**, Canteloupe-type	0	1000	0	0.10	0.04	0.02	0.6	Tr	0.11	0	5	0.13	N	26
920	*weighed with skin*	0	660	0	0.10	0.03	0.01	0.4	Tr	0.07	0	3	0.09	N	17
921	Galia	0	N	0	(0.10)	(0.03)	(0.01)	(0.4)	Tr	(0.09)	0	(3)	(0.17)	N	15
922	*weighed with skin*	0	N	0	(0.07)	(0.02)	(0.01)	(0.3)	Tr	(0.06)	0	(2)	(0.12)	N	10
923	Honeydew	0	48	0	0.10	0.03	0.01	0.3	Tr	0.06	0	2	0.21	N	9
924	*weighed with skin*	0	31	0	0.07	0.02	0.01	0.2	Tr	0.04	0	1	0.14	N	6
925	**watermelon**	0	230	0	(0.10)	0.05	0.01	0.1	Tr	0.14	0	2	0.21	1.0	8
926	**Mixed peel**	0	Tr	0	N	N	N	N	0.1	N	0	N	N	N	Tr
927	**Nectarines**	0	58	0	N	0.02	0.04	0.6	0.3	0.03	0	Tr	0.16	(0.2)	37
928	*weighed with stones*	0	52	0	N	0.02	0.04	0.5	0.3	0.03	0	Tr	0.14	(0.2)	33

[a] Levels ranged from 300 to 3000μg carotene per 100g

Fruit continued

No.	Food	Description and main data sources	Edible proportion	Water g	Protein g	Fat g	Carbo-hydrate g	Energy value kcal	kJ
929	**Olives**, *in brine*	Bottled, drained; flesh and skin, green	1.00	76.5	0.9	11.0	Tr	103	422
930	*in brine, weighed with stones*	Calculated from No. 929	0.80	61.2	0.7	8.8	Tr	82	337
931	**Oranges**	Assorted varieties; flesh only	1.00	86.1	1.1	0.1	8.5	37	158
932	*weighed with peel and pips*[a]	Calculated from No. 931	0.70	60.3	0.8	0.1	5.9	26	112
933	**Passion fruit**	Analysis and literature sources; flesh and pips	1.00	74.9	2.6	0.4	5.8	36	152
934	*weighed with skin*	Calculated from No. 933	0.61	45.7	1.7	0.2	3.5	22	92
935	**Paw-paw**, *raw*	Literature sources; flesh only	1.00	88.5	0.5	0.1	8.8	36	153
936	*raw, weighed with skin and pips*	Calculated from No. 935	0.75	66.4	0.4	0.1	6.6	27	116
937	*canned in juice*	10 samples	1.00	80.4	0.2	Tr	17.0	65	275
938	**Peaches**, *raw*	10 samples; flesh and skin	1.00	88.9	1.0	0.1	7.6	33	142
939	*raw, weighed with stone*	Calculated from No. 938	0.90	80.0	0.9	0.1	6.8	30	128
940	*canned in juice*	10 samples, 7 brands; halves and slices	1.00	86.7	0.6	Tr	9.7	39	165
941	*canned in syrup*	10 samples, 9 brands; halves and slices	1.00	81.1	0.5	Tr	14.0	55	233

[a] Levels ranged from 0.60 to 0.74

Fruit *continued*

Composition of food per 100g

No.	Food	Total nitrogen g	Fatty acids			Cholest- erol mg	Starch g	Total sugars g	Dietary fibre	
			Satd g	Mono unsatd g	Poly unsatd g				Southgate method g	Englyst method g
929	**Olives**, *in brine*	0.14	1.7	5.7	1.3	0	0	Tr	4.0	2.9
930	*in brine, weighed with stones*	0.11	1.4	4.6	1.0	0	0	Tr	3.2	2.3
931	**Oranges**	0.18	Tr	Tr	Tr	0	0	8.5	1.8	1.7
932	*weighed with peel and pips*	0.13	Tr	Tr	Tr	0	0	5.9	1.3	1.2
933	**Passion fruit**	0.45	0.1	0.1	0.1	0	0	5.8	N	3.3
934	*weighed with skin*	0.27	Tr	0.1	0.1	0	0	3.5	N	2.0
935	**Paw-paw**, *raw*	0.08	Tr	Tr	Tr	0	0	8.8	2.3	2.2
936	*raw, weighed with skin and pips*	0.06	Tr	Tr	Tr	0	0	6.6	1.7	1.7
937	*canned in juice*	0.03	Tr	Tr	Tr	0	0	17.0	(0.7)	0.7
938	**Peaches**, *raw*	0.16	Tr	Tr	Tr	0	0	7.6	2.3	1.5
939	*raw, weighed with stone*	0.14	Tr	Tr	Tr	0	0	6.8	2.1	1.3
940	*canned in juice*	0.09	Tr	Tr	Tr	0	0	9.7	(0.9)	0.8
941	*canned in syrup*	0.08	Tr	Tr	Tr	0	0	14.0	0.9	0.9

No.	Food	Na	K	Ca	Mg	P	Fe	Cu	Zn	Cl	Mn	Se	I
						mg						μg	
929	**Olives**, *in brine*	2250	91	61	22	17	1.0	0.23	N	3750	N	N	N
930	*in brine, weighed with stones*	1800	73	49	18	14	0.8	0.18	N	3000	N	N	N
931	**Oranges**	5	150	47	10	21	0.1	0.05	0.1	3	Tr	(1)	2
932	*weighed with peel and pips*	3	110	33	7	15	0.1	0.03	0.1	2	Tr	(1)	1
933	**Passion fruit**	19	200	11	29	64	1.3	N	0.8	N	N	N	N
934	*weighed with skin*	12	120	7	18	39	0.8	N	0.5	N	N	N	N
935	**Paw-paw**, *raw*	5	200	23	11	13	0.5	0.08	0.2	11	0.1	N	N
936	*raw, weighed with skin and pips*	4	170	17	8	10	0.4	0.06	0.1	8	0.1	N	N
937	*canned in juice*	8	110	23	8	6	0.4	0.10	0.3	40	N	N	N
938	**Peaches**, *raw*	1	160	7	9	22	0.4	0.06	0.1	Tr	0.1	(1)	3
939	*raw, weighed with stone*	1	140	6	8	20	0.4	0.05	0.1	Tr	0.1	(1)	3
940	*canned in juice*	12	170	4	7	19	0.4	0.04	0.1	(4)	0.1	Tr	N
941	*canned in syrup*	4	110	3	5	11	0.2	Tr	Tr	4	Tr	Tr	N

Fruit continued

No.	Food	Retinol µg	Carotene µg	Vitamin D µg	Vitamin E mg	Thiamin mg	Ribo-flavin mg	Niacin mg	Trypt 60 mg	Vitamin B6 mg	Vitamin B12 µg	Folate µg	Panto-thenate mg	Biotin µg	Vitamin C mg
929	Olives, in brine	0	180a	0	1.99	Tr	Tr	Tr	0.1	0.02	0	Tr	0.02	Tr	0
930	in brine, weighed with stones	0	145	0	1.59	Tr	Tr	Tr	0.1	0.02	0	Tr	0.02	Tr	0
931	Oranges	0	28b	0	0.24	0.11	0.04	0.4	0.1	0.10	0	31	0.37	1.0	54c
932	weighed with peel and pips	0	20	0	0.17	0.08	0.03	0.3	0.1	0.07	0	22	0.26	0.7	38
933	Passion fruit	0	750	0	N	0.03	0.12	1.5	0.4	N	0	N	N	N	23
934	weighed with skin	0	460	0	N	0.02	0.07	0.9	0.2	N	0	N	N	N	14
935	Paw-paw, raw	0	810	0	N	0.03	0.04	0.3	0.1	0.03	0	1	0.22	N	60
936	raw, weighed with skin and pips	0	585	0	N	0.02	0.03	0.2	0.1	0.02	0	1	0.17	N	45
937	canned in juice	0	(255)	0	N	0.02	0.02	0.2	Tr	(0.01)	0	Tr	(0.20)	N	15
938	Peaches, raw	0	58	0	N	0.02	0.04	0.6	0.2	0.02	0	3	0.17	(0.2)	31
939	raw, weighed with stone	0	53	0	N	0.02	0.04	0.5	0.2	0.02	0	3	0.15	(0.2)	28
940	canned in juice	0	67	0	N	0.01	0.01	0.6	0.1	0.02	0	2	0.06	0.2	6
941	canned in syrup	0	75	0	N	0.01	0.01	0.6	0.1	0.02	0	7	0.05	0.1	5

a Values for green olives. Ripe black olives contain 40µg carotene per 100g

b Blood oranges have been found to contain 155µg carotene per 100g

c Levels ranged from 44 to 79mg vitamin C per 100g

No.	Food	Description and main data sources	Edible proportion	Water g	Protein g	Fat g	Carbo-hydrate g	Energy value kcal	Energy value kJ
942	**Pears**, *average, raw*	Average of Comice, Conference and Williams varieties; flesh and skin	1.00	83.8	0.3	0.1	10.0	40	169
943	*average, raw, weighed with core*	Calculated from No. 942	0.91	76.3	0.3	0.1	9.1	36	155
944	-, *raw, peeled*	Literature sources and calculation from No. 942; flesh only	1.00	83.8	0.3	0.1	10.4	41	175
945	canned in juice	10 samples, 7 brands	1.00	86.8	0.3	Tr	8.5	33	141
946	canned in syrup	10 samples, 8 brands	1.00	82.6	0.2	Tr	13.2	50	215
947	**Pineapple**, *raw*	10 samples; flesh only	1.00	86.5	0.4	0.2	10.1	41	176
948	canned in juice	10 samples, 10 brands; cubes and slices	1.00	86.8	0.3	Tr	12.2	47	200
949	canned in syrup	10 samples, 10 brands; cubes and slices	1.00	82.2	0.5	Tr	16.5	64	273
950	**Plums**, *average, raw*	Assorted varieties; flesh and skin	1.00	83.9	0.6	0.1	8.8	36	155
951	*average, raw, weighed with stones*	Calculated from No. 950	0.94	78.9	0.5	0.1	8.3	34	145
952	-, *stewed with sugar, weighed with stones*	Calculated from Plums stewed with sugar	0.95	70.5	0.5	0.1	19.2	75	319
953	-, *stewed without sugar, weighed with stones*	Calculated from Plums stewed without sugar	0.95	81.0	0.4	0.1	6.9	29	121
954	canned in syrup	10 samples, 7 brands; Red, Golden and Victoria	1.00	81.4	0.3	Tr	15.5	59	253
955	**Prunes**, canned in juice	10 samples; stones removed	0.93	74.1	0.7	0.2	19.7	79	335
956	canned in syrup	11 samples, 6 brands; stones removed	0.92	69.9	0.6	0.2	23.0	90	386
957	ready-to-eat	4 samples; semi-dried	1.00	31.1	2.5	0.4	34.0	141	601

Fruit *continued*

Composition of food per 100g

No.	Food	Total nitrogen g	Fatty acids			Cholest- erol mg	Starch g	Total sugars g	Dietary fibre	
			Satd g	Mono unsatd g	Poly unsatd g				Southgate method g	Englyst method g
942	**Pears**, *average, raw*	0.05	Tr	Tr	Tr	0	0	10.0	N	2.2
943	*average, raw, weighed with core*	0.05	Tr	Tr	Tr	0	0	9.1	N	2.0
944	*-, raw, peeled*	0.05	Tr	Tr	Tr	0	0	10.4	2.1	1.7
945	*canned in juice*	0.04	Tr	Tr	Tr	0	0	8.5	(1.5)	1.4
946	*canned in syrup*	0.04	Tr	Tr	Tr	0	0	13.2	1.5	1.1
947	**Pineapple**, *raw*	0.06	Tr	0.1	0.1	0	0	10.1	1.3	1.2
948	*canned in juice*	0.05	Tr	Tr	Tr	0	0	12.2	(0.8)	0.5
949	*canned in syrup*	0.08	Tr	Tr	Tr	0	0	16.5	0.8	0.7
950	**Plums**, *average, raw*	0.09	Tr	Tr	Tr	0	0	8.8	2.3	1.6
951	*average, raw, weighed with stones*	0.08	Tr	Tr	Tr	0	0	8.3	2.2	1.5
952	*-, stewed with sugar, weighed with stones*	0.08	Tr	Tr	Tr	0	0	19.2	1.8	1.2
953	*-, stewed without sugar, weighed with stones*	0.07	Tr	Tr	Tr	0	0	6.9	1.8	1.2
954	*canned in syrup*	0.05	Tr	Tr	Tr	0	0	15.5	1.0	0.8
955	**Prunes**, *canned in juice*	0.12	Tr	0.1	0.1	0	0	19.7	N	2.4
956	*canned in syrup*	0.10	Tr	0.1	0.1	0	0	23.0	N	2.8
957	*ready-to-eat*	0.40	N	N	N	0	0	34.0	12.8	5.7

Fruit *continued*

No.	Food	Na	K	Ca	Mg	P	Fe	Cu	Zn	Cl	Mn	Se	I
						mg						μg	
942	**Pears**, average, raw	3	150	11	7	13	0.2	0.06	0.1	1	Tr	Tr	1
943	average, raw, weighed with core	3	140	10	6	12	0.2	0.05	0.1	1	Tr	Tr	1
944	-, raw, peeled	3	150	11	7	13	0.2	0.06	0.1	1	Tr	Tr	1
945	canned in juice	3	81	6	5	10	0.2	Tr	0.1	(3)	Tr	Tr	Tr
946	canned in syrup	3	68	6	4	7	0.2	0.02	0.1	3	Tr	Tr	Tr
947	**Pineapple**, raw	2	160	18	16	10	0.2	0.11	0.1	29	0.5	Tr	Tr
948	canned in juice	1	71	8	13	5	0.5	0.08	0.1	(4)	0.9	Tr	Tr
949	canned in syrup	2	79	6	11	5	0.2	0.02	0.1	4	0.9	Tr	Tr
950	**Plums**, average, raw	2	240	13	8	23	0.4	0.10	0.1	Tr	0.1	Tr	Tr
951	average, raw, weighed with stones	2	230	12	7	22	0.4	0.09	0.1	Tr	0.1	Tr	Tr
952	-, stewed with sugar, weighed with stones	2	190	10	7	18	0.3	0.08	0.1	Tr	0.1	Tr	Tr
953	-, stewed without sugar, weighed with stones	2	190	10	7	18	0.3	0.08	0.1	Tr	0.1	Tr	Tr
954	canned in syrup	6	79	9	4	10	N	Tr	Tr	N	Tr	Tr	N
955	**Prunes**, canned in juice	18	340	26	15	30	2.2	0.09	1.0	N	0.1	Tr	N
956	canned in syrup	(18)	(340)	(26)	(15)	(30)	(2.2)	(0.09)	(1.0)	N	(0.1)	Tr	N
957	ready-to-eat	11	760	34	24	73	2.6	0.14	0.4	3	0.3	3	N

No.	Food	Retinol µg	Carotene µg	Vitamin D µg	Vitamin E mg	Thiamin mg	Ribo-flavin mg	Niacin mg	Trypt 60 mg	Vitamin B6 mg	Vitamin B12 µg	Folate µg	Panto-thenate mg	Biotin µg	Vitamin C mg
942	**Pears**, *average, raw*	0	18	0	0.50	0.02	0.03	0.2	Tr	0.02	0	2	0.07	0.2	6
943	*average, raw, weighed with core*	0	16	0	0.45	0.02	0.03	0.2	Tr	0.02	0	2	0.06	0.2	5
944	*-, raw, peeled*	0	19	0	Tr	0.02	0.03	0.2	Tr	0.02	0	2	0.07	0.2	6
945	*canned in juice*	0	Tr	0	Tr	0.01	0.01	0.2	Tr	0.03	0	4	0.04	0.2	3
946	*canned in syrup*	0	Tr	0	Tr	0.01	0.01	0.2	Tr	0.03	0	3	0.04	0.2	2
947	**Pineapple**, *raw*	0	18	0	0.10	0.08	0.03	0.3	0.1	0.09	0	5	0.16	0.3	12
948	*canned in juice*	0	12	0	(0.05)	0.09	0.01	0.2	0.1	0.09	0	1	0.11	0.1	11
949	*canned in syrup*	0	11	0	0.06	0.07	0.01	0.2	0.1	0.07	0	(1)	0.07	0.1	13
950	**Plums**, *average, raw*	0	295	0	0.61	0.05	0.03	1.1	0.1	0.05	0	3	0.15	Tr	4
951	*average, raw, weighed with stones*	0	275	0	0.57	0.05	0.03	1.0	0.1	0.05	0	3	0.14	Tr	4
952	*-, stewed with sugar, weighed with stones*	0	62	0	0.48	0.03	0.02	0.7	0.1	0.03	0	Tr	0.09	Tr	3
953	*-, stewed without sugar, weighed with stones*	0	62	0	0.47	0.03	0.02	0.7	0.1	0.03	0	Tr	0.09	Tr	3
954	*canned in syrup*	0	29	0	0.25	0.01	0.01	0.3	Tr	(0.02)	0	Tr	(0.04)	Tr	1
955	**Prunes**, *canned in juice*	0	140	0	N	0.02	0.02	0.5	0.1	(0.06)	0	5	(0.07)	Tr	Tr
956	*canned in syrup*	0	(140)	0	N	(0.02)	(0.02)	(0.5)	0.1	(0.05)	0	(5)	(0.06)	Tr	Tr
957	*ready-to-eat*	0	140	0	N	0.09	0.18	1.3	0.4	0.21	0	3	0.41	Tr	Tr

Fruit *continued*

Composition of food per 100g

No.	Food	Description and main data sources	Edible proportion	Water g	Protein g	Fat g	Carbo-hydrate g	Energy value kcal	kJ
958	**Raisins**	10 samples, 8 brands. Large stoned variety	1.00	13.2	2.1	0.4	69.3	272	1159
959	**Raspberries**, *raw*	9 samples; whole fruit	1.00	87.0	1.4	0.3	4.6	25	109
960	canned in syrup	Mixed sample	1.00	74.0	0.6	0.1	22.5	88	374
961	**Rhubarb**, *raw*	Stems only	0.87	94.2	0.9	0.1	0.8	7	32
962	*stewed with sugar*	1000g fruit, 100g water, 120g sugar	1.00	84.6	0.9	0.1	11.5	48	203
963	*stewed without sugar*	500g fruit, 50g water and calculation from No. 962	1.00	94.1	0.9	0.1	0.7	7	30
964	canned in syrup	10 samples, 6 brands	1.00	90.6	0.5	Tr	7.6	31	130
965	**Satsumas**	10 samples; flesh only	1.00	87.4	0.9	0.1	8.5	36	155
966	*weighed with peel*	Calculated from No. 965	0.71	62.1	0.6	0.1	6.0	26	110
967	**Strawberries**, *raw*	9 samples; flesh and pips	0.95	89.5	0.8	0.1	6.0	27	113
968	canned in syrup	10 samples	1.00	81.7	0.5	Tr	16.9	65	279
969	**Sultanas**	10 samples, 9 brands; whole fruit	1.00	15.2	2.7	0.4	69.4	275	1171
970	**Tangerines**	Flesh only	1.00	86.7	0.9	0.1	8.0	35	147
971	*weighed with peel and pips*	Calculated from No. 970	0.73	63.3	0.7	0.1	5.8	25	108

Fruit *continued*

Composition of food per 100g

No.	Food	Total nitrogen g	Fatty acids			Cholest- erol mg	Starch g	Total sugars g	Dietary fibre	
			Satd g	Mono unsatd g	Poly unsatd g				Southgate method g	Englyst method g
958	**Raisins**	0.34	N	N	N	0	0	69.3	6.1	2.0
959	**Raspberries,** *raw*	0.22	0.1	0.1	0.1	0	0	4.6	6.7	2.5
960	canned in syrup	0.10	Tr	Tr	Tr	0	0	22.5	(4.5)	1.5
961	**Rhubarb,** *raw*	0.14	Tr	Tr	Tr	0	0	0.8	2.3	1.4
962	*stewed with sugar*	0.14	Tr	Tr	Tr	0	0	11.5	2.0	1.2
963	*stewed without sugar*	0.15	Tr	Tr	Tr	0	0	0.7	2.1	1.3
964	*canned in syrup*	0.08	Tr	Tr	Tr	0	0	7.6	1.3	0.8
965	**Satsumas**	0.14	Tr	Tr	Tr	0	0	8.5	(1.7)	1.3
966	*weighed with peel*	0.10	Tr	Tr	Tr	0	0	6.0	(1.2)	0.9
967	**Strawberries,** *raw*	0.13	Tr	Tr	Tr	0	0	6.0	2.0	1.1
968	*canned in syrup*	0.07	Tr	Tr	Tr	0	0	16.9	0.9	0.7
969	**Sultanas**	0.43	N	N	N	0	0	69.4	6.3	2.0
970	**Tangerines**	0.14	Tr	Tr	Tr	0	0	8.0	1.7	1.3
971	*weighed with peel and pips*	0.10	Tr	Tr	Tr	0	0	5.8	1.2	0.9

Fruit *continued*

Inorganic constituents per 100g

No.	Food	Na	K	Ca	Mg	P	Fe	Cu	Zn	Cl	Mn	Se	I
						mg						μg	
958	**Raisins**	60	1020	46	35	76	3.8	0.39	0.7	9	0.3	(8)	N
959	**Raspberries**, *raw*	3	170	25	19	31	0.7	0.10	0.3	22	0.4	N	N
960	*canned in syrup*	4	100	14	11	14	1.7	0.10	N	5	0.3	N	N
961	**Rhubarb**, *raw*	3	290	93	13	17	0.3	0.07	0.1	87	0.2	Tr	N
962	*stewed with sugar*	1	210	33	6	18	0.1	0.02	Tr	75	0.3	Tr	N
963	*stewed without sugar*	1	230	35	6	19	0.1	0.02	Tr	79	0.3	Tr	N
964	*canned in syrup*	4	89	36	5	8	0.8	Tr	0.1	15	0.1	Tr	N
965	**Satsumas**	4	130	31	10	18	0.1	0.01	0.1	(2)	Tr	N	N
966	*weighed with peel*	3	92	22	7	13	0.1	0.01	0.1	(1)	Tr	N	N
967	**Strawberries**, *raw*	6	160	16	10	24	0.4	0.07	0.1	18	0.3	Tr	9
968	*canned in syrup*	9	87	11	7	15	1.1	Tr	0.1	(5)	0.2	Tr	N[a]
969	**Sultanas**	19	1060	64	31	86	2.2	0.40	0.3	16	0.3	N	N
970	**Tangerines**	2	160	42	11	17	0.3	0.01	0.1	2	Tr	N	N
971	*weighed with peel and pips*	1	120	31	8	12	0.2	0.01	0.1	1	Tr	N	N

[a] Iodine from erythrosine is present but largely unavailable

No.	Food	Retinol µg	Carotene µg	Vitamin D µg	Vitamin E mg	Thiamin mg	Ribo-flavin mg	Niacin mg	Trypt 60 mg	Vitamin B6 mg	Vitamin B12 µg	Folate µg	Panto-thenate mg	Biotin µg	Vitamin C mg
958	**Raisins**	0	12	0	N	0.12	0.05	0.6	0.2	0.25	0	10	0.15	2.0	1
959	**Raspberries**, *raw*	0	6	0	0.48	0.03	0.05	0.5	0.3	0.06	0	33	0.24	1.9	32
960	*canned in syrup*	0	3	0	0.15	0.01	0.03	0.3	0.1	0.04	0	(10)	0.17	(0.7)	7
961	**Rhubarb**, *raw*	0	60	0	0.20	0.03	0.03	0.3	0.1	0.02	0	7	0.09	N	6
962	*stewed with sugar*	0	28	0	0.17	0.03	0.02	0.2	0.1	0.02	0	4	0.08	N	5
963	*stewed without sugar*	0	30	0	0.18	0.03	0.02	0.2	0.1	0.02	0	4	0.08	N	5
964	*canned in syrup*	0	(18)	0	(0.11)	0.02	0.01	0.1	0.1	0.01	0	3	0.05	N	3
965	**Satsumas**	0	75	0	N	0.09	0.04	0.3	0.1	(0.07)	0	33	(0.20)	N	27
966	*weighed with peel*	0	53	0	N	0.06	0.03	0.2	0.1	(0.05)	0	23	(0.14)	N	19
967	**Strawberries**, *raw*	0	8	0	0.20	0.03	0.03	0.6	0.1	0.06	0	20	0.34	1.1	77
968	*canned in syrup*	0	4	0	N	0.01	0.02	0.3	0.1	0.03	0	6	0.21	(1.0)	29
969	**Sultanas**	0	12	0	0.70	0.09	0.05	0.8	0.2	0.25	0	27	0.09	4.8	Tr
970	**Tangerines**	0	97	0	N	0.07	0.02	0.2	0.1	0.07	0	21	0.20	N	30
971	*weighed with peel and pips*	0	71	0	N	0.05	0.01	0.1	0.1	0.05	0	15	0.15	N	22

Nuts

Data in this section of the Tables have been completely revised and now reflects the increased range of nuts and nut products available.

It should be noted that the levels for biotin from new analyses are significantly higher than those quoted in earlier editions.

Taxonomic names for foods in this part of the Tables can be found in Section 3.6.

Nuts and seeds

Composition of food per 100g

No.	Food	Description and main data sources	Edible proportion	Water g	Protein g	Fat g	Carbo-hydrate g	Energy value kcal	kJ
972	**Almonds**	10 blanched samples, flaked and ground	1.00	4.2	21.1	55.8	6.9	612	2534
973	*weighed with shells*	Calculated from No. 972	0.37	1.5	7.8	20.6	2.5	229	935
974	**Brazil nuts**	10 samples, kernel only	1.00	2.8	14.1	68.2	3.1	682	2813
975	*weighed with shells*	Calculated from No. 974	0.46	1.3	6.5	31.4	1.4	314	1295
976	**Cashew nuts**, *roasted and salted*	10 samples, kernels only	1.00	2.4	20.5	50.9	18.8	611	2533
977	**Chestnuts**	Analysis and literature sources; kernel only	1.00	51.7	2.0	2.7	36.6	170	719
978	**Coconut**, *creamed block*	7 samples, 2 brands; block of dried kernel	1.00	2.5	6.0	68.8	7.0	669	2760
979	*desiccated*	Analytical and literature sources	1.00	2.3	5.6	62.0	6.4	604	2492
980	**Hazelnuts**	10 samples, kernel only	1.00	4.6	14.1	63.5	6.0	650	2685
981	*weighed with shells*	Calculated from No. 980	0.38	1.7	5.4	24.1	2.3	247	1020
982	**Macadamia nuts**, salted	8 samples	1.00	1.3	7.9	77.6	4.8	748	3082
983	**Marzipan**, *homemade*	Recipe	1.00	10.2	10.4	25.8	50.2	461	1933
984	*retail*	10 samples, white and yellow	1.00	7.9	5.3	14.4	67.6	404	1705
985	**Mixed nuts**[a]	Calculated from recipe proportions	1.00	2.5	22.9	54.1	7.9	607	2515

[a] Calculated as peanuts 67%, almonds 17%, cashews 8% and hazelnuts 7%

No.	Food	Total nitrogen g	Fatty acids Satd g	Mono unsatd g	Poly unsatd g	Cholest-erol mg	Starch g	Total sugars g	Dietary fibre Southgate method g	Englyst method g
972	**Almonds**	4.07	4.7	34.4	14.2	0	2.7	4.2	(12.9)	(7.4)
973	*weighed with shells*	1.51	1.7	12.7	5.3	0	1.0	1.5	(4.8)	(2.7)
974	**Brazil nuts**	2.61	16.4	25.8	23.0	0	0.7	2.4	8.1	4.3
975	*weighed with shells*	1.20	7.5	11.9	10.6	0	0.3	1.1	3.7	2.0
976	**Cashew nuts**, *roasted and salted*	3.87	10.1	29.4	9.1	0	13.2	5.6	N	3.2
977	**Chestnuts**	0.37	0.5	1.0	1.1	0	29.6	7.0	6.1	4.1
978	**Coconut**, *creamed block*	1.14	59.3	3.9	1.6	0	0	7.0	N	N
979	*desiccated*	1.05	53.4	3.5	1.5	0	0	6.4	21.1	13.7
980	**Hazelnuts**	2.66	4.7	50.0	5.9	0	2.0	4.0	8.9	6.5
981	*weighed with shells*	1.01	1.8	18.9	2.3	0	0.8	1.5	3.4	2.5
982	**Macadamia nuts**, salted	1.49	11.2	60.8	1.6	0	0.8	4.0	N	5.3
983	**Marzipan**, *homemade*	1.98	2.3	15.7	6.5	28	1.2	48.9	(5.8)	(3.3)
984	*retail*	1.02	1.2	8.9	3.7	0	0	67.6	(3.2)	(1.9)
985	**Mixed nuts**	4.27	8.4	28.2	14.8	0	3.9	4.0	7.5	6.0

Nuts and seeds

No.	Food	Na	K	Ca	Mg	P	Fe	Cu	Zn	Cl	Mn	Se	I
						mg						μg	
972	**Almonds**	14	780	240	270	550	3.0	1.00	3.2	18	1.7	4	2
973	*weighed with shells*	5	290	89	100	200	1.1	0.37	1.2	7	0.6	1	1
974	**Brazil nuts**	3	660	170	410	590	2.5	1.76	4.2	57	1.2	1530[a]	20
975	*weighed with shells*	1	300	78	190	270	1.1	0.81	1.9	26	0.5	700	9
976	**Cashew nuts**, *roasted and salted*	290	730	35	250	510	6.2	2.04	5.7	490	1.8	34	11
977	**Chestnuts**	11	500	46	33	74	0.9	0.23	0.5	15	0.5	Tr	N
978	**Coconut**, *creamed block*	30	650	23	73	170	3.7	0.56	0.9	190	1.8	(2)	(2)
979	*desiccated*	28	660	23	90	160	3.6	0.55	0.9	200	1.8	(3)	(3)
980	**Hazelnuts**	6	730	140	160	300	3.2	1.23	2.1	18	4.9	Tr	17
981	*weighed with shells*	2	280	53	61	110	1.2	0.47	0.8	7	1.9	Tr	6
982	**Macadamia nuts**, salted	280	300	47	100	200	1.6	0.43	1.1	390	5.5	7	N
983	**Marzipan**, *homemade*	16	360	110	120	260	1.5	0.46	1.6	20	0.8	3	5
984	*retail*	20	160	66	68	130	0.9	0.24	0.8	23	0.4	1	Tr
985	**Mixed nuts**	300	790	78	200	430	2.1	0.79	3.1	490	2.1	5	12

a Selenium can range from 230 to 5300μg per 100g

No.	Food	Retinol µg	Carotene µg	Vitamin D µg	Vitamin E mg	Thiamin mg	Ribo- flavin mg	Niacin mg	Trypt 60 mg	Vitamin B6 mg	Vitamin B12 µg	Folate µg	Panto- thenate mg	Biotin µg	Vitamin C mg
972	**Almonds**	0	0	0	23.98	0.21	0.75	3.1	3.4	0.15	0	48	0.44	64.0	0
973	*weighed with shells*	0	0	0	8.87	0.08	0.28	1.1	1.3	0.05	0	18	0.16	24.0	0
974	**Brazil nuts**	0	0	0	7.18	0.67	0.03	0.3	3.0	0.31	0	21	0.41	11.0	0
975	*weighed with shells*	0	0	0	3.30	0.31	0.01	0.1	1.4	0.14	0	10	0.19	5.0	0
976	**Cashew nuts**, *roasted and salted*	0	6	0	1.30	0.41	0.16	1.3	5.2	0.43	0	68	1.08	13.0	0
977	**Chestnuts**	0	0	0	1.20	0.14	0.02	0.5	0.4	0.34	0	N	0.49	1.4	Tr
978	**Coconut**, *creamed block*	0	0	0	1.40	(0.03)	(0.05)	(0.9)	1.2	N	0	(9)	(0.50)	N	0
979	*desiccated*	0	0	0	1.26	0.03	0.05	0.9	1.1	(0.09)	0	9	0.50	N	0
980	**Hazelnuts**	0	0	0	24.98	0.43	0.16	1.1	4.0	0.59	0	72	1.51	76.0	0
981	*weighed with shells*	0	0	0	9.49	0.16	0.06	0.4	1.5	0.22	0	27	0.57	29.0	0
982	**Macadamia nuts**, *salted*	0	0	0	1.49	0.28	0.06	1.6	1.7	0.28	0	N	0.61	6.0	0
983	**Marzipan**, *homemade*	14	Tr	0.1	10.82	0.10	0.37	1.4	1.8	0.08	0.2	25	0.33	30.2	1
984	*retail*	0	0	0	6.18	(0.05)	(0.19)	(0.7)	0.9	(0.04)	0	(12)	(0.11)	(16.0)	0
985	**Mixed nuts**	0	Tr	0	6.44	0.22	0.22	9.9	4.9	0.53	0	54	1.42	86.4	0

Nuts and seeds continued

Composition of food per 100g

No.	Food	Description and main data sources	Edible proportion	Water g	Protein g	Fat g	Carbo-hydrate g	Energy value kcal	kJ
986	**Peanut butter**, smooth	10 samples, 3 brands	1.00	1.1	22.6	53.7	13.1	623	2581
987	**Peanuts**, plain	10 samples, kernel only	1.00	6.3	25.6	46.1	12.5	564	2341
988	plain, weighed with shells	Calculated from No. 987	0.69	4.3	17.7	31.8	8.6	389	1615
989	dry roasted	10 samples, 5 brands	1.00	1.8	25.5	49.8	10.3	589	2441
990	roasted and salted	20 samples	1.00	1.9	24.5	53.0	7.1	602	2491
991	**Pecan nuts**	9 samples, kernel only	1.00	3.7	9.2	70.1	5.8	689	2843
992	**Pine nuts**	20 samples, pine kernels	1.00	2.7	14.0	68.6	4.0	688	2840
993	**Pistachio nuts**, weighed with shells	Calculated from Pistachio nuts without shells	0.55	1.1	9.9	30.5	4.6	331	1370
994	**Sesame seeds**	10 samples, with and without hulls	1.00	4.6	18.2	58.0	0.9	598	2470
995	**Sunflower seeds**	Analysis and literature sources	1.00	4.4	19.8	47.5	18.6[a]	581	2410
996	**Tahini paste**	Ref. McCarthy and Matthews (1984) and calculation from No. 994	1.00	3.1	18.5	58.9	0.9	607	2508
997	**Walnuts**	10 samples, kernel only	1.00	2.8	14.7	68.5	3.3	688	2837
998	weighed with shells	Calculated from No. 997	0.43	1.2	6.3	29.4	1.4	295	1217

[a] Including oligosaccharides

Composition of food per 100g

No.	Food	Total nitrogen	Fatty acids			Cholest-erol	Starch	Total sugars	Dietary fibre	
			Satd	Mono unsatd	Poly unsatd				Southgate method	Englyst method
		g	g	g	g	mg	g	g	g	g
986	**Peanut butter, smooth**	4.17	11.7	21.3	18.4	0	6.4	6.7	6.8	5.4
987	**Peanuts, plain**	4.73	8.2	21.1	14.3	0	6.3	6.2	7.3	6.2
988	*plain, weighed with shells*	3.26	6.5	14.5	9.9	0	4.3	4.3	5.0	4.3
989	*dry roasted*	4.71	8.9	22.8	15.5	0	6.5	3.8	7.4	6.4
990	*roasted and salted*	4.53	9.5	24.2	16.5	0	3.3	3.8	6.9	6.0
991	**Pecan nuts**	1.74	5.7	42.5	18.7	0	1.5	4.3	N	4.7
992	**Pine nuts**	2.64	4.6	19.9	41.1	0	0.1	3.9	N	1.9
993	**Pistachio nuts,** *weighed with shells*	1.86	4.1	15.2	9.8	0	1.4	3.2	N	3.3
994	**Sesame seeds**	3.44	8.3	21.7	25.5	0	0.5	0.4	N	7.9
995	**Sunflower seeds**	3.74	4.5	9.8	31.0	0	16.3	1.7[a]	N	6.0
996	**Tahini paste**	3.49	8.4	22.0	25.8	0	0.5	0.4	N	8.0
997	**Walnuts**	2.77	5.6	12.4	47.5	0	0.7	2.6	5.9	3.5
998	*weighed with shells*	1.19	2.4	5.3	20.4	0	0.3	1.1	2.5	1.5

[a] Not including oligosaccharides

Nuts and seeds continued

Inorganic constituents per 100g

No.	Food	Na	K	Ca	Mg	P	Fe	Cu	Zn	Cl	Mn	Se	I
						mg						μg	
986	**Peanut butter**, smooth	350	700	37	180	330	2.1	0.70	3.0	500	1.7	3	N
987	**Peanuts**, *plain*	2	670	60	210	430	2.5	1.02	3.5	7	2.1	3	20
988	*plain, weighed with shells*	1	460	41	140	300	1.7	0.70	2.4	5	1.4	2	14
989	*dry roasted*	790	730	52	190	420	2.1	0.64	3.3	1140	2.2	3	19
990	*roasted and salted*	400	810	37	180	410	1.3	0.54	2.9	660	1.9	4	19
991	**Pecan nuts**	1	520	61	130	310	2.2	1.07	5.3	15	4.6	12	N
992	**Pine nuts**	1	780	11	270	650	5.6	1.32	6.5	41	7.9	N	N
993	**Pistachio nuts**, *weighed with shells*	290	570	61	71	230	1.7	0.46	1.2	450	0.5	(3)	N
994	**Sesame seeds**	20	570	670	370	720	10.4	1.46	5.3	10	1.5	N	N
995	**Sunflower seeds**	3	710	110	390	640	6.4	2.27	5.1	N	2.2	(49)	N
996	**Tahini paste**	20	580	680	380	730	10.6	1.48	5.4	10	1.5	N	N
997	**Walnuts**	7	450	94	160	380	2.9	1.34	2.7	24	3.4	19	9
998	*weighed with shells*	3	190	40	69	160	1.2	0.58	1.2	10	1.5	8	4

Nuts and seeds continued

No.	Food	Retinol μg	Carotene μg	Vitamin D μg	Vitamin E mg	Thiamin mg	Ribo-flavin mg	Niacin mg	Trypt 60 mg	Vitamin B6 mg	Vitamin B12 μg	Folate μg	Panto-thenate mg	Biotin μg	Vitamin C mg
986	**Peanut butter**, smooth	0	0	0	4.99	0.17	0.09	12.5	4.9	0.58	0	53	1.56	94.0	0
987	**Peanuts**, plain	0	0	0	10.09	1.14	0.10	13.8	5.5	0.59	0	110	2.66	72.0	0
988	plain, weighed with shells	0	0	0	6.97	0.79	0.07	9.5	3.8	0.41	0	76	1.83	49.7	0
989	dry roasted	0	0	0	1.11	0.18	0.13	13.1	5.5	0.54	0	66	1.59	130.0	0
990	roasted and salted	0	0	0	0.66	0.18	0.10	13.6	5.3	0.63	0	52	1.70	102.0	0
991	**Pecan nuts**	0	50	0	4.34	0.71	0.15	1.4	4.1	0.19	0	39	1.71	N	0
992	**Pine nuts**	0	10	0	13.65	0.73	0.19	3.8	3.1	N	0	N	N	N	Tr
993	**Pistachio nuts**, *weighed with shells*	0	71	0	2.28	0.39	0.13	0.9	2.2	N	0	32	N	N	0
994	**Sesame seeds**	0	6	0	2.53	0.93	0.17	5.0	5.4	0.75	0	97	2.14	11.0	0
995	**Sunflower seeds**	0	15	0	37.77	1.60	0.19	4.1	5.0	N	0	N	N	N	0
996	**Tahini paste**	0	6	0	2.57	0.94	0.17	5.1	4.1	0.76	0	99	2.17	11.0	0
997	**Walnuts**	0	0	0	3.83	0.40	0.14	1.2	2.8	0.67	0	66	1.60	19.0	0[a]
998	*weighed with shells*	0	0	0	1.64	0.17	0.06	0.5	1.2	0.29	0	28	0.69	8.0	0

[a] Value for ripe dried walnuts. Unripe dried walnuts contain 1300 to 3000mg vitamin C per 100g

Sugars, preserves and snacks

This section of the Tables contains evaluated and revised data for sugar, syrups, preserves, confectionery and savoury snacks. Several new types of sweets and savoury snacks have been added with data based on recent analytical studies.

Sugar, preserves and snacks

999 to 1013
Composition of food per 100g

Sugars, syrups and preserves

No.	Food	Description and main data sources	Edible proportion	Water g	Protein g	Fat g	Carbo-hydrate g	Energy value kcal	kJ
999	**Chocolate nut spread**	8 samples, 5 brands	1.00	Tr	6.2	33.0	60.5	549	2294
1000	**Glucose liquid**, BP	1 sample	1.00	20.4	Tr	0	84.7	318	1355
1001	**Honey**	2 samples	1.00	23.0	0.4	0	76.4	288	1229
1002	**Honeycomb**	2 samples, honey and comb together	1.00	20.2	0.6	4.6[a]	74.4	281	1201
1003	**Jaggery**	5 assorted samples	1.00	3.4	0.5	0	97.2	367	1564
1004	**Jam**, *fruit with edible seeds*	10 samples, 5 flavours	1.00	29.8	0.6	0	69.0	261	1114
1005	*stone fruit*	8 samples, 4 flavours	1.00	29.6	0.4	0	69.3	261	1116
1006	*reduced sugar*	9 samples, 5 brands; assorted flavours	1.00	65.3	0.5	0.1	31.9	123	523
1007	**Lemon curd**, *starch base*	10 jars, 4 brands	1.00	30.1	0.6	5.1	62.7	283	1202
1008	**Marmalade**	4 brands	1.00	28.0	0.1	0	69.5	261	1114
1009	**Mincemeat**	10 samples of the same brand	1.00	27.5	0.6	4.3	62.1	274	1163
1010	**Sugar**, demerara	5 samples	1.00	Tr	0.5	0	104.5[b]	394	1681
1011	white	Granulated and loaf sugar	1.00	Tr	Tr	0	105.0[c]	394	1680
1012	**Syrup**, golden	3 samples of the same brand	1.00	20.0	0.3	0	79.0	298	1269
1013	**Treacle**, black	3 samples	1.00	28.5	1.2	0	67.2	257	1096

[a] Waxy material, probably not available as fat; disregarded in calculating energy values
[b] 99.3g per 100g expressed as sucrose
[c] 99.9g per 100g expressed as sucrose

Sugars, preserves and snacks

Composition of food per 100g

No.	Food	Total nitrogen g	Fatty acids			Cholest- erol mg	Starch g	Total sugars g	Dietary fibre	
			Satd g	Mono unsatd g	Poly unsatd g				Southgate method g	Englyst method g

Sugars, syrups and preserves

No.	Food	Total nitrogen g	Satd g	Mono unsatd g	Poly unsatd g	Cholest- erol mg	Starch g	Total sugars g	Southgate method g	Englyst method g
999	**Chocolate nut spread**	0.99	10.1	16.8	4.6	2	0.8	59.7	1.2	0.8
1000	**Glucose liquid**, BP	Tr	0	0	0	0	44.5[a]	40.2	0	0
1001	**Honey**	0.06	0	0	0	0	0	76.4	0	0
1002	**Honeycomb**	0.09	0	0	0	0	0	74.4	0	0
1003	**Jaggery**	0.08	0	0	0	0	7.9	89.3	0	0
1004	**Jam**, *fruit with edible seeds*	0.10	0	0	0	0	0	69.0	1.0	N
1005	*stone fruit*	0.06	0	0	0	0	0	69.3	0.9	N
1006	*reduced sugar*	0.08	Tr	Tr	Tr	0	0	31.9	N	N
1007	**Lemon curd**, *starch base*	0.09	N	N	N	N	22.3	40.4	0.2	(0.2)
1008	**Marmalade**	0.01	0	0	0	0	0	69.5	0.6	(0.6)
1009	**Mincemeat**	0.10	2.4	1.6	0.1	4	Tr	62.1	3.0	1.3
1010	**Sugar**, *demerara*	0.08	0	0	0	0	0	104.5[b]	0	0
1011	*white*	Tr	0	0	0	0	0	105.0[c]	0	0
1012	**Syrup**, *golden*	0.05	0	0	0	0	0	79.0	0	0
1013	**Treacle**, *black*	0.19	0	0	0	0	0	67.2	0	0

[a] Including oligosaccharides

[b] 99.3g per 100g expressed as sucrose

[c] 99.9g per 100g expressed as sucrose

325

Sugar, preserves and snacks

Inorganic constituents per 100g

No.	Food	mg										µg	
		Na	K	Ca	Mg	P	Fe	Cu	Zn	Cl	Mn	Se	I
	Sugars, syrups and preserves												
999	**Chocolate nut spread**	50	390	130	65	180	2.2	0.48	1.0	60	1.1	N	N
1000	**Glucose liquid**, BP	150	3	8	2	11	0.5	0.09	N	190	Tr	Tr	Tr
1001	**Honey**	11	51	5	2	17	0.4	0.05	0.9	18	0.3	(1)	Tr
1002	**Honeycomb**	7	35	8	2	32	0.2	0.04	N	26	N	(1)	Tr
1003	**Jaggery**	79	285	92	117	72	1.6	0.75	0.1	250	0.5	Tr	Tr
1004	**Jam**, *fruit with edible seeds*	16	110	24	10	18	1.5	0.23	0.2	9	(0.1)	Tr	7
1005	*stone fruit*	12	100	12	5	18	1.0	0.12	0.2	4	(0.1)	Tr	7
1006	*reduced sugar*	17	120	19	7	15	0.4	0.05	Tr	Tr	0.1	Tr	N
1007	**Lemon curd**, *starch base*	65	11	9	2	15	0.5	(0.30)	1.3	150	N	N	N
1008	**Marmalade**	18	44	35	4	13	0.6	0.12	0.2	7	N	(1)	(7)
1009	**Mincemeat**	140	190	30	10	17	1.5	0.20	0.2	200	0.2	N	N
1010	**Sugar**, demerara	6	89	53	15	20	0.9	0.06	N	35	0.2	Tr	Tr
1011	white	Tr	2	2	Tr	Tr	Tr	0.02	0.2	Tr	Tr	Tr	Tr
1012	**Syrup**, golden	270	240	26	10	20	1.5	0.09	N	42	N	Tr	Tr
1013	**Treacle**, black	96	1470	500	140	31	9.2	0.43	0.9	820	N	N	Tr

Sugar, preserves and snacks

Vitamins per 100g

No.	Food	Retinol µg	Carotene µg	Vitamin D µg	Vitamin E mg	Thiamin mg	Ribo-flavin mg	Niacin mg	Trypt 60 mg	Vitamin B6 mg	Vitamin B12 µg	Folate µg	Panto-thenate mg	Biotin µg	Vitamin C mg
	Sugars, syrups and preserves														
999	**Chocolate nut spread**	Tr	Tr	Tr	N	0.03	0.10	0.5	1.5	0.10	Tr	N	N	N	Tr
1000	**Glucose liquid**, BP	0	0	0	0	0	0	0	0	0	0	0	0	0	0
1001	**Honey**	0	0	0	0	Tr	0.05	0.2	Tr	N	0	N	N	N	0
1002	**Honeycomb**	0	0	0	N	Tr	0.05	0.2	Tr	N	0	N	N	N	0
1003	**Jaggery**	0	0	0	0	Tr	0.04	Tr	Tr	Tr	0	Tr	Tr	Tr	0
1004	**Jam**, *fruit with edible seeds*	0	Tr	0	0	Tr	Tr	Tr	Tr	Tr	0	Tr	Tr	Tr	10[a]
1005	*stone fruit*	0	Tr	0	0	Tr	Tr	Tr	Tr	Tr	0	Tr	Tr	Tr	0
1006	*reduced sugar*	0	Tr	0	Tr	Tr	Tr	Tr	Tr	Tr	0	Tr	Tr	Tr	25
1007	**Lemon curd**, *starch base*	(10)	Tr	(0.1)	N	Tr	(0.02)	Tr	(0.1)	Tr	Tr	Tr	(0.10)	(1.0)	Tr
1008	**Marmalade**	0	50	0	Tr	Tr	Tr	Tr	Tr	Tr	0	5	Tr	Tr	10
1009	**Mincemeat**	0	9	Tr	N	0.04	0.02	0.4	0.1	(0.10)	Tr	8	0.03	Tr	Tr
1010	**Sugar**, demerara	0	0	0	0	Tr	Tr	Tr	Tr	Tr	0	Tr	Tr	Tr	0
1011	white	0	0	0	0	0	0	0	0	0	0	0	0	0	0
1012	**Syrup**, golden	0	0	0	0	Tr	Tr	Tr	Tr	Tr	0	Tr	Tr	Tr	0
1013	**Treacle**, black	0	0	0	0	Tr	Tr	Tr	Tr	Tr	0	Tr	Tr	Tr	0

[a] Blackcurrant jam, included in the sample, contains 24mg vitamin C per 100g

Sugar, preserves and snacks continued

Composition of food per 100g

No.	Food	Description and main data sources	Edible proportion	Water g	Protein g	Fat g	Carbo-hydrate g	Energy value kcal	Energy value kJ
Chocolate confectionery									
1014	**Bounty bar**	8 samples; Mars	1.00	7.6	4.8	26.1	58.3	473	1980
1015	**Chocolate**, milk	10 samples of the same brand	1.00	2.2	8.4	30.3	59.4	529	2214
1016	plain	10 samples of the same brand	1.00	0.6	4.7	29.2	64.8	525	2197
1017	white	14 samples; buttons ánd bars	1.00	0.6	8.0	30.9	58.3	529	2212
1018	**Chocolates**, fancy and filled	8 samples of different brands, mixed, milk and plain	1.00	5.7	4.1	18.8	73.3	460	1938
1019	**Creme eggs**	10 samples; Cadbury's	1.00	5.3	4.1	16.8	58.0	385	1619
1020	**Kit Kat**	12 samples; Rowntree Mackintosh	1.00	1.3	8.2	26.6	60.5	499	2092
1021	**Mars bar**	8 samples: Mars	1.00	6.9	5.3	18.9	66.5	441	1853
1022	**Milky Way**	10 samples; Mars	1.00	6.6	4.4	15.8	63.4	397	1674
1023	**Smartie-type sweets**	10 samples; Smarties and M and M's	1.00	1.5	5.4	17.5	73.9	456	1921
1024	**Twix**	10 samples; Mars	1.00	3.5	5.6	24.5	63.2	480	2013
Non chocolate confectionery									
1025	**Boiled sweets**	6 samples	1.00	N	Tr	Tr	87.3	327	1397
1026	**Fruit gums**	8 samples of the same brand	1.00	12.0	1.0	0	44.8	172	734
1027	**Liquorice allsorts**	6 samples	1.00	6.6	3.9	2.2	74.1	313	1333
1028	**Pastilles**	6 samples of different brands	1.00	10.2	5.2	0	61.9	253	1079
1029	**Peppermints**	Several samples of 6 different brands	1.00	0.2	0.5	0.7	102.2	392	1670
1030	**Popcorn**, candied	Recipe	1.00	2.6	2.1	20.0	77.6	480	2018
1031	plain	Recipe	1.00	0.9	6.2	42.8	48.6	592	2467
1032	**Toffees**, *mixed*	8 samples of different brands	1.00	4.8	2.1	17.2	71.1	430	1810
1033	**Turkish delight**, *without nuts*	7 assorted samples	1.00	16.1	0.6	0	77.9	295	1257

Sugars, preserves and snacks continued

No.	Food	Total nitrogen g	Fatty acids			Cholest-erol mg	Starch g	Total sugars g	Dietary fibre	
			Satd g	Mono unsatd g	Poly unsatd g				Southgate method g	Englyst method g
Chocolate confectionery										
1014	**Bounty bar**	0.77	21.2	3.2	0.4	10	4.6	53.7	N	N
1015	**Chocolate**, milk	1.35	17.8	9.5	1.5	30	2.9	56.5	Tr	Tr
1016	plain	0.75	16.9	9.3	1.2	9	5.3	59.5	N	N
1017	white	1.28	18.2	9.9	1.1	N	Tr	58.3	0	0
1018	**Chocolates**, fancy and filled	0.66	N	N	N	N	7.5	65.8	N	N
1019	**Creme eggs**	0.66	N	N	N	N	Tr	58.0	Tr	Tr
1020	**Kit Kat**	1.31	13.8	9.7	1.6	N	13.7	46.8	N	N
1021	**Mars bar**	0.84	10.0	7.2	0.8	25	0.7	65.8	Tr	Tr
1022	**Milky Way**	0.70	8.3	5.8	0.9	N	0.8	62.6	Tr	Tr
1023	**Smartie-type sweets**	0.86	10.3	5.6	0.6	(17)	3.1	70.8	Tr	Tr
1024	**Twix**	0.90	12.0	9.5	1.6	N	15.5	47.7	N	N
Non chocolate confectionery										
1025	**Boiled sweets**	Tr	0	0	0	0	0.4	86.9	0	0
1026	**Fruit gums**	0.16	0	0	0	0	2.2	42.6	0	0
1027	**Liquorice allsorts**	0.63	0.6	0.5	0.7	0	6.9	67.2	N	N
1028	**Pastilles**	0.84	0	0	0	0	0	61.9	0	0
1029	**Peppermints**	0.08	N	N	N	0	0	102.2	0	0
1030	**Popcorn**, candied	0.33	2.0	6.8	9.2	18	15.5	62.1	N	N
1031	plain	0.99	4.3	14.5	19.7	0	47.6	1.0	N	N
1032	**Toffees**, *mixed*	0.34	13.7	2.2	0.2	17	1.0	70.1	0	0
1033	**Turkish delight**, *without nuts*	0.10	0	0	0	0	9.3	68.6	0	0

Sugar, preserves and snacks *continued*

No.	Food	Na	K	Ca	Mg	P	Fe	Cu	Zn	Cl	Mn	Se	I
		mg										µg	
Chocolate confectionery													
1014	**Bounty bar**	180	320	110	43	140	1.3	0.47	N	400	N	N	N
1015	**Chocolate**, milk	120	420	220	55	240	1.6	0.30	0.2	270	0.5	(4)	N
1016	plain	11	300	38	100	140	2.4	0.70	0.2	100	0.5	(2)	N
1017	white	110	350	270	26	230	0.2	Tr	0.9	240	Tr	N	N
1018	**Chocolates**, fancy and filled	60	240	92	51	120	1.8	0.45	N	180	(0.5)	(2)	120
1019	**Creme eggs**	55	210	120	27	130	0.8	0.10	0.6	110	0.1	Tr	N
1020	**Kit Kat**	110	330	200	52	190	1.5	0.28	1.1	210	0.3	N	N
1021	**Mars bar**	150	250	160	35	150	1.1	0.31	N	300	N	(2)	N
1022	**Milky Way**	100	230	120	42	130	1.7	0.13	0.8	160	0.3	N	N
1023	**Smartie-type sweets**	58	270	150	48	160	1.5	0.25	0.9	110	0.3	N	N
1024	**Twix**	190	190	110	28	130	1.1	0.08	0.7	250	0.2	N	N
Non chocolate confectionery													
1025	**Boiled sweets**	25	8	5	2	12	0.4	0.09	N	68	N	Tr	N
1026	**Fruit gums**	64	360	360	110	4	4.2	1.43	N	160	N	Tr	N
1027	**Liquorice allsorts**	75	220	63	38	29	8.1	0.39	N	120	N	N	N
1028	**Pastilles**	77	40	40	12	Tr	1.4	0.32	N	120	N	Tr	N
1029	**Peppermints**	9	Tr	7	3	Tr	0.2	0.04	N	22	N	Tr	Tr
1030	**Popcorn**, candied	56	75	6	26	58	0.4	N	0.7	101	0.1	N	3
1031	plain	4	220	10	81	170	1.1	N	1.7	8	0.3	N	2
1032	**Toffees**, mixed	320	210	95	25	64	1.5	0.40	N	480	N	Tr	N
1033	**Turkish delight**, *without nuts*	31	4	10	2	7	0.2	0.12	0.7	110	Tr	Tr	Tr

No.	Food	Retinol µg	Carotene µg	Vitamin D µg	Vitamin E mg	Thiamin mg	Ribo-flavin mg	Niacin mg	Trypt 60 mg	Vitamin B6 mg	Vitamin B12 µg	Folate µg	Panto-thenate mg	Biotin µg	Vitamin C mg
	Chocolate confectionery														
1014	**Bounty bar**	Tr	(40)	Tr	N	(0.04)	(0.10)	(0.3)	0.8	(0.03)	Tr	N	(0.59)	(2)	0
1015	**Chocolate**, milk	Tr	(40)	Tr	0.74	0.10	0.23	0.2	1.4	(0.07)	Tr	(10)	(1.08)	(3)	0
1016	plain	0	(40)	0	0.85	0.07	0.08	0.4	0.8	(0.07)	0	(10)	(1.08)	(3)	0
1017	white	Tr	75	Tr	1.14	0.08	0.49	0.2	2.6	0.07	Tr	(10)	(0.59)	(2)	0
1018	**Chocolates**, fancy and filled	Tr	(40)	Tr	N	(0.08)	(0.35)	(0.3)	0.7	(0.05)	Tr	(10)	(0.73)	(3)	0
1019	**Creme eggs**	47	(55)	0.6	1.07	0.06	0.34	0.2	1.3	0.03	1.4	12	(0.59)	(3)	0
1020	**Kit Kat**	Tr	47	Tr	1.03	0.11	0.44	0.5	2.6	0.06	Tr	N	0.70	4	0
1021	**Mars bar**	Tr	(40)	Tr	N	(0.05)	(0.20)	(0.2)	0.9	(0.03)	Tr	N	(0.59)	(2)	0
1022	**Milky Way**	Tr	Tr	Tr	1.91	0.05	0.20	0.2	1.1	0.03	Tr	(10)	0.59	2	0
1023	**Smartie-type sweets**	Tr	28	Tr	0.80	0.08	0.79	0.3	1.7	0.03	Tr	(10)	0.67	2	0
1024	**Twix**	Tr	7	Tr	3.72	0.06	0.22	0.6	0.9	0.05	Tr	N	0.61	3	0
	Non chocolate confectionery														
1025	**Boiled sweets**	0	0	0	0	0	0	0	0	0	0	0	0	0	0
1026	**Fruit gums**	0	0	0	0	0	0	0	0	0	0	0	0	0	0
1027	**Liquorice allsorts**	0	0	0	0	0	0	0	0.7	0	0	0	0	0	0
1028	**Pastilles**	0	0	0	0	0	0	0	0	0	0	0	0	0	0
1029	**Peppermints**	0	0	0	0	0	0	0	0	0	0	0	0	0	0
1030	**Popcorn**, candied	52	98	0	3.75	0.06	0.04	0.3	0.2	0.07	0	3	0.1	1	0
1031	plain	0	230	0	11.03	0.18	0.11	1.0	0.7	0.20	0	9	0.3	4	0
1032	**Toffees**, *mixed*	0	N	0	N	0	0	0	0.4	0	0	0	0	0	0
1033	**Turkish delight**, *without nuts*	0	0	0	0	0.13	N	N	N	N	N	N	N	N	0

Sugar, preserves and snacks *continued*

1034 to 1042

Composition of food per 100g

No.	Food	Description and main data sources	Edible proportion	Water g	Protein g	Fat g	Carbo-hydrate g	Energy value kcal	kJ
	Savoury snacks								
1034	Bombay Mix	20 samples; savoury mix of gram flour, assorted peas, lentils, nuts and seeds	1.00	3.5	18.8	32.9	35.1	503	2099
1035	Corn snacks	20 samples, assorted types e.g. Wotsits, Monster Munch, Nik-Naks	1.00	3.3	7.0	31.9	54.3	519	2168
1036	Peanuts and raisins[a]	Calculated from recipe proportions	1.00	9.3	15.3	26.0	37.5	435	1820
1037	Potato crisps	20 samples, mixed plain and flavoured	1.00	1.9	5.6	37.6	49.3	546	2275
1038	low fat	20 samples, mixed plain and flavoured	1.00	1.1	6.6	21.5	63.0	456	1916
1039	Potato hoops	18 samples, assorted flavours; Hula Hoop type	1.00	2.8	3.9	32.0	58.5	523	2186
1040	Tortilla chips	20 samples, 6 brands, maize chips	1.00	0.9	7.6	22.6	60.1	459	1927
1041	Trail Mix	10 samples, mix of nuts and dried fruit	1.00	8.9	9.1	28.5	37.2	432	1804
1042	Twiglets	20 samples, savoury wholewheat sticks	1.00	3.2	11.3	11.7	62.0	383	1617

[a] Calculated as peanuts 56% and raisins 44%

Sugars, preserves and snacks *continued*

Composition of food per 100g

No.	Food	Total nitrogen g	Fatty acids Satd g	Fatty acids Mono unsatd g	Fatty acids Poly unsatd g	Cholest- erol mg	Starch g	Total sugars g	Dietary fibre Southgate method g	Dietary fibre Englyst method g
Savoury snacks										
1034	**Bombay Mix**	3.01	4.0	16.2	11.3	0	32.8	2.3	N	6.2
1035	**Corn snacks**	1.12	11.8	12.9	5.8	0	49.7	4.6	N	1.0
1036	**Peanuts and raisins**	2.80	4.6	11.8	8.0	0	3.5	34.0	6.8	4.4
1037	**Potato crisps**	0.89	9.2	12.0	9.9	0	48.6	0.7	10.7	4.9
1038	low fat	1.06	6.2	8.0	6.6	0	62.0	1.0	13.7	6.3
1039	**Potato hoops**	0.62	N	N	N	0	58.0	0.5	N	2.6
1040	**Tortilla chips**	1.22	4.0	10.6	6.7	0	58.9	1.2	N	4.9
1041	**Trail Mix**	1.45	N	N	N	0	0.1	37.1	N	4.3
1042	**Twiglets**	1.98	4.9	4.4	1.8	0	60.9	1.1	N	10.3

Sugar, preserves and snacks continued

Inorganic constituents per 100g

No.	Food	Na	K	Ca	Mg	P	Fe	Cu	Zn	Cl	Mn	Se	I
						mg						µg	
Savoury snacks													
1034	**Bombay Mix**	770	770	58	100	290	3.8	0.62	2.5	1410	1.4	N	N
1035	**Corn snacks**	1130	200	68	18	130	0.8	0.04	0.5	1840	0.1	N	N
1036	**Peanuts and raisins**	27	820	53	130	270	2.1	0.74	2.3	7	1.3	2	11
1037	**Potato crisps**	1070[a]	1060	37	45	120	1.8	0.13	0.7	1460	0.3	N	N
1038	low fat	(1070)	(1060)	36	48	130	1.8	0.38	0.9	(1460)	0.4	N	N
1039	**Potato hoops**	1070	540	22	28	100	1.0	0.16	0.7	(1650)	0.2	N	N
1040	**Tortilla chips**	850	220	150	89	230	1.6	0.09	1.2	1400	0.4	N	N
1041	**Trail Mix**	27	620	69	110	210	3.7	0.55	1.5	N	1.6	N	N
1042	**Twiglets**	1330	460	45	81	370	2.9	0.32	2.0	2520	1.6	N	N

[a] The sodium content of samples ranged from 600 to 1500mg per 100g

Sugar, preserves and snacks *continued*

Savoury snacks

No.	Food	Retinol µg	Carotene µg	Vitamin D µg	Vitamin E mg	Thiamin mg	Ribo-flavin mg	Niacin mg	Trypt 60 mg	Vitamin B6 mg	Vitamin B12 µg	Folate µg	Panto-thenate mg	Biotin µg	Vitamin C mg
1034	**Bombay Mix**	0	Tr	0	4.71	0.38	0.10	4.3	3.5	0.54	0	N	1.19	24	Tr
1035	**Corn snacks**	0	460	0	5.85	0.19	0.16	0.9	0.7	0.13	0	N	N	N	Tr
1036	**Peanuts and raisins**	0	5	0	5.65	0.69	0.08	8.0	3.1	0.44	0	65	1.56	41	Tr
1037	**Potato crisps**	0	2	0	3.10	0.11	0.07	4.6	1.3	0.32	0	41	N	N	27
1038	low fat	0	(2)	0	(1.77)	0.19	0.14	5.0	1.6	0.53	0	(41)	N	N	14
1039	**Potato hoops**	0	Tr	0	N	N	N	N	0.1	N	0	N	N	N	3
1040	**Tortilla chips**	0	455	0	1.94	0.17	0.09	1.8	0.8	0.31	0	N	N	N	Tr
1041	**Trail Mix**	0	47	0	4.53	0.23	0.09	2.0	1.5	N	0	25	N	N	Tr
1042	**Twiglets**	0	Tr	0	2.47	0.37	0.48	7.8	2.3	0.38	0	N	1.54	15	Tr

Section 2.12

This section of the Tables has been extended by including data from the *'Milk Products and Eggs'* supplement and additional data for an increased number of fruit juices.

This section includes beverages which are made up and drunk with milk as well as carbonated drinks, squash, cordials and fruit juices. Most beverages which are commonly drunk with milk have been given made up with either whole or semi-skimmed milk, although Build-Up and Complan are also included made up with skimmed milk. As it is difficult to cover the range of strengths in which instant coffee, squash and cordials are consumed, only one entry, for the undiluted form, is given.

The vitamin composition of beverages may be different from that quoted in these Tables if manufacturers have added to or changed the fortification of products; where these have recently changed in specification the data have been updated in this edition. Concentrations of vitamin C in many fruit-based drinks can vary widely depending on fortification practices. The user should check the label of any beverage of this type to establish its vitamin C composition.

Losses of labile vitamins assigned to made up drinks have been estimated from figures in Section 3.4.

As many beverages may be sold or measured by volume, typical specific gravities (densities) of some of these products are given below.

Specific gravities of selected beverages

Coca-cola	1.039
Lemonade, bottled	1.015
Lime juice cordial undiluted	1.102
Lucozade	1.074
Orange drink undiluted	1.116
Ribena undiluted	1.283

337

Beverages

Composition of food per 100g

No.	Food	Description and main data sources	Edible proportion	Water g	Protein g	Fat g	Carbo-hydrate g	Energy value kcal	Energy value kJ
Powdered drinks and essences									
1043	**Bournvita powder**	6 samples	1.00	1.5	7.7	1.5	79.0	341	1450
1044	*made up with whole milk*	Calculated from 8g powder to 200ml milk	1.00	84.6	3.4	3.8	7.6	76	320
1045	*made up with semi-skimmed milk*	Calculated from 8g powder to 200ml milk	1.00	86.5	3.5	1.6	7.8	58	243
1046	**Build-up powder**	Manufacturer's data (Nestlé)	1.00	3.0	24.5	0.9	65.0[a]	350	1490
1047	*made up with whole milk*	Made up according to packet directions	1.00	78.1	5.6	3.6	11.7	98	414
1048	*made up with semi-skimmed milk*	Made up according to packet directions	1.00	79.8	5.7	1.5	11.9	80	343
1049	*made up with skimmed milk*	Made up according to packet directions	1.00	81.0	5.7	0.2	11.9	69	294
1050	**Cocoa powder**	10 samples, 2 brands	1.00	3.4	18.5[b]	21.7	11.5	312	1301
1051	*made up with whole milk*	Calculated from 4g powder, 4g sugar and 200ml milk	1.00	84.6	3.4	4.2	6.8	76	320
1052	*made up with semi-skimmed milk*	Calculated from 4g powder, 4g sugar and 200ml milk	1.00	86.5	3.5	1.9	7.0	57	243
1053	**Coffee**, *infusion, 5 minutes*	60g ground coffee from mixed sample; boiled in percolator with 900ml water and strained	1.00	N	0.2	Tr	0.3	2	8
1054	*instant*	10 jars, 2 brands	1.00	3.4	14.6[b]	0	11.0	100	424
1055	**Coffee and chicory essence**	7 bottles of the same brand	1.00	36.9	1.6[b]	0.2	56.0	218	931
1056	**Coffeemate**	Analysis and manufacturer's data (Nestlé)	1.00	3.0	2.7	34.9	57.3[a]	540	2254

[a] Including oligosaccharides from the glucose syrup/maltodextrins in the product

[b] (Total N – purine N) x 6.25

338

Beverages

Powdered drinks and essences

No.	Food	Total nitrogen g	Fatty acids Satd g	Mono unsatd g	Poly unsatd g	Cholest-erol mg	Starch g	Total sugars g	Dietary fibre Southgate method g	Englyst method g
1043	**Bournvita powder**	1.23	N	N	N	N	27.0	52.0	N	N
1044	*made up with whole milk*	0.53	N	N	N	N	1.0	6.6	Tr	Tr
1045	*made up with semi-skimmed milk*	0.55	N	N	N	N	1.0	6.8	Tr	Tr
1046	**Build-up powder**	3.84	0.6	0.3	Tr	12	Tr	54.7[a]	Tr	Tr
1047	*made up with whole milk*	0.88	2.2	1.0	0.1	13	Tr	10.5	Tr	Tr
1048	*made up with semi-skimmed milk*	0.90	0.9	0.4	Tr	7	Tr	10.7	Tr	Tr
1049	*made up with skimmed milk*	0.90	0.1	0.1	Tr	3	Tr	10.7	Tr	Tr
1050	**Cocoa powder**	3.70[b]	12.8	7.2	0.6	0	11.5	Tr	N	12.1
1051	*made up with whole milk*	0.55	2.6	1.2	0.1	13	0.2	6.6	(0.2)	0.2
1052	*made up with semi-skimmed milk*	0.57	1.2	0.6	0.1	6	0.2	6.8	(0.2)	0.2
1053	**Coffee**, *infusion, 5 minutes*	1.28	Tr	Tr	Tr	0	0	0.3	0	0
1054	*instant*	3.26[c]	Tr	Tr	Tr	0	4.5	6.5	0	0
1055	**Coffee and chicory essence**	0.33[d]	Tr	Tr	Tr	0	2.2	53.8	0	0
1056	**Coffeemate**	0.42	32.1	1.1	Tr	2	Tr	9.8[a]	0	0

[a] Not including oligosaccharides from the glucose syrup/maltodextrins in the product
[c] Includes 0.93g purine nitrogen

[b] Includes 0.74g purine nitrogen
[d] Includes 0.08g purine nitrogen

Beverages

No.	Food	mg										µg	
		Na	K	Ca	Mg	P	Fe	Cu	Zn	Cl	Mn	Se	I
	Powdered drinks and essences												
1043	**Bournvita powder**	190	330	62	110	250	1.9	0.64	1.4	70	N	N	N
1044	*made up with whole milk*	59	150	110	14	98	0.1	0.02	0.4	99	N	N	N
1045	*made up with semi-skimmed milk*	59	160	120	14	100	0.1	0.02	0.4	99	N	N	N
1046	**Build-up powder**	380	1100	800	190	650	11.0	4.90	7.6	700	2.6	N	150
1047	*made up with whole milk*	92	250	190	32	160	1.3	0.56	1.1	170	0.3	N	31
1048	*made up with semi-skimmed milk*	92	260	200	32	160	1.3	0.56	1.2	170	0.3	N	31
1049	*made up with skimmed milk*	92	260	200	32	160	1.3	0.56	1.2	170	0.3	N	31
1050	**Cocoa powder**	950	1500	130	520	660	10.5	3.90	6.9	460	N	N	N
1051	*made up with whole milk*	70	160	110	20	100	0.2	0.07	0.5	100	N	N	N
1052	*made up with semi-skimmed milk*	70	170	120	20	100	0.2	0.07	0.5	100	N	N	N
1053	**Coffee**, *infusion, 5 minutes*	Tr	66	2	6	2	Tr	Tr	N	Tr	Tr	Tr	Tr
1054	*instant*	41	4000	160	390	350	4.40	0.05	0.50	50	Tr	Tr	Tr
1055	**Coffee and chicory essence**	65	750	30	39	90	0.70	0.60	N	85	N	Tr	N
1056	**Coffeemate**	200	900	4	N	350	N	N	N	N	N	N	N

No.	Food	Retinol µg	Carotene µg	Vitamin D µg	Vitamin E mg	Thiamin mg	Ribo- flavin mg	Niacin mg	Trypt 60 mg	Vitamin B6 mg	Vitamin B12 µg	Folate µg	Panto- thenate mg	Biotin µg	Vitamin C mg
	Powdered drinks and essences														
1043	**Bournvita powder**	Tr	Tr	Tr	Tr	N	N	N	1.7	N	Tr	N	N	N	0
1044	*made up with whole milk*	50	20	0.03	0.07	N	N	N	0.80	N	0.4	N	N	N	Tr
1045	*made up with semi-skimmed milk*	20	8	0.01	0.02	N	N	N	0.80	N	0.4	N	N	N	1
1046	**Build-up powder**	660	Tr	10.0	10.00	1.50	1.50	20.00	5.80	3.00	3.0	400	13.00	55	60
1047	*made up with whole milk*	120	18	1.17	1.23	0.21	0.32	2.40	1.30	0.40	0.7	51	1.80	8	8
1048	*made up with semi-skimmed milk*	94	7	1.16	1.17	0.21	0.33	2.40	1.30	0.40	0.7	51	1.78	8	8
1049	*made up with skimmed milk*	76	Tr	1.15	1.15	0.21	0.33	2.40	1.30	0.40	0.7	51	1.78	8	8
1050	**Cocoa powder**	0	(40)	0	0.68	0.16	0.06	1.70	3.90	0.07	0	38	N	N	0
1051	*made up with whole milk*	50	(20)	0.03	0.08	0.04	0.15	0.10	0.80	0.05	0.4	5	0.30	2	Tr
1052	*made up with semi-skimmed milk*	20	(9)	0.01	0.04	0.04	0.16	0.10	0.80	0.05	0.4	5	0.28	2	1
1053	**Coffee**, *infusion, 5 minutes*	0	N	0	N	Tr	0.01	0.70	0	Tr	0	Tr	Tr	Tr	0
1054	*instant*	0	N	0	N	0	0.11	22.00[a]	2.90	0.03	0	Tr	0.40	Tr	0
1055	**Coffee and chicory essence**	0	N	0	N	0	0.03	2.80	N	N	0	N	N	N	0
1056	**Coffeemate**	0	200	0	N	0	1.00	0	0.6	0	0	0	0	0	0

[a] Can be as high as 39mg per 100g. Decaffinated instant coffee contains about the same amount

Beverages *continued*

Powdered drinks and essences

No.	Food	Description and main data sources	Edible proportion	Water g	Protein g	Fat g	Carbo-hydrate g	Energy value kcal	kJ
1057	**Complan powder**, savoury	Chicken flavour, manufacturer's data (Farley's)	1.00	3.8	22.0	16.0	55.0[a]	438	1846
1058	savoury, *made up with water*	Made up according to packet directions	1.00	78.7	4.9	3.5	12.2	97	409
1059	sweet	3 flavours, manufacturer's data (Farley's)	1.00	3.5	20.0	14.0	59.7[a]	430	1813
1060	sweet, *made up with water*	Made up according to packet directions	1.00	78.3	4.5	3.1	13.4	96	407
1061	sweet, *made up with whole milk*	Made up according to packet directions	1.00	69.3	6.9	6.1	16.9	145	612
1062	sweet, *made up with semi-skimmed milk*	Made up according to packet directions	1.00	70.8	7.0	4.3	17.0	130	550
1063	sweet, *made up with skimmed milk*	Made up according to packet directions	1.00	71.9	7.0	3.1	17.0	120	507
1064	**Drinking chocolate powder**	10 tins, 3 brands	1.00	2.1	5.5[b]	6.0	77.4	366	1554
1065	*made up with whole milk*	Calculated from 18g powder to 200ml milk	1.00	80.9	3.4	4.1	10.6	90	377
1066	*made up with semi-skimmed milk*	Calculated from 18g powder to 200ml milk	1.00	82.8	3.5	1.9	10.8	71	304
1067	**Horlicks LowFat Instant powder**	Manufacturer's data (SmithKline Beechams)	1.00	N	17.4	3.3	72.9	373	1584
1068	*made up with water*	Calculated from 32g powder to 200ml water	1.00	(86.2)	2.4	0.5	10.1	51	218
1069	**Horlicks powder**	Manufacturer's data (SmithKline Beechams)	1.00	2.5	12.4	4.0	78.0	378	1607
1070	*made up with whole milk*	Calculated from 25g powder to 200ml milk	1.00	78.6	4.2	3.9	12.7	99	419
1071	*made up with semi-skimmed milk*	Calculated from 25g powder to 200ml milk	1.00	80.4	4.3	1.9	12.9	81	347

[a] Including oligosaccharides from the glucose syrup/maltodextrins in the product

[b] (Total N − purine N) x 6.25

Beverages *continued*

1057 to 1071
Composition of food per 100g

Powdered drinks and essences

No.	Food	Total nitrogen g	Fatty acids			Cholesterol mg	Starch g	Total sugars g	Dietary fibre	
			Satd g	Mono unsatd g	Poly unsatd g				Southgate method g	Englyst method g
1057	**Complan powder**, savoury	3.45	7.5	6.0	1.8	N	8.8	13.7[a]	0	0
1058	savoury, *made up with water*	0.77	1.7	1.3	0.4	N	2.0	3.0	0	0
1059	sweet	3.13	5.9[b]	6.0[b]	1.5[b]	N	Tr	46.6[a]	Tr	Tr
1060	sweet, *made up with water*	0.70	1.3	1.3	0.3	N	Tr	10.5	Tr	Tr
1061	sweet, *made up with whole milk*	1.08	3.2	2.2	0.4	N	Tr	14.0	Tr	Tr
1062	sweet, *made up with semi-skimmed milk*	1.09	2.1	1.7	0.4	N	Tr	14.1	Tr	Tr
1063	sweet, *made up with skimmed milk*	1.09	1.3	1.3	0.3	N	Tr	14.1	Tr	Tr
1064	**Drinking chocolate powder**	1.04[c]	3.5	2.0	0.2	0	3.6	73.8	N	N
1065	*made up with whole milk*	0.54	2.5	1.2	0.1	12	0.3	10.3	Tr	Tr
1066	*made up with semi-skimmed milk*	0.56	1.2	0.6	0.1	6	0.3	10.5	Tr	Tr
1067	**Horlicks LowFat Instant powder**	2.78	N	N	N	Tr	N	N	N	N
1068	*made up with water*	0.38	N	N	N	Tr	N	N	N	N
1069	**Horlicks powder**	1.98	N	N	N	N	25.0	53.0	N	N
1070	*made up with whole milk*	0.66	N	N	N	N	2.7	10.0	Tr	Tr
1071	*made up with semi-skimmed milk*	0.68	N	N	N	N	2.7	10.2	Tr	Tr

[a] Not including oligosaccharides from the glucose syrup/maltodextrins in the product
[b] Chocolate flavour has a slightly different fatty acid composition
[c] Includes 0.16g purine nitrogen

No.	Food	Na	K	Ca	Mg	P	Fe	Cu	Zn	Cl	Mn	Se	I
						mg						µg	
												Se	I

Powdered drinks and essences

No.	Food	Na	K	Ca	Mg	P	Fe	Cu	Zn	Cl	Mn	Se	I
1057	**Complan powder**, savoury	1800	650	310	5	360	6.5	0.54	6.5	1700	0.9	N	61
1058	*savoury, made up with water*	400	140	68	1	79	1.4	0.12	1.4	380	0.2	N	13
1059	sweet	290	950	710	76	590	6.5	0.54	6.5	640	0.9	N	61
1060	*sweet, made up with water*	65	210	160	17	130	1.5	0.12	1.5	140	0.2	N	14
1061	*sweet, made up with whole milk*	110	320	250	25	200	1.5	0.12	1.7	220	0.2	N	25
1062	*sweet, made up with semi-skimmed milk*	110	330	250	25	200	1.5	0.12	1.7	220	0.2	N	25
1063	*sweet, made up with skimmed milk*	110	330	250	26	200	1.5	0.12	1.7	220	0.2	N	25
1064	**Drinking chocolate powder**	250	410	33	150	190	2.4	1.10	1.9	130	N	N	N
1065	*made up with whole milk*	70	160	110	22	99	0.2	0.09	0.5	100	N	N	N
1066	*made up with semi-skimmed milk*	70	170	110	22	100	0.2	0.09	0.5	100	N	N	N
1067	**Horlicks LowFat Instant powder**	590	860	580	N	N	N	N	N	N	N	N	N
1068	*made up with water*	81	120	79	N	N	N	N	N	N	N	N	N
1069	**Horlicks powder**	460	670	430	50	300	1.4	0.30	1.3	N	N	N	N
1070	*made up with whole milk*	98	200	150	15	110	0.2	0.03	0.5	N	N	N	N
1071	*made up with semi-skimmed milk*	98	210	150	15	120	0.2	0.03	0.5	N	N	N	N

Powdered drinks and essences

No.	Food	Retinol µg	Carotene µg	Vitamin D µg	Vitamin E mg	Thiamin mg	Ribo-flavin mg	Niacin mg	Trypt/60 mg	Vitamin B6 mg	Vitamin B12 µg	Folate µg	Panto-thenate mg	Biotin µg	Vitamin C mg
1057	**Complan powder**, savoury	430	Tr	2.2	5.20	0.77	0.87	8.70	5.20	0.95	2.2	150	3.00	33	21
1058	savoury, *made up with water*	95	Tr	0.5	1.15	0.17	0.18	1.90	1.10	0.21	0.5	33	0.67	7	5
1059	sweet	430	Tr	2.2	5.20	0.77	0.87	8.70	4.70	0.95	2.2	170	3.00	33	43
1060	sweet, *made up with water*	96	Tr	0.5	1.17	0.18	0.20	1.90	1.10	0.21	0.5	38	0.67	7	10
1061	sweet, *made up with whole milk*	135	16	0.51	1.21	0.20	0.32	2.00	1.60	0.26	0.8	42	0.93	9	10
1062	sweet, *made up with semi-skimmed milk*	110	7	0.49	1.17	0.20	0.33	2.00	1.60	0.26	0.8	42	0.91	9	11
1063	sweet, *made up with skimmed milk*	95	Tr	0.48	1.14	0.20	0.33	2.00	1.60	0.26	0.8	42	0.91	9	11
1064	**Drinking chocolate powder**	0	N	0	0.18	0.06	0.04	0.50	1.20	0.02	0	10	N	N	0
1065	*made up with whole milk*	47	N	0.03	0.08	0.04	0.14	0.10	0.80	0.05	0.3	5	N	N	Tr
1066	*made up with semi-skimmed milk*	19	N	0.01	0.04	0.04	0.15	0.10	0.80	0.05	0.3	5	N	N	1
1067	**Horlicks LowFat Instant powder**	470	Tr	1.6	N	0.75	1.00	11.30	4.20	N	1.3	190	N	N	19
1068	*made up with water*	64	Tr	0.2	N	0.10	0.14	1.50	0.60	N	0.2	25	N	N	3
1069	**Horlicks powder**	600	0	2.0	N	0.96	1.28	14.40	3.00	N	N	N	N	N	0
1070	*made up with whole milk*	115	18	0.25	N	0.14	0.29	1.70	1.00	N	N	N	N	N	Tr
1071	*made up with semi-skimmed milk*	85	8	0.23	N	0.14	0.30	1.70	1.00	N	N	N	N	N	1

Powdered drinks and essences

No.	Food	Description and main data sources	Edible proportion	Water g	Protein g	Fat g	Carbo-hydrate g	Energy value kcal	kJ
1072	**Milk shake**, *purchased*	21 samples, thick take-away type	1.00	80.0	2.9	3.2	13.2	90	379
1073	**Milk shake powder**	6 samples (Nesquik), 3 flavours	1.00	0.5	1.3	1.6	98.3	388	1654
1074	*made up with whole milk*	Calculated from 15g powder to 200ml milk	1.00	81.9	3.1	3.7	11.1	87	368
1075	*made up with semi-skimmed milk*	Calculated from 15g powder to 200ml milk	1.00	83.7	3.2	1.6	11.3	69	294
1076	**Ovaltine powder**	Manufacturer's data (Wander)	1.00	2.0	9.0	2.7	79.4	358	1523
1077	*made up with whole milk*	Calculated from 25g powder to 200ml milk	1.00	78.5	3.8	3.8	12.9	97	410
1078	*made up with semi-skimmed milk*	Calculated from 25g powder to 200ml milk	1.00	80.3	3.9	1.7	13.0	79	338
1079	**Tea**, Indian, *infusion*	10g leaves from mixed sample; infused with 1000ml boiling water 2-10 minutes and strained	1.00	N	0.1	Tr	Tr	Tr	2

No.	Food	Total nitrogen g	Fatty acids			Cholest- erol mg	Starch g	Total sugars g	Dietary fibre	
			Satd g	Mono unsatd g	Poly unsatd g				Southgate method g	Englyst method g
	Powdered drinks and essences									
1072	**Milk shake**, *purchased*	0.46	(2.0)	(0.9)	(0.1)	10	Tr	13.2	Tr	Tr
1073	**Milk shake powder**	0.21	N	N	N	Tr	Tr	98.3	Tr	Tr
1074	*made up with whole milk*	0.48	N	N	N	13	Tr	11.1	Tr	Tr
1075	*made up with semi-skimmed milk*	0.50	N	N	N	6	Tr	11.3	Tr	Tr
1076	**Ovaltine powder**	1.41	N	N	N	N	N	N	N	N
1077	*made up with whole milk*	0.60	N	N	N	N	N	N	Tr	Tr
1078	*made up with semi-skimmed milk*	0.62	N	N	N	N	N	N	Tr	Tr
1079	**Tea**, Indian, *infusion*	Tr	Tr	Tr	Tr	0	0	Tr	0	0

No.	Food	Na	K	Ca	Mg	P	Fe	Cu	Zn	Cl	Mn	Se	I
						mg						μg	
												Se	I
	Powdered drinks and essences												
1072	**Milk shake,** *purchased*	55	130	100	9	86	1.0	0.01	0.4	91	Tr	N	N
1073	**Milk shake powder**	20	150	8	27	53	2.0	0.10	0.4	27	0.2	N	32
1074	*made up with whole milk*	52	140	110	12	89	0.2	0.01	0.4	95	Tr	N	16
1075	*made up with semi-skimmed milk*	52	150	110	12	92	0.2	0.01	0.4	95	Tr	N	16
1076	**Ovaltine powder**	160	640	83	96	430	1.9	1.00	1.2	210	N	N	N
1077	*made up with whole milk*	66	190	110	20	130	0.3	0.11	0.5	110	N	N	N
1078	*made up with semi-skimmed milk*	66	200	110	20	130	0.3	0.11	0.5	110	N	N	N
1079	**Tea,** *Indian, infusion*	Tr	17	Tr	1	1	Tr	Tr	Tr	Tr	0.14	Tr	Tr

Powdered drinks and essences

No.	Food	Retinol µg	Carotene µg	Vitamin D µg	Vitamin E mg	Thiamin mg	Ribo- flavin mg	Niacin mg	Trypt 60 mg	Vitamin B6 mg	Vitamin B12 µg	Folate µg	Panto- thenate mg	Biotin µg	Vitamin C mg
1072	**Milk shake, purchased**	21	51	Tr	0.05	0.03	0.18	0.10	0.70	0.05	0.3	4	0.31	2	1
1073	**Milk shake powder**	Tr	Tr	0	0.15	Tr	0.02	0.20	0.30	0.01	0	3	N	N	0
1074	*made up with whole milk*	48	19	0.03	0.09	0.04	0.16	0.10	0.70	0.06	0.4	5	N	N	1
1075	*made up with semi-skimmed milk*	19	8	0.01	0.04	0.04	0.17	0.10	0.70	0.06	0.4	5	N	N	1
1076	**Ovaltine powder**	625	Tr	2.10	N	1.00	1.30	15.00	2.10	N	1.7	N	N	N	0
1077	*made up with whole milk*	115	18	0.25	N	0.14	0.28	1.70	0.90	N	0.5	N	N	N	Tr
1078	*made up with semi-skimmed milk*	86	8	0.24	N	0.14	0.29	1.70	0.90	N	0.5	N	N	N	1
1079	**Tea**, Indian, *infusion*	0	0	0	N	Tr	0.01	0.10	0	Tr	0	Tr	Tr	Tr	0

Beverages *continued*

1080 to 1093
Composition of food per 100g

No.	Food	Description and main data sources	Edible proportion	Water g	Protein g	Fat g	Carbohydrate g	Energy value kcal	Energy value kJ
	Carbonated drinks								
1080	Coca-cola	8 cans and 5 bottles	1.00	89.8	Tr	0	10.5	39	168
1081	Lemonade, *bottled*	7 bottles of the same brand	1.00	94.6	Tr	0	5.6	21	90
1082	Lucozade	Mixed sample and manufacturer's data (SmithKline Beechams)	1.00	81.7	Tr	0	18.0[a]	67	288
	Squash and cordials								
1083	Lime juice cordial, *undiluted*	6 bottles of the same brand	1.00	70.5	0.1	0	29.8	112	479
1084	Orange drink, *undiluted*	Mixed sample; orange squash	1.00	71.2	Tr	0	28.5	107	456
1085	Ribena, *undiluted*	Mixed sample and manufacturer's data (SmithKline Beechams); blackcurrant juice drink	1.00	40.3	0.1	0	60.8[a]	228	975
1086	Rosehip syrup, *undiluted*	9 bottles, 4 brands	1.00	32.5	Tr	0	61.9	232	990
	Juices								
1087	Apple juice, *unsweetened*	10 samples; bottles and cartons	1.00	88.0	0.1	0.1	9.9	38	164
1088	Grape juice, *unsweetened*	10 samples, 6 brands; red and white juice	1.00	85.4	0.3	0.1	11.7	46	196
1089	Grapefruit juice, *unsweetened*	50 samples; fresh, canned, bottled and frozen	1.00	89.4	0.4	0.1	8.3	33	140
1090	Lemon juice	Analysis and literature sources; juice from fresh lemon	1.00	91.4	0.3	Tr	1.6	7	31
1091	Orange juice, *unsweetened*	60 samples; fresh, canned, bottled and frozen	1.00	89.2	0.5	0.1	8.8	36	153
1092	Pineapple juice, *unsweetened*	18 samples; fresh juice	1.00	87.8	0.3	0.1	10.5	41	177
1093	Tomato juice	10 samples, 9 brands	1.00	93.8	0.8	Tr	3.0	14	62

[a] Including oligosaccharides

Beverages *continued*

Composition of food per 100g

No.	Food	Total nitrogen g	Fatty acids			Cholest-erol mg	Starch g	Total sugars g	Dietary fibre	
			Satd g	Mono unsatd g	Poly unsatd g				Southgate method g	Englyst method g
Carbonated drinks										
1080	**Coca-cola**	Tr	0	0	0	0	Tr	10.5	0	0
1081	**Lemonade**, *bottled*	Tr	0	0	0	0	0	5.6	0	0
1082	**Lucozade**	Tr	0	0	0	0	Tr	8.6[a]	0	0
Squash and cordials										
1083	**Lime juice cordial**, *undiluted*	0.01	0	0	0	0	Tr	29.8	0	0
1084	**Orange drink**, *undiluted*	Tr	0	0	0	0	0	28.5	0	0
1085	**Ribena**, *undiluted*	0.02	0	0	0	0	Tr	59.1[a]	0	0
1086	**Rosehip syrup**, *undiluted*	Tr	0	0	0	0	0.1	61.8	0	0
Juices										
1087	**Apple juice**, *unsweetened*	0.01	Tr	Tr	0.1	0	0	9.9	Tr	Tr
1088	**Grape juice**, *unsweetened*	0.05	Tr	Tr	Tr	0	0	11.7	0	0
1089	**Grapefruit juice**, *unsweetened*	0.07	Tr	Tr	Tr	0	0	8.3	Tr	Tr
1090	**Lemon juice**	0.05	Tr	Tr	Tr	0	0	1.6	0.1	0.1
1091	**Orange juice**, *unsweetened*	0.08	Tr	Tr	Tr	0	0	8.8	0.1	0.1
1092	**Pineapple juice**, *unsweetened*	0.05	Tr	Tr	Tr	0	0	10.5	Tr	Tr
1093	**Tomato juice**	0.13	Tr	Tr	Tr	0	Tr	3.0	N	0.6

[a] Not including oligosaccharides

No.	Food	Na	K	Ca	Mg	P (mg)	Fe	Cu	Zn	Cl	Mn	Se (µg)	I (µg)
	Carbonated drinks												
1080	**Coca-cola**	8	1	4	1	15	Tr	(0.03)	Tr	(10)	Tr	Tr	Tr
1081	**Lemonade**, *bottled*	7	1	5	Tr	Tr	Tr	0.01	Tr	Tr	Tr	Tr	Tr
1082	**Lucozade**	28	2	4	2	4	Tr	0.01	Tr	37	Tr	Tr	Tr
	Squash and cordials												
1083	**Lime juice cordial**, *undiluted*	8	49	9	4	5	0.3	0.07	N	4	Tr	Tr	Tr
1084	**Orange drink**, *undiluted*	21	17	8	3	2	0.1	0.01	N	4	Tr	Tr	Tr
1085	**Ribena**, *undiluted*	26	65	5	3	6	0.4	0.01	0.1	16	N	Tr	Tr
1086	**Rosehip syrup**, *undiluted*	280	26	N	N	N	N	N	N	N	N	Tr	Tr
	Juices												
1087	**Apple juice**, *unsweetened*	2	110	7	5	6	0.1	Tr	Tr	3	Tr	Tr	Tr
1088	**Grape juice**, *unsweetened*	7	55	19	7	14	0.9	Tr	0.1	6	0.1	(1)	N
1089	**Grapefruit juice**, *unsweetened*	7	100	14	8	11	0.2	0.01	Tr	4	0.2	(1)	N
1090	**Lemon juice**	1	130	7	7	8	0.1	0.03	Tr	3	Tr	(1)	N
1091	**Orange juice**, *unsweetened*	10	150	10	8	13	0.2	Tr	Tr	9	0.1	(1)	(2)
1092	**Pineapple juice**, *unsweetened*	8	53	8	6	1	0.2	0.02	0.1	15	0.7	Tr	Tr
1093	**Tomato juice**	230	230	10	10	19	0.4	0.06	0.1	400	0.1	Tr	(2)

Beverages *continued*

No.	Food	Retinol µg	Carotene µg	Vitamin D µg	Vitamin E mg	Thiamin mg	Riboflavin mg	Niacin mg	$\frac{Trypt}{60}$ mg	Vitamin B6 mg	Vitamin B12 µg	Folate µg	Pantothenate mg	Biotin µg	Vitamin C mg
	Carbonated drinks														
1080	**Coca-cola**	0	0	0	0	0	0	0	0	0	0	0	0	0	0
1081	**Lemonade**, *bottled*	0	Tr	0	Tr	Tr	Tr	Tr	Tr	Tr	0	Tr	Tr	Tr	Tr[a]
1082	**Lucozade**	0	0	0	0	Tr	Tr	Tr	Tr	Tr	0	Tr	Tr	Tr	3
	Squash and cordials														
1083	**Lime juice cordial**, *undiluted*	0	Tr	0	Tr	Tr	Tr	Tr	Tr	Tr	0	Tr	Tr	Tr	Tr
1084	**Orange drink**, *undiluted*	0	N	0	Tr	Tr	Tr	Tr	Tr	Tr	0	Tr	Tr	Tr	Tr[b]
1085	**Ribena**, *undiluted*	0	N	0	N	N	N	7.8[c]	Tr	1.01[c]	2.6[c]	N	N	N	78[c]
1086	**Rosehip syrup**, *undiluted*	0	(500)	0	Tr	0	Tr	Tr	Tr	Tr	0	Tr	Tr	Tr	295
	Juices														
1087	**Apple juice**, *unsweetened*	0	Tr	0	Tr	0.01	0.01	0.1	Tr	0.02	0	4	0.04	0.8	14
1088	**Grape juice**, *unsweetened*	0	Tr	0	Tr	Tr	0.01	0.1	Tr	0.04	0	1	0.03	0.7	Tr
1089	**Grapefruit juice**, *unsweetened*	0	1	0	0.19	0.04	0.01	0.2	Tr	0.02	0	6	0.08	1.0	31
1090	**Lemon juice**	0	12	0	N	0.03	0.01	0.1	Tr	0.05	0	13	0.10	0.3	36
1091	**Orange juice**, *unsweetened*	0	17	0	0.17	0.08	0.02	0.2	0.1	0.07	0	20	0.13	1.0	39
1092	**Pineapple juice**, *unsweetened*	0	8	0	0.03	0.06	0.01	0.1	0.1	0.05	0	8	0.07	Tr	11
1093	**Tomato juice**	0	200	0	1.01	0.02	0.02	0.7	0.1	0.06	0	10	(0.20)	(1.5)	8

[a] Vitamin C may be added to some brands and the content may range from 5 to 15 mg per 100g

[b] Vitamin C is added to some brands and the content may range from 20 to 60 mg per 100g. There is a loss on storage of opened bottles, particularly when exposed to light

[c] These are declared amounts and represent the levels present at the end of shelf life

Section 2.13

Alcoholic beverages

The values for wines were obtained on typical examples but due to the variety of alcoholic strengths available, these should only be used as a guide, not as the definitive source for the composition of wines.

In contrast to foods in other parts of the Tables the data here represent composition per 100ml.

For information regarding the specific gravity of drinks, see below.

Specific gravities of alcoholic beverages

Beers		Ciders		Fortified wines	
Beer, bitter, *canned*	1.008	**Cider**, dry	1.007	**Port**	1.026
draught	1.004	sweet	1.012	**Sherry**, dry	0.988
keg	1.001	vintage	1.017	medium	0.998
mild, *draught*	1.009	*Wines*		sweet	1.009
Brown ale, *bottled*	1.008	**Red wine**	0.998	*Vermouths*	
Lager, *bottled*	1.005	**Rosé wine**, medium	1.003	**Vermouth**, dry	1.005
Pale ale, *bottled*	1.003	**White wine**, dry	0.995	sweet	1.046
Stout, *bottled*	1.014	medium	1.005	*Liqueurs*	
extra	1.002	sparkling	0.995	**Advocaat**	1.093
Strong ale	1.018	sweet	1.016	**Cherry brandy**	1.093
				Curacao	1.052
				Spirits	
				40% volume	0.950

Alcoholic beverages

No.	Food	Description and main data sources	Alcohol g	Solids g	Protein g	Fat g	Carbo-hydrate g	Energy value kcal	kJ
Beers									
1094	**Beer**, bitter, *canned*	6 samples	3.1	3.3	0.3	Tr	2.3	32	132
1095	*draught*	5 samples from different brewers	3.1	3.3	0.3	Tr	2.3	32	132
1096	*keg*	6 samples from different brewers	3.0	3.6	0.3	Tr	2.3	31	129
1097	mild, *draught*	5 samples from different brewers	2.6	2.5	0.2	Tr	1.6	25	104
1098	**Brown ale**, *bottled*	6 samples from different brewers	2.2	4.2	0.3	Tr	3.0	28	117
1099	**Lager**, *bottled*	6 samples	3.2	2.4	0.2	Tr	1.5	29	120
1100	**Pale ale**, *bottled*	6 samples from different brewers	3.3	3.3	0.3	Tr	2.0	32	133
1101	**Stout**, *bottled*	4 samples from different brewers	2.9	5.8	0.3	Tr	4.2	37	156
1102	extra	6 samples of the same brand	4.3	3.6	0.3	Tr	2.1	39	163
1103	**Strong ale**	6 samples from different brewers, barley wine type	6.6	8.0	0.7	Tr	6.1	72	301
Ciders									
1104	**Cider**, dry	3 samples of different brands	3.8	3.7	Tr	0	2.6	36	152
1105	sweet	3 samples of different brands	3.7	5.1	Tr	0	4.3	42	176
1106	vintage	3 samples of the same brand	10.5	8.9	Tr	0	7.3	101	421
Wines									
1107	**Red wine**	3 samples, Beaujolais, Burgundy, claret	9.5	2.3	0.2	0	0.3	68	284
1108	**Rosé wine**, medium	5 samples from different vintners	8.7	4.0	0.1	0	2.5	71	294
1109	**White wine**, dry	5 samples from different vintners	9.1	1.8	0.1	0	0.6	66	275
1110	medium	1 sample, Graves	8.8	5.4	0.1	0	3.4	75	311
1111	sparkling	1 sample, Champagne	9.9	3.3	0.3	0	1.4	76	315
1112	sweet	1 sample, Sauternes	10.2	9.2	0.2	0	5.9	94	394

Alcoholic beverages

No.	Food	Total nitrogen g	Fatty acids Satd g	Fatty acids Mono unsatd g	Fatty acids Poly unsatd g	Cholest-erol mg	Starch g	Total sugars g	Dietary fibre Southgate method g	Dietary fibre Englyst method g
Beers										
1094	**Beer, bitter**, canned	0.04	Tr	Tr	Tr	0	0	2.3	0	0
1095	draught	0.04	Tr	Tr	Tr	0	0	2.3	0	0
1096	keg	0.04	Tr	Tr	Tr	0	0	2.3	0	0
1097	mild, draught	0.03	Tr	Tr	Tr	0	0	1.6	0	0
1098	**Brown ale**, bottled	0.04	Tr	Tr	Tr	0	0	3.0	0	0
1099	**Lager**, bottled	0.03	Tr	Tr	Tr	0	0	1.5	0	0
1100	**Pale ale**, bottled	0.05	Tr	Tr	Tr	0	0	2.0	0	0
1101	**Stout**, bottled	0.05	Tr	Tr	Tr	0	0	4.2	0	0
1102	extra	0.05	Tr	Tr	Tr	0	0	2.1	0	0
1103	**Strong ale**	0.11	Tr	Tr	Tr	0	0	6.1	0	0
Ciders										
1104	**Cider**, dry	Tr	0	0	0	0	0	2.6	0	0
1105	sweet	Tr	0	0	0	0	0	4.3	0	0
1106	vintage	Tr	0	0	0	0	0	7.3	0	0
Wines										
1107	**Red wine**	0.03	0	0	0	0	0	0.3	0	0
1108	**Rosé wine**, medium	0.01	0	0	0	0	0	2.5	0	0
1109	**White wine**, dry	0.02	0	0	0	0	0	0.6	0	0
1110	medium	0.02	0	0	0	0	0	3.4	0	0
1111	sparkling	0.04	0	0	0	0	0	1.4	0	0
1112	sweet	0.03	0	0	0	0	0	5.9	0	0

Alcoholic beverages

Inorganic constituents per 100ml

No.	Food	Na	K	Ca	Mg	P	Fe	Cu	Zn	Cl	Mn	Se	I
		mg										µg	
Beers													
1094	**Beer**, bitter, *canned*	9	37	8	7	11	0.01	Tr	Tr	N	Tr	Tr	8
1095	*draught*	12	38	11	9	13	0.01	0.08	0.05	32	Tr	Tr	8
1096	*keg*	8	35	8	7	9	0.01	0.01	0.02	30	Tr	Tr	8
1097	mild, *draught*	11	33	10	8	12	0.02	0.05	0.03	34	Tr	Tr	N
1098	**Brown ale**, *bottled*	16	33	7	6	11	0.03	0.07	0.03	37	Tr	Tr	N
1099	**Lager**, *bottled*	4	34	4	6	12	Tr	Tr	0.01	19	Tr	Tr	N
1100	**Pale ale**, *bottled*	10	49	9	10	15	0.02	0.04	0.03	31	Tr	Tr	N
1101	**Stout**, *bottled*	23	45	8	8	17	0.05	0.08	0.01	48	Tr	Tr	N
1102	extra	4	86	5	9	28	0.02	0.03	0.01	24	Tr	Tr	N
1103	**Strong ale**	15	110	14	20	40	0.03	0.08	0.01	57	Tr	Tr	N
Ciders													
1104	**Cider**, dry	7	72	8	3	3	0.49	0.04	Tr	6	Tr	Tr	N
1105	sweet	7	72	8	3	3	0.49	0.04	Tr	6	Tr	Tr	N
1106	vintage	2	97	5	4	9	0.31	0.02	Tr	5	Tr	Tr	N
Wines													
1107	**Red wine**	10	130	7	11	14	0.90	0.12	0.05	18	Tr	Tr	N
1108	**Rosé wine**, medium	4	75	12	7	6	0.95	0.02	0.04	7	0.1	Tr	N
1109	**White wine**, dry	4	61	9	8	6	0.50	0.01	0.02	10	0.1	Tr	N
1110	medium	21	88	14	9	8	1.21	0.01	0.02	4	0.1	Tr	N
1111	sparkling	4	57	3	6	7	0.50	0.01	0.03	7	Tr	Tr	N
1112	sweet	13	110	14	11	13	0.58	0.05	0.02	7	0.1	Tr	N

Alcoholic beverages

No.	Food	Retinol µg	Carotene µg	Vitamin D µg	Vitamin E mg	Thiamin mg	Riboflavin mg	Niacin mg	Trypt 60 mg	Vitamin B6 mg	Vitamin B12 µg	Folate µg	Pantothenate mg	Biotin µg	Vitamin C mg
Beers															
1094	**Beer, bitter**, canned	0	Tr	0	N	Tr	(0.03)	(0.30)	0.13	(0.02)	(0.15)	(4)	(0.10)	(0.5)	0
1095	draught	0	Tr	0	N	Tr	0.04	0.47	0.13	0.02	0.2	9	(0.10)	(0.5)	0
1096	keg	0	Tr	0	N	Tr	0.03	0.32	0.13	0.02	0.2	5	(0.10)	(0.5)	0
1097	mild, draught	0	Tr	0	N	Tr	(0.03)	(0.30)	0.10	(0.02)	(0.2)	(5)	(0.10)	(0.5)	0
1098	**Brown ale**, bottled	0	Tr	0	N	Tr	0.02	0.26	0.13	0.01	0.1	4	0.10	0.5	0
1099	**Lager**, bottled	0	Tr	0	N	Tr	0.02	0.33	0.21	0.02	0.1	4	(0.10)	(0.5)	0
1100	**Pale ale**, bottled	0	Tr	0	N	Tr	0.02	0.35	0.17	0.01	0.1	4	(0.10)	(0.5)	0
1101	**Stout**, bottled	0	Tr	0	N	Tr	0.03	0.26	0.17	0.01	0.1	4	(0.10)	(0.5)	0
1102	extra	0	Tr	0	N	Tr	0.04	0.51	0.17	0.02	0.2	6	N	N	0
1103	**Strong ale**	0	Tr	0	N	Tr	0.06	0.83	0.37	0.04	0.4	9	N	N	0
Ciders															
1104	**Cider, dry**	0	Tr	0	N	Tr	Tr	0.01	Tr	0.01	Tr	N	0.04	0.6	0
1105	sweet	0	Tr	0	N	Tr	Tr	0.01	Tr	0.01	Tr	N	0.03	0.6	0
1106	vintage	0	Tr	0	N	Tr	Tr	0.01	Tr	(0.01)	Tr	N	(0.03)	(0.6)	0
Wines															
1107	**Red wine**	0	Tr	0	N	Tr	0.02	0.09	Tr	0.02	Tr	Tr	(0.04)	N	0
1108	**Rosé wine**, medium	0	Tr	0	N	Tr	0.01	0.07	Tr	0.02	Tr	Tr	0.04	N	0
1109	**White wine**, dry	0	Tr	0	N	Tr	0.01	0.06	Tr	0.02	Tr	Tr	0.03	N	0
1110	medium	0	Tr	0	N	Tr	0.01	0.08	Tr	0.01	Tr	Tr	0.03	N	0
1111	sparkling	0	Tr	0	N	Tr	0.01	0.07	Tr	0.02	Tr	Tr	0.03	N	0
1112	sweet	0	Tr	0	N	Tr	0.01	0.08	Tr	0.01	Tr	Tr	0.03	N	0

Alcoholic beverages *continued*

No.	Food	Description and main data sources	Alcohol g	Solids g	Protein g	Fat g	Carbo-hydrate g	Energy value kcal	Energy value kJ
	Fortified wines								
1113	**Port**	2 samples	15.9	13.0	0.1	0	12.0	157	655
1114	**Sherry,** dry	1 sample	15.7	3.3	0.2	0	1.4	116	481
1115	medium	5 samples from different importers	14.8	4.7	0.1	0	3.6	118	489
1116	sweet	1 sample	15.6	9.6	0.3	0	6.9	136	568
	Vermouths								
1117	**Vermouth,** dry	5 samples of different brands	13.9	6.6	0.1	0	5.5	118	493
1118	sweet	5 samples of different brands	13.0	16.4	Tr	0	15.9	151	631
	Liqueurs								
1119	**Advocaat**	4 samples of different brands	12.8	39.6	4.7	6.3	28.4	272	1139
1120	**Cherry brandy**	6 samples of different brands	19.0	33.3	Tr	0	32.6	255	1073
1121	**Curacao**	4 samples of different brands	29.3	27.8	Tr	0	28.3	311	1303
	Spirits								
1122	**40% volume**	Mean of brandy, gin, rum, whisky	31.7	Tr	Tr	0	Tr	222	919

Alcoholic beverages continued

No.	Food	Total nitrogen g	Fatty acids Satd g	Fatty acids Mono unsatd g	Fatty acids Poly unsatd g	Cholest-erol mg	Starch g	Total sugars g	Dietary fibre Southgate method g	Dietary fibre Englyst method g
Fortified wines										
1113	**Port**	0.02	0	0	0	0	0	12.0	0	0
1114	**Sherry,** dry	0.03	0	0	0	0	0	1.4	0	0
1115	medium	0.02	0	0	0	0	0	3.6	0	0
1116	sweet	0.05	0	0	0	0	0	6.9	0	0
Vermouths										
1117	**Vermouth,** dry	0.01	0	0	0	0	0	5.5	0	0
1118	sweet	Tr	0	0	0	0	0	15.9	0	0
Liqueurs										
1119	**Advocaat**	0.75	1.9	3.0	0.9	242	0	28.4	0	0
1120	**Cherry brandy**	Tr	0	0	0	0	0	32.6	0	0
1121	**Curacao**	Tr	0	0	0	0	0	28.3	0	0
Spirits										
1122	**40% volume**	Tr	0	0	0	0	0	Tr	0	0

Alcoholic beverages *continued*

No.	Food	mg										µg	
		Na	K	Ca	Mg	P	Fe	Cu	Zn	Cl	Mn	Se	I
Fortified wines													
1113	**Port**	4	97	4	11	12	0.40	0.10	N	8	Tr	Tr	N
1114	**Sherry, dry**	10	57	7	13	11	0.39	0.03	N	12	Tr	Tr	N
1115	**medium**	6	89	9	8	7	0.53	0.10	0.27	7	Tr	Tr	N
1116	**sweet**	13	110	7	11	10	0.37	0.11	N	14	Tr	Tr	N
Vermouths													
1117	**Vermouth, dry**	17	40	7	5	7	0.34	0.06	0.04	9	Tr	Tr	N
1118	**sweet**	28	30	6	4	6	0.36	0.04	0.03	16	Tr	Tr	N
Liqueurs													
1119	**Advocaat**	N	N	N	N	N	N	N	N	N	N	N	N
1120	**Cherry brandy**	N	N	N	N	N	N	N	N	N	N	N	N
1121	**Curacao**	N	N	N	N	N	N	N	N	N	N	N	N
Spirits													
1122	**40% volume**	Tr	Tr	Tr	Tr	Tr	Tr	Tr	Tr	Tr	Tr	Tr	Tr

Alcoholic beverages continued

No.	Food	Retinol µg	Carotene µg	Vitamin D µg	Vitamin E mg	Thiamin mg	Ribo-flavin mg	Niacin mg	Trypt 60 mg	Vitamin B6 mg	Vitamin B12 µg	Folate µg	Panto-thenate mg	Biotin µg	Vitamin C mg
Fortified wines															
1113	**Port**	0	Tr	0	0	Tr	0.01	0.06	Tr	0.01	Tr	Tr	N	N	0
1114	**Sherry**, dry	0	Tr	0	0	Tr	0.01	0.10	Tr	0.01	Tr	Tr	N	N	0
1115	medium	0	Tr	0	0	Tr	0.01	0.08	Tr	0.01	Tr	Tr	N	N	0
1116	sweet	0	Tr	0	0	Tr	0.01	0.07	Tr	0.01	Tr	Tr	N	N	0
Vermouths															
1117	**Vermouth**, dry	0	Tr	0	0	Tr	Tr	0.04	Tr	0.01	Tr	Tr	N	N	0
1118	sweet	0	Tr	0	0	Tr	Tr	0.04	Tr	Tr	Tr	Tr	N	N	0
Liqueurs															
1119	**Advocaat**	N	N	Tr	N	N	N	N	1.40	N	N	N	N	N	0
1120	**Cherry brandy**	0	Tr	0	0	Tr	Tr	Tr	Tr	Tr	Tr	Tr	Tr	Tr	0
1121	**Curacao**	0	Tr	0	0	Tr	Tr	Tr	Tr	Tr	Tr	Tr	Tr	Tr	0
Spirits															
1122	**40% volume**	0	0	0	0	0	0	0	0	0	0	0	0	0	0

Soups, sauces and miscellaneous foods

The foods in this group cover homemade, canned and packet soups; dairy sauces; salad sauces, dressings and pickles; non salad sauces; and a selection of miscellaneous food items.

An entry for water has been included in the miscellaneous foods section, mainly for use in recipe calculations. There is considerable variation in the composition of tap water both by area of the country and source of supply. The local water board will be able to provide information on the composition of tap water from a specific area.

Dried soups as made up were corrected for evaporative loss.

The fatty acid profile of the margarine used in the recipe calculations is an average of hard, soft and polyunsaturated varieties.

Losses of labile vitamins assigned to recipes were estimated from figures in Section 3.4.

Soups, sauces and miscellaneous foods

Composition of food per 100g

No.	Food	Description and main data sources	Edible proportion	Water g	Protein g	Fat g	Carbo-hydrate g	Energy value kcal	kJ
	Homemade soups								
1123	**Lentil soup**	Recipe	1.00	77.7	4.4	3.8	12.7[a]	99	418
	Canned soups								
1124	**Cream of chicken soup**, canned, *ready to serve*	10 cans, 3 brands	1.00	87.9	1.7	3.8	4.5	58	242
1125	**condensed**, canned	7 cans of the same brand	1.00	82.2	2.6	7.2	6.0	98	407
1126	*ready to serve*	Diluted with an equal volume of water	1.00	91.1	1.3	3.6	3.0	49	203
1127	**Cream of mushroom soup**, canned, *ready to serve*	10 cans, 3 brands	1.00	89.2	1.1	3.8	3.9	53	222
1128	**Cream of tomato soup**, canned, *ready to serve*	10 cans, 3 brands	1.00	84.2	0.8	3.3	5.9	55	230
1129	**condensed**, canned	7 cans, 2 brands	1.00	70.6	1.7	6.8	14.6	123	514
1130	*ready to serve*	Diluted with an equal volume of water	1.00	85.3	0.9	3.4	7.3	62	258
1131	**Low calorie soup**, canned	7 cans, 3 brands; tomato, vegetable and minestrone varieties	1.00	93.3	0.9	0.2	4.0	20	87
1132	**Oxtail soup**, canned, *ready to serve*	10 cans, 3 brands	1.00	88.5	2.4	1.7	5.1	44	185
1133	**Vegetable soup**, canned, *ready to serve*	10 cans, 4 brands	1.00	86.4	1.5	0.7	6.7	37	159

a Including oligosaccharides

Soups, sauces and miscellaneous foods

Composition of food per 100g

No.	Food	Total nitrogen g	Fatty acids			Cholest-erol mg	Starch g	Total sugars g	Dietary fibre	
			Satd g	Mono unsatd g	Poly unsatd g				Southgate method g	Englyst method g
	Homemade soups									
1123	**Lentil soup**	0.71	1.3	1.3	0.9	7	9.8	2.3ᵃ	2.0	1.1
	Canned soups									
1124	**Cream of chicken soup**, canned, ready to serve	0.27	N	N	N	N	3.4	1.1		
1125	**condensed**, canned	0.41	N	N	N	N	4.6	1.4	N	N
1126	ready to serve	0.20	N	N	N	N	2.3	0.7	N	N
1127	**Cream of mushroom soup**, canned, ready to serve	0.17	N	N	N	N	3.1	0.8	N	N
1128	**Cream of tomato soup**, canned, ready to serve	0.13	N	N	N	N	3.3	2.6	N	N
1129	**condensed**, canned	0.27	N	N	N	N	3.4	11.2	N	N
1130	ready to serve	0.14	N	N	N	N	1.7	5.6	N	N
1131	**Low calorie soup**, canned	0.14	Tr	Tr	Tr	0	2.0	2.0	N	N
1132	**Oxtail soup**, canned, *ready to serve*	0.38	N	N	N	N	4.2	0.9	N	N
1133	**Vegetable soup**, canned, *ready to serve*	0.24	N	N	N	N	4.2	2.5	N	1.5

ᵃ Not including oligosaccharides

Soups, sauces and miscellaneous foods

Inorganic constituents per 100g

No.	Food	Na	K	Ca	Mg	P	Fe	Cu	Zn	Cl	Mn	Se	I
		mg										µg	
Homemade soups													
1123	**Lentil soup**	45	170	38	15	71	1.2	0.09	0.5	76	0.2	1	4
Canned soups													
1124	**Cream of chicken soup**, canned, *ready to serve*	460	41	27	5	27	0.4	0.02	0.3	700	Tr	N	(16)
1125	**condensed**, canned	710	(62)	(41)	(7)	(41)	(0.5)	(0.03)	(0.5)	1070	Tr	N	(5)
1126	*ready to serve*	350	(31)	(20)	(4)	(20)	(0.3)	(0.02)	(0.3)	530	Tr	N	(3)
1127	**Cream of mushroom soup**, canned, *ready to serve*	470	55	30	4	30	0.3	0.04	0.3	750	Tr	N	(16)
1128	**Cream of tomato soup**, canned, *ready to serve*	460	190	17	8	20	0.4	0.06	0.2	740	0.1	N	(16)
1129	**condensed**, canned	830	(360)	(32)	(15)	(38)	(0.7)	(0.11)	0.3	1320	0.1	N	(5)
1130	*ready to serve*	410	(180)	(16)	(8)	(19)	(0.3)	(0.06)	0.2	660	0.1	N	(3)
1131	**Low calorie soup**, canned	370	130	13	7	17	0.3	0.01	0.1	580	0.1	N	N
1132	**Oxtail soup**, canned, *ready to serve*	440	93	40	6	37	1.0	0.04	0.4	660	N	N	(16)
1133	**Vegetable soup**, canned, *ready to serve*	500	140	17	10	27	0.6	0.06	0.3	750	0.1	N	(16)

Soups, sauces and miscellaneous foods

No.	Food	Retinol µg	Carotene µg	Vitamin D µg	Vitamin E mg	Thiamin mg	Ribo-flavin mg	Niacin mg	Trypt 60 mg	Vitamin B6 mg	Vitamin B12 µg	Folate µg	Panto-thenate mg	Biotin µg	Vitamin C mg
Homemade soups															
1123	**Lentil soup**	36	325	0.28	0.35	0.07	0.04	0.3	0.8	0.07	Tr	4	0.21	N	1
Canned soups															
1124	**Cream of chicken soup**, canned, ready to serve	0	0	0	N	0.01	0.03	0.2	0.3	0.01	0	120	N	N	0
1125	**condensed**, canned	0	0	0	N	(0.02)	0.04	0.6	0.5	N	0	N	N	N	0
1126	ready to serve	0	0	0	N	(0.01)	0.02	0.3	0.2	N	0	N	N	N	0
1127	**Cream of mushroom soup**, canned, ready to serve	0	0	0	N	Tr	0.05	0.3	0.2	0.01	0	N	N	N	0
1128	**Cream of tomato soup**, canned, ready to serve	0	210	0	N	0.03	0.02	0.5	0.1	0.06	0	12	N	N	Tr
1129	**condensed**, canned	0	(400)	0	N	(0.06)	0.05	1.0	0.2	N	0	N	N	N	Tr
1130	ready to serve	0	(200)	0	N	(0.03)	0.03	0.5	0.1	N	0	N	N	N	Tr
1131	**Low calorie soup**, canned	0	N	0	N	0.35	0.14	2.0	0.1	0.20	0	(10)	N	N	Tr
1132	**Oxtail soup**, canned, ready to serve	0	0	0	N	0.02	0.03	0.7	0.5	0.03	0	N	N	N	0
1133	**Vegetable soup**, canned, ready to serve	0	18	0	N	0.03	0.02	0.4	0.2	0.05	0	10	N	N	Tr

No.	Food	Description and main data sources	Edible proportion	Water g	Protein g	Fat g	Carbo-hydrate g	Energy value kcal	kJ
Packet soups									
1134	**Chicken noodle soup**, dried	10 packets, 5 brands	1.00	4.8	13.8	5.0	60.9	329	1394
1135	dried, *ready to serve*	Calculated from 35g soup powder to 570ml water	1.00	94.2	0.8	0.3	3.7	20	84
1136	**Instant soup powder**, dried	10 packets, 3 brands; assorted flavours	1.00	4.1	6.5	14.0	64.4	393	1659
1137	dried, *made up with water*	Calculated from 37g powder to 190ml water	1.00	84.4	1.1	2.3	10.5	64	272
1138	**Minestrone soup**, dried	10 packets, 3 brands	1.00	3.9	10.1	8.8	47.6	298	1259
1139	dried, *ready to serve*	Calculated from 45g soup powder to 570ml water	1.00	92.6	0.8	0.7	3.7	23	99
1140	**Oxtail soup**, dried	10 packets, 5 brands	1.00	3.0	17.6	10.5	51.0	356	1504
1141	dried, *ready to serve*	Calculated from 45g soup powder to 570ml water	1.00	92.5	1.4	0.8	3.9	27	116
1142	**Tomato soup**, dried	10 packets, 4 brands	1.00	2.8	6.6	5.6	65.0	321	1359
1143	dried, *ready to serve*	Calculated from 58g soup powder to 570ml water	1.00	90.6	0.6	0.5	6.3	31	130

No.	Food	Total nitrogen	Fatty acids			Cholest-erol	Starch	Total sugars	Dietary fibre	
			Satd	Mono unsatd	Poly unsatd				Southgate method	Englyst method
		g	g	g	g	mg	g	g	g	g
Packet soups										
1134	**Chicken noodle soup**, dried	2.20	N	N	N	N	50.7	10.2	N	4.3
1135	dried, *ready to serve*	0.13	Tr	Tr	Tr	N	3.1	0.6	N	0.2
1136	**Instant soup powder**, dried	1.04	N	N	N	Tr	34.1[a]	30.3[a]	N	N
1137	dried, *made up with water*	0.17	N	N	N	Tr	5.6[a]	4.9[a]	N	N
1138	**Minestrone soup**, dried	1.62	N	N	N	Tr	32.6	15.0	5.9	N
1139	dried, *ready to serve*	0.12	N	N	N	Tr	2.5	1.2	0.5	N
1140	**Oxtail soup**, dried	2.81	N	N	N	N	41.8	9.2	3.4	N
1141	dried, *ready to serve*	0.22	N	N	N	N	3.2	0.7	0.3	N
1142	**Tomato soup**, dried	1.05	N	N	N	0	28.9	36.1	3.0	N
1143	dried, *ready to serve*	0.10	N	N	N	0	2.8	3.5	0.3	N

[a] Including maltodextrins

No.	Food	Na	K	Ca	Mg	P	Fe	Cu	Zn	Cl	Mn	Se	I
						mg						μg	
Packet soups													
1134	**Chicken noodle soup**, dried	6120	270	45	44	160	2.7	0.26	1.2	9030	0.7	N	(74)
1135	dried, *ready to serve*	370	16	3	3	10	0.2	0.02	0.1	550	Tr	Tr	(4)
1136	**Instant soup powder**, dried	3440	610	48	27	600	1.7	0.17	0.7	4780	0.3	N	N
1137	dried, *made up with water*	560	99	7	4	97	0.3	0.03	0.1	780	0.1	N	N
1138	**Minestrone soup**, dried	5600	800	120	48	160	2.8	0.29	1.0	7670	0.6	Tr	(74)
1139	dried, *ready to serve*	430	62	9	7	12	0.2	0.02	0.1	590	Tr	Tr	(5)
1140	**Oxtail soup**, dried	5250	700	140	44	260	4.3	0.25	2.4	7670	N	N	(74)
1141	dried, *ready to serve*	400	54	11	3	20	0.3	0.02	0.2	590	Tr	Tr	(5)
1142	**Tomato soup**, dried	4040	920	140	39	130	1.8	0.33	0.8	6620	0.3	N	(74)
1143	dried, *ready to serve*	390	89	14	4	13	0.2	0.03	0.1	640	Tr	Tr	(7)

Soups, sauces and miscellaneous foods *continued*

No.	Food	Retinol µg	Carotene µg	Vitamin D µg	Vitamin E mg	Thiamin mg	Ribo-flavin mg	Niacin mg	Trypt 60 mg	Vitamin B6 mg	Vitamin B12 µg	Folate µg	Panto-thenate mg	Biotin µg	Vitamin C mg
Packet soups															
1134	**Chicken noodle soup**, dried	0	0	0	N	0.23	0.08	2.2	2.6	N	0	N	N	N	0
1135	dried, *ready to serve*	0	0	0	N	0.01	Tr	0.1	0.2	N	0	N	N	N	0
1136	**Instant soup powder**, dried	0	N	0	N	(0.05)	(0.02)	(0.4)	1.0	N	0	N	N	N	0
1137	dried, *made up with water*	0	N	0	N	(0.01)	Tr	(0.1)	0.2	N	0	N	N	N	0
1138	**Minestrone soup**, dried	0	N	0	N	0.21	0.15	3.1	1.9	N	0	N	N	N	0
1139	dried, *ready to serve*	0	N	0	N	0.02	0.01	0.2	0.1	N	0	N	N	N	0
1140	**Oxtail soup**, dried	0	0	0	N	10.40[a]	0.30	3.5	3.8	N	0	N	N	N	0
1141	dried, *ready to serve*	0	0	0	N	0.80[a]	0.02	0.3	0.3	N	0	N	N	N	0
1142	**Tomato soup**, dried	0	N	0	N	0.23	0.18	1.9	0.9	N	0	52	N	N	Tr
1143	dried, *ready to serve*	0	N	0	N	0.02	0.02	0.2	0.1	N	0	N	N	N	Tr

[a] This remarkably high content is derived from the flavouring agent

No.	Food	Description and main data sources	Edible proportion	Water g	Protein g	Fat g	Carbo-hydrate g	Energy value kcal	Energy value kJ
Dairy sauces									
1144	**Bread sauce,** *made with whole milk*	Recipe	1.00	76.3	4.2	5.1	12.6	110	463
1145	*made with semi-skimmed milk*	Recipe	1.00	78.1	4.3	3.1	12.8	93	393
1146	**Cheese sauce,** *made with whole milk*	Recipe	1.00	66.9	8.0	14.6	9.0	197	819
1147	*made with semi-skimmed milk*	Recipe	1.00	68.7	8.1	12.6	9.1	179	750
1148	**Cheese sauce packet mix,** *made up with whole milk*	Recipe	1.00	77.2	5.3	6.1	9.3	110	462
1149	*made up with semi-skimmed milk*	Recipe	1.00	79.2	5.4	3.8	9.5	90	383
1150	**Onion sauce,** *made with whole milk*	Recipe	1.00	80.4	2.8	6.5	8.3	99	414
1151	*made with semi-skimmed milk*	Recipe	1.00	81.7	2.9	5.0	8.4	86	361
1152	**White sauce,** *savoury, made with whole milk*	Recipe	1.00	73.7	4.1	10.3	10.9	150	624
1153	*made with semi-skimmed milk*	Recipe	1.00	75.8	4.2	7.8	11.1	128	539
1154	*sweet, made with whole milk*	Recipe	1.00	68.3	3.8	9.5	18.6	170	712
1155	*made with semi-skimmed milk*	Recipe	1.00	70.2	3.9	7.2	18.8	150	634

Soups, sauces and miscellaneous foods *continued*

Composition of food per 100g

No.	Food	Total nitrogen g	Fatty acids Satd g	Mono unsatd g	Poly unsatd g	Cholest-erol mg	Starch g	Total sugars g	Dietary fibre Southgate method g	Englyst method g
Dairy sauces										
1144	**Bread sauce**, *made with whole milk*	0.69	2.6	1.6	0.6	14	8.0	4.7	0.6	0.3
1145	*made with semi-skimmed milk*	0.71	1.4	1.0	0.5	8	8.0	4.8	0.6	0.3
1146	**Cheese sauce**, *made with whole milk*	1.26	7.5	4.7	1.7	39	4.6	4.3	0.2	0.2
1147	*made with semi-skimmed milk*	1.28	6.3	4.1	1.6	33	4.6	4.5	0.2	0.2
1148	**Cheese sauce packet mix**, *made up with whole milk*	0.84	N	N	N	N	3.9	5.4	N	N
1149	*made up with semi-skimmed milk*	0.86	N	N	N	N	3.9	5.6	N	N
1150	**Onion sauce**, *made with whole milk*	0.45	2.8	2.2	1.1	16	3.5	4.7	0.6	0.4
1151	*made with semi-skimmed milk*	0.46	1.8	1.8	1.1	11	3.5	4.9	0.6	0.4
1152	**White sauce**, *savoury, made with whole milk*	0.66	4.4	3.5	1.8	26	5.6	5.3	0.3	0.2
1153	*made with semi-skimmed milk*	0.68	2.9	2.8	1.7	18	5.6	5.5	0.3	0.2
1154	*sweet, made with whole milk*	0.61	4.1	3.3	1.7	24	5.2	13.5	0.2	0.2
1155	*made with semi-skimmed milk*	0.62	2.7	2.6	1.6	17	5.2	13.7	0.2	0.2

No.	Food	mg										µg	
		Na	K	Ca	Mg	P	Fe	Cu	Zn	Cl	Mn	Se	I
Dairy sauces													
1144	**Bread sauce**, *made with whole milk*	480	140	120	16	96	0.3	0.03	0.4	760	0.1	6	15
1145	*made with semi-skimmed milk*	480	150	120	16	99	0.3	0.03	0.5	760	0.1	6	15
1146	**Cheese sauce**, *made with whole milk*	450	150	240	17	180	0.2	0.02	0.8	710	Tr	3	23
1147	*made with semi-skimmed milk*	450	150	240	17	180	0.2	0.02	0.8	710	Tr	3	23
1148	**Cheese sauce packet mix**, *made up with whole milk*	460	170	160	14	170	0.1	Tr	0.6	650	Tr	N	N
1149	*made up with semi-skimmed milk*	460	190	170	14	170	0.1	Tr	0.7	650	Tr	N	N
1150	**Onion sauce**, *made with whole milk*	430	140	90	12	73	0.3	0.02	0.3	690	0.1	1	13
1151	*made with semi-skimmed milk*	435	150	93	12	75	0.3	0.02	0.3	690	0.1	1	13
1152	**White sauce**, *savoury, made with whole milk*	400	160	130	15	110	0.2	0.01	0.4	640	Tr	1	19
1153	*made with semi-skimmed milk*	400	170	140	15	110	0.2	0.01	0.5	640	Tr	1	19
1154	*sweet, made with whole milk*	110	150	120	12	98	0.2	0.01	0.4	190	Tr	1	17
1155	*made with semi-skimmed milk*	110	160	130	12	100	0.2	0.01	0.5	190	Tr	1	17

Soups, sauces and miscellaneous foods *continued*

No.	Food	Retinol µg	Carotene µg	Vitamin D µg	Vitamin E mg	Thiamin mg	Ribo-flavin mg	Niacin mg	Trypt 60 mg	Vitamin B6 mg	Vitamin B12 µg	Folate µg	Panto-thenate mg	Biotin µg	Vitamin C mg
Dairy sauces															
1144	**Bread sauce,** *made with whole milk*	58	31	0.16	0.20	0.05	0.14	0.30	0.90	0.05	0.3	4	0.27	1.8	Tr
1145	*made with semi-skimmed milk*	31	20	0.14	0.16	0.05	0.14	0.30	1.00	0.05	0.3	4	0.25	1.9	1
1146	**Cheese sauce,** *made with whole milk*	150	105	0.55	0.68	0.05	0.22	0.20	1.90	0.06	0.5	6	0.31	2.3	1
1147	*made with semi-skimmed milk*	125	93	0.54	0.62	0.05	0.23	0.20	1.90	0.06	0.5	6	0.29	2.3	1
1148	**Cheese sauce packet mix,** *made up with whole milk*	67	37	0.09	0.29	0.04	0.23	0.20	1.20	0.07	0.5	5	N	N	1
1149	*made up with semi-skimmed milk*	36	25	0.07	0.23	0.04	0.24	0.20	1.20	0.07	0.5	5	N	N	1
1150	**Onion sauce,** *made with whole milk*	71	78	0.39	0.45	0.06	0.12	0.30	0.60	0.08	0.3	6	0.25	1.5	1
1151	*made with semi-skimmed milk*	50	70	0.38	0.41	0.06	0.12	0.30	0.70	0.08	0.3	6	0.23	1.6	2
1152	**White sauce, savoury,** *made with whole milk*	115	77	0.62	0.71	0.05	0.18	0.20	0.90	0.06	0.4	4	0.32	2.1	Tr
1153	*made with semi-skimmed milk*	79	64	0.60	0.64	0.05	0.19	0.20	1.00	0.06	0.4	4	0.29	2.2	1
1154	*sweet, made with whole milk*	105	71	0.57	0.65	0.05	0.17	0.20	0.90	0.06	0.4	3	0.29	1.9	Tr
1155	*made with semi-skimmed milk*	73	59	0.55	0.59	0.05	0.18	0.20	0.90	0.06	0.4	3	0.27	2.0	1

Soups, sauces and miscellaneous foods *continued*

Composition of food per 100g

No.	Food	Description and main data sources	Edible proportion	Water g	Protein g	Fat g	Carbo-hydrate g	Energy value kcal	kJ
Salad sauces, dressings and pickles									
1156	**Apple chutney**	Recipe	1.00	43.5	0.9	0.2	52.2[a]	201	858
1157	**French dressing**	Recipe	1.00	22.8	0.3	72.1	0.1	651	2674
1158	**Mango chutney**, oily	10 assorted samples	1.00	34.8	0.4	10.9	49.5	285	1202
1159	**Mayonnaise**, retail	6 samples, 5 brands	1.00	18.8	1.1	75.6	1.7	691	2843
1160	**Pickle**, sweet	10 jars, 3 brands, including Branston, Pan Yan	1.00	58.9	0.6	0.3	34.4	134	572
1161	**Salad cream**	3 samples, different brands	1.00	47.2	1.5	31.0	16.7	348	1440
1162	reduced calorie	Analysis and manufacturer's data	1.00	N	1.0	17.2	9.4	194	804
1163	**Tomato chutney**	Recipe	1.00	54.0	1.2	0.4	40.9[a]	162	690
Non salad sauces									
1164	**Barbecue sauce**	Ref. Marsh (1980)	1.00	80.9	1.8	1.8	12.2	75	314
1165	**Brown sauce**, *bottled*	6 bottles of different brands	1.00	64.0	1.1	0	25.2	99	422
1166	**Cook-in-sauces**, canned	9 samples, 3 brands; assorted flavours	1.00	87.4	1.1	0.8	8.3	43	181
1167	**Curry sauce**, canned	10 samples, 4 brands; assorted flavours	1.00	81.4	1.5	5.0	7.1	78	324
1168	**Horseradish sauce**	8 samples, 5 brands; creamed and plain samples	1.00	64.0	2.5	8.4	17.9	153	640
1169	**Mint sauce**	8 samples, 4 brands	1.00	68.7	1.6	Tr	21.5	87	371
1170	**Pasta sauce**, tomato based	9 samples, 4 brands; assorted types	1.00	83.9	2.0	1.5	6.9	47	200
1171	**Soy sauce**	Ref. Marsh (1980); dark, thick variety	1.00	67.6	8.7	0	8.3	64	266
1172	**Tomato ketchup**	6 bottles of different brands	1.00	64.8	2.1	Tr	24.0	98	420
1173	**Tomato sauce**	Recipe	1.00	80.3	2.2	5.5	8.6[a]	91	379

[a] including oligosaccharides

No.	Food	Total nitrogen	Fatty acids			Cholesterol	Starch	Total sugars	Dietary fibre	
			Satd	Mono unsatd	Poly unsatd				Southgate method	Englyst method
		g	g	g	g	mg	g	g	g	g
Salad sauces, dressings and pickles										
1156	**Apple chutney**	0.16	Tr	Tr	Tr	0	0.2	51.1[a]	1.8	1.2
1157	**French dressing**	0.06	10.0	49.8	8.0	0	0	0.1	0	0
1158	**Mango chutney**, oily	0.06	N	N	N	0	0.4	49.1	1.4	0.9
1159	**Mayonnaise**, retail	0.18	11.1	17.3	43.9	75	0.4	1.3	0	0
1160	**Pickle**, sweet	0.09	Tr	Tr	Tr	0	1.8	32.6	1.5	1.2
1161	**Salad cream**	0.23	3.9	6.2	19.4	43	Tr	16.7	N	N
1162	reduced calorie	0.16	2.5	4.7	9.1	7	0.2	9.2	N	N
1163	**Tomato chutney**	0.19	0.1	0.1	0.1	0	0.1	40.1[a]	1.9	1.4
Non salad sauces										
1164	**Barbecue sauce**	0.29	0.3	0.8	0.7	0	N	N	N	N
1165	**Brown sauce**, *bottled*	0.18	Tr	Tr	Tr	0	2.1	23.1	N	0.7
1166	**Cook-in-sauces**, canned	0.18	0.1	0.4	0.2	Tr	3.3	5.0	N	N
1167	**Curry sauce**, canned	0.24	N	N	N	0	3.4	3.7	N	N
1168	**Horseradish sauce**	0.40	1.1	3.8	3.2	14	3.0	14.9	N	2.5
1169	**Mint sauce**	0.26	Tr	Tr	Tr	0	0	21.5	N	N
1170	**Pasta sauce**, tomato based	0.32	0.2	0.3	0.8	0	1.2	5.7	N	N
1171	**Soy sauce**	1.39	0	0	0	0	N	N	0	0
1172	**Tomato ketchup**	0.34	Tr	Tr	Tr	0	1.1	22.9	N	0.9
1173	**Tomato sauce**	0.36	1.8	2.2	1.1	10	4.3	4.0[a]	1.7	1.4

[a] Not including oligosaccharides

Soups, sauces and miscellaneous foods *continued*

Inorganic constituents per 100g

No.	Food	Na	K	Ca	Mg	P	Fe	Cu	Zn	Cl	Mn	Se	I
						mg						µg	
	Salad sauces, dressings and pickles												
1156	**Apple chutney**	180	220	23	15	34	0.8	0.11	0.3	280	0.2	Tr	2
1157	**French dressing**	930	45	12	14	11	0.3	0.03	Tr	1440	0.1	Tr	1
1158	**Mango chutney,** oily	1090	57	23	27	10	2.3	0.10	0.1	1720	0.1	N	N
1159	**Mayonnaise,** retail	450	16	8	1	27	0.3	0.02	0.1	750	Tr	N	35
1160	**Pickle,** sweet	1700	110	19	10	11	2.0	0.10	1.4	2600	0.1	N	N
1161	**Salad cream**	1040	40	18	9	48	0.5	0.02	0.3	1620	0.1	N	11
1162	reduced calorie	N	N	N	N	N	N	N	N	N	N	N	N
1163	**Tomato chutney**	130	310	24	16	40	1.0	0.08	0.2	240	0.2	Tr	2
	Non salad sauces												
1164	**Barbecue sauce**	810	170	19	N	20	0.9	N	N	N	N	N	N
1165	**Brown sauce,** *bottled*	980	390	43	29	36	3.1	0.33	0.2	1550	0.3	N	N
1166	**Cook-in-sauces,** canned	940	130	7	5	20	0.4	0.03	0.1	620	0.1	N	N
1167	**Curry sauce,** canned	980	180	30	18	31	1.1	0.05	0.2	760	0.2	N	N
1168	**Horseradish sauce**	910	220	43	18	42	0.5	0.05	0.4	1710	0.2	N	N
1169	**Mint sauce**	700	210	120	46	27	7.4	0.30	0.2	1100	0.9	Tr	Tr
1170	**Pasta sauce,** tomato based	410	490	23	21	42	0.7	0.16	0.2	830	0.1	N	N
1171	**Soy sauce**	5720	360	19	43	210	2.7	0.1	0.2	N	Tr	N	N
1172	**Tomato ketchup**	1120	590	25	19	43	1.2	0.40	0.1	1810	0.1	N	N
1173	**Tomato sauce**	340	280	19	10	39	0.6	0.03	0.3	560	0.1	1	5

Soups, sauces and miscellaneous foods *continued*

No.	Food	Retinol µg	Carotene µg	Vitamin D µg	Vitamin E mg	Thiamin mg	Ribo-flavin mg	Niacin mg	Trypt 60 mg	Vitamin B6 mg	Vitamin B12 µg	Folate µg	Panto-thenate mg	Biotin µg	Vitamin C mg
	Salad sauces, dressings and pickles														
1156	**Apple chutney**	0	14	0	0.17	0.05	0.01	0.3	0.1	0.08	0	4	0.06	0.4	3
1157	**French dressing**	0	2	0	3.64	0	0	0	0	0	0	0	0	0	0
1158	**Mango chutney,** oily	0	130	0	N	0.02	0.03	0.1	Tr	N	0	N	N	N	1
1159	**Mayonnaise,** retail	86	100	0.33	18.94	0.02	0.07	Tr	0.3	0.01	0.5	4	N	N	N
1160	**Pickle,** sweet	0	250	0	N	0.03	0.01	0.2	0.1	0.01	0	Tr	N	Tr	Tr
1161	**Salad cream**	9	17	0.19	10.47	N	N	N	0.3	0.03	0.5	3	N	N	0
1162	reduced calorie	N	N	N	N	N	N	N	0.2	N	N	N	N	N	0
1163	**Tomato chutney**	0	390	0	0.88	0.08	0.01	0.8	0.2	0.12	0	8	0.15	1.1	6
	Non salad sauces														
1164	**Barbecue sauce**	0	520	0	N	0.03	0.02	0.9	0.3	N	0	N	N	N	7
1165	**Brown sauce,** *bottled*	0	40	0	N	0.13	0.09	0.1	0.2	0.10	0	8	N	N	Tr
1166	**Cook-in-sauces,** canned	Tr	N	0	N	Tr	0.01	0.1	0.1	0.03	0	1	N	N	Tr
1167	**Curry sauce,** canned	0	N	0	N	Tr	0.03	0.1	0.2	0.02	0	N	N	N	Tr
1168	**Horseradish sauce**	Tr	Tr	Tr	N	N	N	N	N	N	Tr	N	N	N	Tr
1169	**Mint sauce**	0	Tr	0	Tr	Tr	Tr	Tr	0.3	Tr	0	Tr	Tr	Tr	Tr
1170	**Pasta sauce,** tomato based	0	N	0	N	0.06	0.50	0.1	0.3	0.06	0	10	N	N	0
1171	**Soy sauce**	0	0	0	N	0.05	0.13	3.4	1.4	N	0	11	N	N	0
1172	**Tomato ketchup**	0	230	0	N	1.00	0.09	2.1	0.3	0.03	0	1	N	N	2
1173	**Tomato sauce**	26	1070	0.27	1.47	0.11	0.02	1.2	0.4	0.14	0	9	0.22	1.2	8

Soups, sauces and miscellaneous foods *continued*

Composition of food per 100g

No.	Food	Description and main data sources	Edible proportion	Water g	Protein g	Fat g	Carbo-hydrate g	Energy value kcal	kJ
	Miscellaneous foods								
1174	**Baking powder**	6 samples of the same brand	1.00	6.3	5.2	Tr	37.8	163	693
1175	**Bovril**	9 jars	1.00	38.7	38.0[a]	0.7	2.9	169	718
1176	**Gelatin**	Literature sources	1.00	13.0	84.4	0	0	338	1435
1177	**Gravy instant granules**	7 samples, 3 brands	1.00	4.0	4.4	32.5	40.6	462	1927
1178	*made up with water*	Calculated from 23.5g granules and 300ml water	1.00	93.0	0.3	2.4	2.9	33	139
1179	**Marmite**	7 jars	1.00	25.4	39.7[a]	0.7	1.8	172	730
1180	**Mustard**, smooth	10 samples, 7 types; English and French	1.00	63.7	7.1	8.2	9.7	139	579
1181	wholegrain	9 samples, 5 brands	1.00	65.0	8.2	10.2	4.2	140	584
1182	**Oxo cubes**	10 samples	1.00	9.1	38.3[a]	3.4	12.0	229	969
1183	**Salt**, block	2 samples	1.00	0.2	0	0	0	0	0
1184	table	2 samples	1.00	Tr	0	0	0	0	0
1185	**Vinegar**[b]	4 samples	1.00	N	0.4	0	0.6	4	16
1186	**Water**	Included for recipe calculation	1.00	100.0	0	0	0	0	0
1187	**Yeast**, bakers, *compressed*	Literature sources	1.00	70.0	11.4[a]	0.4	1.1	53	226
1188	dried	Literature sources	1.00	5.0	35.6[a]	1.5	3.5	169	717

[a] (Total N – purine N) × 6.25

[b] Contains 4.8ml acetic acid per 100ml

Soups, sauces and miscellaneous foods *continued*

Composition of food per 100g

No.	Food	Total nitrogen g	Fatty acids			Cholesterol mg	Starch g	Total sugars g	Dietary fibre	
			Satd g	Mono unsatd g	Poly unsatd g				Southgate method g	Englyst method g
Miscellaneous foods										
1174	**Baking powder**	0.91	0	0	0	0	37.8	Tr	0	0
1175	**Bovril**	6.25[a]	N	N	N	N	2.9	0	0	0
1176	**Gelatin**	15.20	0	0	0	0	0	0	0	0
1177	**Gravy instant granules**	0.70	N	N	N	N	39.3	1.3	Tr	Tr
1178	*made up with water*	0.05	N	N	N	N	2.9	0.1	Tr	Tr
1179	**Marmite**	6.62[b]	N	N	N	0	1.8	0	0	0
1180	**Mustard**, smooth	1.14	0.5	5.8	1.6	0	1.9	7.8	N	N
1181	wholegrain	1.31	0.6	7.2	1.9	0	0.3	3.9	N	4.9
1182	**Oxo cubes**	6.29[a]	N	N	N	Tr	12.0	0	0	0
1183	**Salt**, block	0	0	0	0	0	0	0	0	0
1184	table	0	0	0	0	0	0	0	0	0
1185	**Vinegar**	0.07	0	0	0	0	0	0.6	0	0
1186	**Water**	0	0	0	0	0	0	0	0	0
1187	**Yeast**, bakers, *compressed*	2.02[c]	N	N	N	0	1.1	Tr	6.2	N
1188	dried	6.32[c]	N	N	N	0	(3.5)	Tr	(19.7)	N

[a] Includes 0.17g purine nitrogen

[b] Includes 0.27g purine nitrogen

[c] Purine nitrogen forms about 10 per cent of the total nitrogen

Soups, sauces and miscellaneous foods *continued*

1174 to 1188

Inorganic constituents per 100g

No.	Food	mg										μg	
		Na	K	Ca	Mg	P	Fe	Cu	Zn	Cl	Mn	Se	I
	Miscellaneous foods												
1174	**Baking powder**	11800[a]	49	1130[a]	9	8430[a]	Tr	Tr	2.8	29	Tr	Tr	Tr
1175	**Bovril**	4800	1200	40	61	590	14.0	0.45	1.8	6800	0.1	N	N
1176	**Gelatin**	N	N	N	N	N	N	0.05	0.2	N	N	N	N
1177	**Gravy instant granules**	6330	150	22	15	71	0.5	0.24	0.3	10000	0.4	N	N
1178	*made up with water*	460	10	1	1	5	Tr	0.02	Tr	730	Tr	Tr	Tr
1179	**Marmite**	4500	2600	95	180	1700	3.7	0.30	2.1	6600	0.2	N	49
1180	**Mustard, smooth**	2950	200	70	82	190	2.9	0.19	1.0	3550	0.7	N	N
1181	wholegrain	1620	220	120	93	200	2.8	0.21	1.2	2210	0.7	N	N
1182	**Oxo cubes**	10300	730	180	59	360	24.5	0.71	N	16000	N	N	44
1183	**Salt, block**	38700	Tr	230	140	Tr	0.3	0.39	N	59600	N	N	N
1184	table	38850	Tr	29	290	8	0.2	0.10	N	59900	N	N	44[b]
1185	**Vinegar**	20	89	15	22	32	0.5	0.04	0.1	47	N	(1)	N
1186	**Water**	N	N	N[c]	N	N	N	N	N	N	Tr	Tr	Tr
1187	**Yeast**, bakers, *compressed*	16	610	25	59	390	5.0	1.60	3.2	20	Tr	N	Tr
1188	dried	50	2000	80	230	1290	20.0	5.00	8.0	N	N	N	N

[a] The sodium, calcium and phosphorus content will depend on the brand

[b] Iodised salt contains 3100μg iodine per 100g

[c] The calcium content can range from Tr to 16mg per 100g

Soups, sauces and miscellaneous foods *continued*

No.	Food	Retinol	Carotene	Vitamin D	Vitamin E	Thiamin	Ribo-flavin	Niacin	Trypt 60	Vitamin B6	Vitamin B12	Folate	Panto-thenate	Biotin	Vitamin C
		µg	µg	µg	mg	mg	mg	mg	mg	mg	µg	µg	mg	µg	mg
Miscellaneous foods															
1174	**Baking powder**	0	0	0	Tr	Tr	Tr	Tr	1.0	Tr	0	Tr	Tr	Tr	0
1175	**Bovril**	0	0	0	N	9.10	7.40	82.0	3.0	0.53	8.3	1040	N	N	0
1176	**Gelatin**	0	0	0	0	Tr	Tr	Tr	Tr	Tr	0	Tr	Tr	Tr	0
1177	**Gravy instant granules**	N	Tr	Tr	N	N	N	N	0.8	N	Tr	Tr	N	N	0
1178	*made up with water*	N	Tr	Tr	N	Tr	Tr	Tr	0.1	Tr	Tr	Tr	Tr	Tr	0
1179	**Marmite**	0	0	0	N	3.10	11.00	58.0	9.0	1.30	0.5	1010	N	N	0
1180	**Mustard**, smooth	0	N	0	N	N	N	N	2.1	N	0	0	N	N	0
1181	wholegrain	0	N	0	N	N	N	N	2.4	N	0	0	N	N	0
1182	**Oxo cubes**	0	0	0	N	N	N	N	N	N	N	N	N	N	0
1183	**Salt**, block	0	0	0	0	0	0	0	0	0	0	0	0	0	0
1184	table	0	0	0	0	0	0	0	0	0	0	0	0	0	0
1185	**Vinegar**	0	0	0	0	0	0	0	0	0	0	0	0	0	0
1186	**Water**	0	0	0	0	0	0	0	0	0	0	0	0	0	0
1187	**Yeast**, bakers, *compressed*	0	Tr	0	Tr	0.71	1.70	11.0	2.0	0.60	Tr	1250	3.50	60	Tr
1188	*dried*	0	Tr	0	Tr	2.33[a]	4.00	8.5	7.0	2.00	Tr	4000	11.00	200	Tr

[a] Value for bakers' yeast. Brewers' yeast contains 15.6mg thiamin per 100g

The

Appendices

ORGANIC ACIDS

Organic acids are minor constituents of many foods, but in some the concentrations of individual acids may be nutritionally significant. The organic acids most frequently found in foods are listed below. This list is not exclusive and many hydroxy and other organic acids occur in foods, usually in low concentrations. For plant foods the concentrations of organic acids present can be extremely variable depending on the conditions under which the plant has been grown, the state of maturity of the food and changes occurring during the interval between harvest and consumption (Hulme, 1971).

Major organic acids in foods

Acetic acid is present in vinegar and consequently all pickled foods. It is also found in fermented food products.

Citric and malic acids are the major organic acids in most fruits, vegetables and their products. The relative proportions of these acids vary according to the variety of fruit or vegetable. Many manufactured products contain added citric acid.

Lactic acid is present in fermented foods such as yogurts and sauerkraut. Measurable levels are also present in cheeses.

Oxalic acid, often as the calcium salt, is found in many vegetables as well as in some fruits and nuts. Inclusions of oxalate crystals can frequently be seen in sections of plant tissue. It has no nutritional value, indeed it renders calcium unavailable for absorption and at higher concentrations is toxic.

Tartaric acid is the major organic acid in wines and in the fruit of the tamarind.

Amounts in foods

Fruit

Lemons and limes contain the highest concentrations at between 3 and 4.5g/100g. Lemon juice may contain up to 6g/100ml of mainly citric acid. Levels of between 1 and 3g/100g have been reported in apricots, berries, cherries, citrus fruit (excluding lemons and limes), currants, nectarines, kiwifruit, plums, pomegranates and rhubarb. Most other fruits contain less than 1g/100g total organic acids.

Vegetables

The total organic acids in vegetables range between 0.1 and 0.8g/100g (Hartman and Hillig, 1934). In most vegetables recently examined, the concentration did not exceed 0.5g/100g (Cashel et al., 1989). For practical nutritional purposes the organic acids in vegetables can be discounted in energy calculations.

Cheeses

In general soft and mould-covered varieties of cheese contain less than 1g/100g and hard cheeses and cheese spreads contain from 1 to 1.5g/100g.

Beverages

Fresh orange juice contains citric acid at levels of 1g/100ml with malic acid present at lower levels of up to 0.2g/100ml. In most other fruit and fruit-based drinks the concentrations of organic acids are less than 1g/100ml. Beverages designated as 'low calorie' may have only minor amounts of energy from organic acids (usually less than 1 kcal/100ml) but some have citric acid concentrations which would provide up to 7 kcal/100ml.

Vinegar

Vinegar contains about 5g acetic acid/100ml providing 17 kcal per 100ml, or 65kJ/15kcal per 100ml for labelling purposes.

Energy value

The organic acids that are normal constituents of metabolic pathways will provide energy to the body. The energy conversion factors in Table 14 are the heats of combustion of the major acids. It is reasonable to use these as the acids listed are completely absorbed and metabolised to carbon dioxide and water. Oxalic and tartaric acids are not included in this table since they are probably not metabolised to any significant extent to provide energy.

Table 14 *Energy value of the major organic acids in foods*

Acid	Energy value (per g)	
	kcal	kJ
Acetic acid	3.5	14.6
Citric acid	2.4	10.3
Malic acid	2.4	10.0
Lactic acid	3.6	15.1

The methods which have been used for the analysis of foods in the Tables are shown; usually the first reference given is the most recent.

The nutrient values quoted in the Tables have been determined by a variety of methods. Although most give results of the same order of accuracy, with new methods merely improving the efficiency of analysis, some methods give different results and these have been documented in the Tables only where they appear to be substantial.

The following abbreviations are used in the text:—
>   GLC   Gas liquid chromatography
>   HPLC   High performance liquid chromatography
>   ICPOES Inductively coupled plasma optical emission spectrophotometry

Nutrient	Method
Water	Freeze drying Vacuum drying at 70°C Air oven at 100°C
Nitrogen	Kjeldahl procedure.
Fat	Werner Schmidt Weibuhl Stoldt } Egan *et al.* (1981) Rose-Gottlieb
Fatty acids	GLC of methyl esters (IUPAC, 1976)
Cholesterol	GLC
Alcohol	Standard Inland Revenue distillation method
Carbohydrates	
Total sugars (as monosaccharides)	Boehringer enzyme kit (Egan *et al.*, 1981) HPLC (Southgate *et al.*, 1978; Dean, 1978) Colorimetry (Southgate, 1976)
Starch	Enzymatic hydrolysis and measurement of glucose (Dean, 1978) Polarimetry (Egan *et al.*, 1981)
Fibre	
Southgate dietary fibre	Southgate (1969) Wenlock *et al.* (1985)
Non-starch polysaccharides	Englyst and Cummings (1988) Englyst and Cummings (1984) Englyst *et al.* (1982)

Nutrient	Method
Inorganics	
Sodium	Emission spectrometry (Moxon, 1983)
	Atomic absorption spectrophotometry
	Flame photometry
Potassium	Emission spectrometry
	Atomic absorption spectrophotometry
	Flame photometry
Calcium	ICPOES
	Atomic absorption spectrophotometry
	Titrimetry
Magnesium	ICPOES
Copper	Atomic absorption spectrophotometry
Iron	Colorimetry
Zinc	
Phosphorus	ICPOES
	Colorimetry
Chloride	Colorimetry
	Titrimetry
Manganese	ICPOES
	Atomic absorption spectrophotometry
Selenium	Fluorimetry (Michie *et al.*, 1978)
Iodine	Spectrophotometry (Moxon, 1980)
Vitamins	
Fat soluble vitamins	
Retinol	Reverse phase HPLC
β-Carotene	Chromatographic separation and
Other carotenoids	absorption spectrophotometry
Vitamin D	Reverse phase HPLC.
	Biological assay and spectrophotometry
	GLC
Vitamin E	Normal phase HPLC
	Reverse phase HPLC
	Colorimetry combined with GLC (Christie *et al.*, 1973)

Nutrient	Method
Water soluble vitamins	
Thiamin	HPLC with fluorimetric detection (Finglas and Faulks, 1984)
	Fluorimetry (Society of Public Analysts and Other Analytical Chemists: Analytical Methods Committee, 1951)
	Microbiological assay (Bell, 1974)
Riboflavin	Fluorimetry (Finglas and Faulks, 1984)
	Microbiological assay (Bell, 1974)
Niacin	HPLC (Kwiatowska *et al.*, 1989)
	Microbiological assay (Bell, 1974)
Vitamin B_6	HPLC with fluorimetric detection (Brubacher *et al.*, 1985)
	Microbiological assay (Bell, 1974)
Vitamin B_{12} Pantothenate Biotin	Microbiological assay (Bell, 1974)
Folate	Microbiological assay (Phillips and Wright, 1983; Bell, 1974)
Vitamin C	
Ascorbic acid	Titrimetry (AOAC, 1975)
Ascorbic acid and dehydroascorbic acid	Fluorimetry (AOAC, 1975)

WEIGHT CHANGES ON PREPARATION OF FOODS

The figures below show the percentage changes in weight recorded during the cooking of foods included in this edition. The values were obtained by Holland *et al.* (1991), Wiles *et al.* (1980), Paul and Southgate (1977), McCance and Shipp (1933), and from previously unpublished determinations where a measure of weight change was available. The values should be used for information only; for more accurate figures users should make their own determinations. The weight changes during cooking of recipe dishes have been included with each recipe in Appendix 3.5.

The majority of changes result from the loss or gain of water, but for many meats and fried foods there will also have been a loss or gain of fat. The values have been calculated as:

$$\frac{\text{Weight of cooked food or dish} - \text{Weight of raw food(s)}}{\text{Weight of raw food(s)}} \times 100$$

A value of +200 thus means <u>not</u> that the food doubled its weight, but that it gained twice its original weight on cooking (i.e. <u>tripled</u> in weight), because:

$$\frac{300 - 100}{100} \times 100 = +200$$

A plus sign (+) indicates that the food or dish gained weight on cooking while a minus sign (−) shows that it lost weight.

For root and leafy vegetables boiled in water there is little difference in weight between the raw and cooked food.

CEREALS AND CEREAL PRODUCTS

Rice and pasta

Brown rice, boiled	+153
Savoury rice, cooked	+197
White rice, easy cook, boiled	+177
Macaroni, boiled	+312
Egg noodles, boiled	+479
Spaghetti, white, boiled	+244
wholemeal, boiled	+190

Bread

Brown bread, toasted	−22
Currant bread, toasted	−12
Hovis, toasted	−22
White bread, fried	−29
toasted	−18
'with added fibre', toasted	−16
Wholemeal bread, toasted	−15

Buns and pastries

Crumpets, toasted	−11
Teacakes, toasted	−10

MEAT AND MEAT PRODUCTS

Bacon

Collar, boiled	−29
Gammon, boiled	−33
slices, grilled	−36
Rashers, lean and fat, fried, back	−41
fried, middle	−43
fried, streaky	−40
grilled, back	−41
grilled, middle	−44
grilled, streaky	−45

Beef

Brisket, boiled	−34
Forerib, roast	−20
Mince, stewed	−28
Rump steak, fried	−29
grilled	−24
Silverside, boiled	−41
Sirloin, roast	−22
Stewing steak, stewed	−36
Topside, roast	−24

Lamb

Breast, roast	−19
Chops, loin, grilled	−31
Cutlets, grilled	−24
Leg, roast	−29
Scrag and neck, stewed	−25
Shoulder, roast	−23

Pork

Belly, grilled	−26
Chops, grilled	−37
Leg, roast	−28

Poultry and game

Chicken, roast	−14
boiled	−22
wing quarter, roast	−27
leg quarter, roast	−33
breaded, fried	−5
Duck, roast	−33
Goose, roast	−33
Grouse, roast	−23
Partridge, roast	−35
Pheasant, roast	−29
Pigeon, roast	−36
Turkey, roast	−18
Hare, stewed	−40
Rabbit, stewed	−31

Offal

Heart, ox, stewed	−29
Kidney, lamb, fried	−37
ox, stewed	−39
pig, stewed	−39
Liver, calf, fried	−16
chicken, fried	−16
lamb, fried	−12
ox, stewed	−18
pig, stewed	−26
Sweetbread, boiled and fried	−40
Tripe, boiled	−44

Meat products

Haggis, boiled	−5
Sausages beef, fried	−25
grilled	−25
Sausages pork, fried	−30
grilled	−28
low fat, fried	−18
grilled	−27
Beefburgers, fried	−24
Grillsteaks, grilled	−25

Meat, cooked dishes

Beef curry, retail, heated	−7
Chicken curry, retail, heated	−11

FISH AND FISH PRODUCTS

White fish

Cod, baked fillets	−19
poached fillets	−14
frozen, grilled steaks	−15
dried, salted, boiled	+19
Haddock, steamed	−22
in crumbs, fried	+25
smoked, steamed	−15
Halibut, steamed	−13

Lemon sole, steamed	−13
fried	+15
Plaice, steamed	−9
fried in crumbs	+21
Saithe, steamed	−24
Whiting, steamed	+16
in crumbs, fried	+13

Fatty fish

Herring, fried	−13
grilled	−9
Kipper, baked	−17
Mackerel, fried	−16
Salmon, steamed	−10
Trout, steamed	−18
Whitebait, fried	−23

Crustacea

Crab, boiled	−20
Lobster, boiled	−19
Scampi in breadcrumbs, fried	−23

Molluscs

Mussels, boiled	−33

Fish products and dishes

Fish cakes, fried	+2
Fish fingers, fried	−10
grilled	−7
Roe, cod, hard, fried	−7
herring, soft, fried	−20

VEGETABLES AND VEGETABLE PRODUCTS

Beans and lentils

Aduki beans, soaked and boiled	+115
Black gram, urad gram, soaked and boiled	+208
Blackeye beans, soaked and boiled	+164
Chick peas, whole, soaked and boiled	+163
Lentils, green and brown, boiled	+139
red split, boiled	+227
Mung beans, whole, soaked and boiled	+199
Red kidney beans, soaked and boiled	+161
Soya beans, soaked and boiled	+156

Calculation of cooked edible matter from raw foods

Although the Tables show the amount of edible material in raw foods and cooked foods, it is sometimes necessary to estimate the amount of cooked edible material that would be obtained from a known weight of the raw food 'as purchased'. This is done by combining the percentage weight loss with the edible matter as a proportion of the cooked food, as follows:

Cooked edible matter as a proportion of raw 'as purchased' food =

$$\frac{\text{Edible matter of cooked food} \times (100 - \% \text{ weight loss on cooking})}{100}$$

For example, the weight loss on grilling lamb chops is 31% and the edible proportion of grilled lamb chops weighed with bone (food no. 396) is 0.79. 200g of raw lamb chops with lean, fat and bone will therefore yield:

$$200 \times \frac{0.79 \times (100 - 31)}{100} = 110g \quad \text{cooked lamb to eat}$$

3.4   COOKED FOODS AND DISHES

Calculation of the composition of dishes prepared from recipes

The composition of cooked dishes in this book has been calculated, as in previous editions, from the recipes, the composition of the ingredients and the changes in weight on cooking. The change in weight on cooking is usually only due to the evaporation of water or to its gain by absorption. The composition of dishes where the method of preparation also involves a change in fat content cannot be calculated directly in this way. In these cases the cooked dishes were either analysed for fat and water before the calculations were made or the weight change corrected for fat uptake measured after preparation.

The method of calculation was as follows. The weights of the raw ingredients were used to calculate the total amounts of nutrients in the dish. A correction for any wastage due to ingredients left on utensils and in the vessels used in preparation was made at this stage. The weight of the raw dish was then measured, using a scale weighing to about 1g . The dish was then cooked and the dish reweighed. (A minor correction to allow for the difference between weighing the dish hot and at room temperature is not usually necessary.) Where the difference in weight was accounted for by water, the composition of the cooked dish was calculated as follows:

$$\text{Composition of cooked dish per 100g} = \frac{\text{Total nutrients in raw ingredients}}{\text{weight of the cooked dish}} \times 100$$

$$\text{Water content of cooked dish per 100g} = \frac{\text{Water in raw ingredients} - \text{weight loss on cooking}}{\text{weight of the cooked dish}} \times 100$$

An example of this calculation is shown in Table 15.

If a recipe is to be calculated from the ingredients, but the weight of the cooked dish is not known, this may be estimated by using the % weight change from a similar recipe as follows (provided that all the weight change can be attributed to water):

$$\text{Wt of cooked dish} = \frac{\text{Wt of raw ingredients} \times (100 - \% \text{ wt loss of a similar dish})}{100}$$

For recipes which gain weight on cooking, for example dumplings:

$$\text{Wt of cooked dish} = \frac{\text{Wt of raw ingredients} \times (100 + \% \text{ wt gain of a similar dish})}{100}$$

Table 15 *Custard, made up with whole milk*

		Amount in recipe	Amounts contributed			
			Protein	Fat	Carbohydrate	Etc.
Ingredient		g	g	g	g	
Milk		515(500ml)	16.5	20.1	24.7	
Custard powder		25	0.1	0.2	23.0	
Sugar		25	0	0	26.3	
Total in recipe	(a)	565	16.6	20.3	74.0	
Cooked weight	(b)	447				
Composition of cooked dish (per 100g)	(c)		3.7	4.5	16.6	

i.e. $c = \dfrac{a}{b} \times 100$

Vitamin loss estimation in foods and recipe calculations

The vitamin losses in cooked foods and dishes were estimated using the values shown in the tables below. For recipes a set of factors was applied to the whole dish depending on the major food ingredient; different factors were not assigned to each ingredient individually. These factors are guidelines for the losses of vitamins and for more accurate information the foods or composite dish should be analysed. Actual losses to the food or dish will be affected by the time of cooking, state of the food and the amount of heat applied.

Table 16 *Cereals and cereal-based recipes: typical percentage losses of vitamins on cooking*

	Boiling	Baking	Toasting
Thiamin	40	25[a]	15
Riboflavin	40	15	
Niacin	40	5	
Vitamin B$_6$	40	25	
Folate	50	50	
Pantothenate	40	25	
Biotin	40	0	
Vitamin C		[b]	

[a] 15% in breadmaking
[b] Losses of vitamin C (for example in fruit pies) were assigned for each individual recipe and ranged from 10% to 70%

Table 17 *Milk and milk-based recipes: typical percentage losses of vitamins on cooking*

	Boiling[a]	Sauces[b]	Baked dishes
Vitamin E	20		
Thiamin	10	20	25
Riboflavin	10		15
Niacin			5
Vitamin B$_6$	10	20	25
Vitamin B$_{12}$	5		
Folate	20	50	50
Pantothenate	10	20	25
Vitamin C	50	50	

[a] In milk-based drinks, custards, etc. [b] For example, for cheese sauce

Table 18 *Eggs and egg dishes: typical percentage losses of vitamins on cooking*

	Scrambled	Omelette	Baked dishes
Thiamin	5	5	15
Riboflavin	20	20	15
Niacin	5	5	5
Vitamin B$_6$	15	15	25
Folate	30	30	50
Pantothenate	15	15	25

Table 19 *Meats and meat-based dishes: typical percentage losses of vitamins on cooking (average loss with range)*

	Roasting, frying and grilling	Stewing and boiling[a]	Cooked dishes
Thiamin	20 (0-40)	60 (40-70)	20
Riboflavin	20 (0-30)	30 (0-40)	20
Niacin	20 (10-30)	50 (30-70)	20
Vitamin B$_6$	20 (0-40)	50 (30-60)	20
Folate	-	30[b]	50
Pantothenate	20	40 (30-50)	20

All methods			
Vitamin A	0	Vitamin B$_{12}$	20 (10-50)
Vitamin E	20 (0-40)	Biotin	10 (0-30)
Vitamin C	20 (0-30) liver only		

[a] These losses refer to the meat alone. Water-soluble vitamins are leached into the gravies and liquors; if these are used the overall losses with these cooking methods are smaller
[b] Liver and kidney only. The content of folate in other meats is too low to make meaningful calculations of losses

Table 20 *Fish: typical percentage losses of vitamins on cooking[a]*

	Poaching	Baking	Frying and grilling
Vitamin E	(0)	0	0
Thiamin	10	30	20
Riboflavin	0	20	20
Niacin	10	20	20
Vitamin B_6	0	10	20
Vitamin B_{12}	0	10	0
Folate	0	20	0
Pantothenate	20	20	20
Biotin	10	10	10
Vitamin C	-	-	20[b]

[a] These losses are based on those seen in a small number of cod samples
[b] Used for roe

Table 21 *Vegetables and vegetable-based dishes: typical percentage losses of vitamins on cooking*

	Boiling	Frying	Cooked dishes
Carotene	-	-	0
Vitamin E	0	0	0
Thiamin	35	20	20
Riboflavin	20	0	20
Niacin	30	0	20
Vitamin B_6	40	25	20
Folate	40	55	50
Pantothenate	-	-	20
Biotin	-	-	20
Vitamin C	45	30	50

Table 22 *Fruit: typical percentage losses of vitamins on stewing*

Thiamin	25
Riboflavin	25
Niacin	25
Vitamin B_6	20
Folate	80
Pantothenate	25
Biotin	25
Vitamin C	25

- In the recipe calculations, an egg has been assumed to weigh 50g. A level teaspoon refers to a standard 5ml spoon and has been taken to hold 5g salt and spices, and 3.5g baking powder, bicarbonate of soda and cream of tartar.

- The fat used in domestic recipes was butter or margarine as indicated. In recipes for commercial products, however, it was taken as unfortified margarine and these products thus contain less vitamin A and vitamin D.

- Unless specified the recipes use whole pasteurised milk, fresh cream, Cheddar cheese, non-dairy vanilla ice cream and plain white flour.

- The baking powder used was a proprietory preparation whose composition is given in the Tables. This contains calcium acid phosphate, sodium bicarbonate, sodium pyrophosphate and flour; the use of another brand could result in a different composition in the cooked dish with respect to sodium, calcium and phosphorus.

- Where the quantity of vegetable oil is in brackets, this represents a measured amount absorbed during deep fat frying.

- For all the recipes, the quantities are reported exactly as in the original recipe source.

- The majority of weight losses were determined experimentally, although a few represent estimations from similar foods.

24 Fried rice

550g boiled rice	2g salt
168g chopped onion	pepper
21g dripping or lard	1g spices
21g clove garlic	

Fry onion and garlic until soft. Add boiled rice and seasoning. Fry until fat has been absorbed and rice is fully coated.

Weight loss: 5.6%

43 Naan bread

336g flour	2 tsp sugar
1 tsp salt	0.5g bakers yeast
112g natural yogurt	56g ghee
150ml milk	

Sift flour with salt and mix in yogurt. Add warmed milk, sugar and yeast. Knead well for about 15 mins then leave for 4-5 hrs. Shape and bake 230°C/mark 8 for 10 mins.

Weight loss: 16.8%

60 Croissants

450g strong flour
170g unfortified margarine
28g dry yeast
28g lard
1 egg
1 tsp salt
240ml water

Glaze:
30ml water
2½ g caster sugar
1 egg

Blend yeast and water, sift flour, salt and rub in lard. Mix together and knead until smooth. Roll out, dot with margarine and fold into three. Repeat twice, cover and rest for 30 mins. Repeat process a further 3 times, then place in fridge for 1 hr. Roll out, trim, cut and shape into crescents. After 30 mins brush with glaze and bake for 20 mins at 220°C/mark 7.

Weight loss: 15%

76 Porridge, made with water

60g oatmeal
7g salt

500ml water

Weight loss: 14%

77 Porridge, made with whole milk

60g oatmeal
7g salt

500ml milk

Weight loss: 14%

98 Flapjacks

120g rolled oats
90g margarine
60g golden syrup

60g brown sugar
2g ginger

Melt fat, add sugar and syrup. Work in the oats. Press into a greased sandwich tin and bake at 170°C/mark 3 for 30 mins.

Weight loss: 5%

100 Homemade biscuits - creaming method

200g plain flour
1 egg

100g margarine
100g caster sugar

Cream fat and sugar. Mix in egg, then flour and knead the dough lightly until smooth. Roll out thinly, prick and shape. Bake 10-15 mins at 180°C/mark 4.

Weight loss: 10%

101 Jaffa cakes

33.1% baked sponge base
39.6% orange jelly
27.3% plain chocolate

Recipe from Flour Milling and Baking Research Association.

106 Shortbread

200g flour   100g butter
50g caster sugar

Beat the butter and sugar to a cream. Mix in the flour and knead until smooth.
Press into a flat tin to about 2cms thick. Bake for about 45 mins at 170°C/mark 3.

Weight loss: 10%

108 Wholemeal crackers/Farmhouse crackers

105g fat   630g plain flour
11.9g salt   210g wholemeal flour
15.4g bakers yeast   2.1g bicarbonate of soda

Recipe from Flour Milling and Baking Research Association.

Weight loss: 11%

109 Battenburg cake

100g flour   67.5g water
75g margarine   8g skimmed milk powder
120g sugar   5g baking powder
90g eggs   2g salt

58g marzipan   8g jam

Recipe from Flour Milling and Baking Research Association.

Weight loss: 10% cake

110 Cake mix - made up

205g powder mix
1 egg
85ml water

Add egg and some of the water to the powder, whisk vigorously. Add remaining
water, repeat whisking until smooth and creamy. Bake at 190°C/mark 5 for 20-25
mins.

Weight loss: 11.2%

111 Crispie cakes

112g plain chocolate
33g crisp rice cereal
33g corn flake type cereal

Melt chocolate in a bowl over hot water. Stir in cereals. Place in cases and allow
to cool and set.

114 Rich fruit cake

200g margarine
200g brown sugar
4 eggs
20g black treacle
20ml brandy

250g flour
¼ tsp salt
750g mixed fruit
150g mixed glacé fruit, chopped
1 tsp mixed spice

Cream the fat and sugar. Beat the eggs, treacle and brandy. Fold in the sifted flour and spices and mix in the fruit. Turn into a 20cm cake tin. Bake for 4 hrs at 150°C/mark 2.

Weight loss: 5%

115 Iced rich fruit cake

1680g fruit cake, rich
70g apricot jam
410g marzipan

Royal icing:
300g icing sugar
1 egg white
1 tsp lemon juice

Make the cake as in Rich Fruit Cake (No. 114) recipe. When cold spread with a thin layer of apricot jam and cover with marzipan. Make the icing by beating the egg whites and icing sugar; finally add the lemon juice.

116 Wholemeal fruit cake

200g margarine
200g brown sugar
3 eggs
200g plain flour
2 tsp baking powder

4g mixed spice
200g wholemeal flour
200g mixed fruit
100ml milk

Cream fat and sugar, beat in eggs. Sift white flour, baking powder and spice, add creamed mixture together with wholemeal flour and fruit. Add milk until soft. Bake for 1½-2 hrs at 180°C/mark 4.

Weight loss: 5%

117 Fresh cream gateau

50% sponge without fat
35% whipping cream
15% sugar

Proportions are derived from recipes and shop bought samples.

119 Sponge cake with fat

150g flour
1 tsp baking powder
150g margarine

150g caster sugar
3 eggs

Cream the fat and sugar until light and fluffy. Add the beaten egg a little at a time and beat well. Fold in the sifted flour and baking powder. Bake for about 20 mins at 190°C/mark 5.

Weight loss: 12.9%

120 **Sponge cake, fatless**

4 eggs
100g caster sugar
100g flour

Whisk the eggs and sugar in a basin over hot water until stiff. Fold in flour. Bake for 25 mins at 190°C/mark 5.

Weight loss: 13.8%

122 **Sponge cake with butter icing**

70% sponge cake with fat
15% margarine
15% sugar

Proportions are derived from recipe review.

124/125 **Flaky pastry**

200g flour   ½ tsp salt
75g margarine   85ml water
75g lard   10ml lemon juice

Divide fat into 4 portions. Sift flour and salt, rub in one portion of fat. Mix with water and lemon juice, then knead until smooth and leave for 15 mins. Roll out, dot two-thirds with another fat portion and fold into 3. Roll out and repeat process with remaining 2 fat portions. Bake at 220°C/mark 7.

Weight loss: 24.3%

126/127 **Shortcrust pastry**

200g flour   ½ tsp salt
50g margarine   30ml water
50g lard

Rub the fat into the flour, mix to a stiff dough with the water, roll out and bake at 200°C/mark 6.

Weight loss: 13.8%

128/129 **Wholemeal pastry**

200g wholemeal flour   2g salt
50g margarine   30ml water
50g lard

Rub the fat into the flour, mix to a stiff dough with the water, roll out and bake at 220°C/mark 7.

Weight loss: 13.6%

130 Chelsea buns

200g strong flour
85g skimmed milk
65g unfortified margarine
45g sugar

35g eggs
15g yeast
55g currants
2g salt

Recipe from Flour Milling and Baking Research Association.

Weight loss: 15%

131 Cream horns

300g cooked flaky pastry (unfortified margarine)
190g whipping cream
35g jam

Recipe from Flour Milling and Baking Research Association.

138 Eccles cakes

200g strong flour
100g water
180g unfortified margarine
20g cooking fat

55g butter
75g sugar
370g currants

Recipe from Flour Milling and Baking Research Association.

Weight loss: 19%

141 Hot cross buns

450g strong flour
28g fresh yeast
1 egg
pinch salt
56g margarine
112g currants
1g cinnamon
1g nutmeg
2g mixed spice

45g peel
150ml milk
60ml water
56g caster sugar

Glaze:
45g sugar
30ml milk
30ml water

Cream yeast with milk and add salt. Add to flour and eggs, and mix. Knead for 10 mins. Sprinkle with sugar, dot with fat and leave for 30 mins. Mix fat, sugar and fruit into mixture and mould. Cut a cross on each bun, glaze and bake for 15 mins at 250°C/mark 9.

Weight loss: 15%

142 Jam tarts

200g raw shortcrust pastry   200g jam

Line about 10 tart tins with thinly rolled pastry. Fill each tart with jam and bake in a hot oven, 200°C/mark 6 for 10-15 mins.

Weight loss: 6.5%

144 Mince pies, individual

300g raw shortcrust pastry
200g mincemeat

Roll out the pastry and cut into rounds. Place half of the rounds in tart tins. Fill with mincemeat and cover with remaining pastry. Bake for about 20 mins at 190°C/mark 5.

Weight loss: 12.6%

146 Plain scones

200g flour   50g margarine
4 tsp baking powder   10g sugar
¼ tsp salt   125ml milk

Sift the flour, sugar and baking powder and rub in fat. Mix in the milk. Roll out and cut into rounds. Bake in a hot oven at 220°C/mark 7 for about 10 mins.

Weight loss: 18.5%

147 Wholemeal scones

200g wholemeal flour   50g margarine
14g baking powder   10g sugar
1g salt   125ml milk

Method as recipe for plain scones (No. 146).

Weight loss: 14%

148 Scotch pancakes

200g flour   25g caster sugar
½ tsp salt   1 egg
1 tsp cream of tartar   200ml milk
50g margarine
½ tsp bicarbonate of soda

Sift flour with salt and raising agent, rub in fat and mix in sugar. Add egg and milk to give a stiff batter. Cook by spoonsful on hot greased griddle.

Weight loss: 9.4%

150 Blackcurrant pie, pastry top and bottom

450g raw shortcrust pastry
450g blackcurrants
80g sugar

Method as for fruit pie, pastry top and bottom (No. 157).

Weight loss: 4.2%

151 Bread pudding

225g bread
275ml milk
50g melted butter
75g demarara sugar

4g mixed spice
1 beaten egg
175g dried fruit

Break bread into pieces, cover with milk and leave for 30 mins. Add remaining ingredients, mix well and bake for 1¼ hrs at 180°C/mark 4.

Weight loss: 24%

152 Christmas pudding

100g flour
300g fresh breadcrumbs
1 tsp mixed spice
½ tsp salt
125g suet
150g raisins
150g sultanas

150g currants
50g chopped mixed peel
30g ground almonds
150g brown sugar
3 eggs
15g treacle
150ml stout

Sift the flour, spices and salt into a basin and mix in all dry ingredients. Whisk the eggs, treacle and stout and stir thoroughly into dry ingredients. Put into well greased basins, cover with greased paper and foil. Boil for 6 hrs. Renew foil and store. Reboil for 2 hrs before serving.

Weight loss: 0%

154/155 Fruit crumble, plain or wholemeal

400g prepared fruit
50g margarine

100g plain or wholemeal flour
100g sugar

Prepare fruit. Arrange in a dish and sprinkle with sugar. Rub together the other ingredients and pile on top. Bake for 40 mins at 190°C/mark 5.

Weight loss: 7.4%

156/159 Fruit pie, one crust, plain or wholemeal

200g raw plain or wholemeal shortcrust pastry
450g prepared fruit
80g sugar

Place fruit in a pie dish and cover with pastry. Bake for 10-15 mins at 200°C/mark 6 to set pastry, then about 20 mins at 180°C/mark 4 to cook fruit.

Weight loss: 4.2%

157/160 Fruit pie, pastry top and bottom, plain or wholemeal

450g raw plain or wholemeal shortcrust pastry
450g fruit (e.g., apple, gooseberry, rhubarb, plum)
80g sugar

Line a pie dish with half the pastry. Fill with prepared fruit and sugar and cover with remaining pastry. Bake for 10-15 mins at 220°C/mark 7 to set the pastry, then for about 20-30 mins at 180°C/mark 4 to cook the fruit.

Weight loss: 4.2%

161 Lemon meringue pie

200g raw shortcrust pastry
2 lemons (80g juice)
2 eggs
125g caster sugar

25g cornflour
15g margarine
125ml water

Boil the cornflour, water, grated rind, lemon juice and 25g of sugar. Cool, stir in egg yolks, and pour mixture into the pastry case. Make a meringue with egg whites and rest of sugar; pile on top of the lemon mixture. Bake for 30 mins at 180°C/mark 4 until crisp and brown on top.

Weight loss: 19.0%

162 Sweet pancakes

100g flour
250ml whole milk
1 egg

50g lard (for pan)
50g sugar

Sieve the flour into a basin, add the egg and about 100ml of the milk, stirring until smooth. Add the rest of the milk and beat to a smooth batter. Heat a little of the lard in a frying pan and pour in enough batter to cover the bottom. Cook both sides and turn onto sugared paper. Dredge lightly with sugar. Repeat until all the batter is used, to give about 10 pancakes.

Weight loss: 20%

163 Pie with pie filling

450g raw shortcrust pastry
450g fruit pie filling

Method as for fruit pie, pastry top and bottom (No. 157).

Weight loss: 4.2%

164 Steamed sponge pudding

100g flour
1 tsp baking powder
50g margarine

50g caster sugar
1 egg
30ml milk

Cream the fat and sugar. Beat in the eggs a little at a time. Fold in the sifted flour and baking powder, adding milk to give a soft dropping consistency. Turn the mixture into a greased basin and steam for about 2½ hrs.

Weight gain: + 3.9%

165 Treacle tart

300g raw shortcrust pastry
250g golden syrup
50g fresh breadcrumbs

Line shallow tins with pastry, pour in the syrup and sprinkle with the breadcrumbs. Bake for 20-30 mins at 200°C/mark 6.

Weight loss: 0%

166 Cauliflower cheese

100g cheese, grated	25g margarine
1 small cauliflower (700g)	25g flour
100ml cauliflower water	250ml milk
½ level tsp salt	pepper

Boil cauliflower until just tender, break into florets. Drain saving 100ml water, place in a dish and keep warm. Make a white sauce from the margarine, flour, milk and cauliflower water. Add 75g cheese and season. Pour over the cauliflower and sprinkle with the remaining cheese. Brown under a grill or in a hot oven, 220°C/mark 7.

Weight loss: 14.6%

167 Dumplings

100g flour	1 tsp baking powder
45g suet	½ tsp salt
75g water	

Mix the dry ingredients together with the cold water to form a dough. Divide into balls, flour them and place in boiling water. Boil for 30 mins.

Weight gain: + 52.7%

168 Macaroni cheese

280g cooked macaroni	25g flour
350ml milk	100g grated cheese
25g margarine	½ tsp salt

Boil the macaroni and drain well. Make a white sauce from the margarine, flour and milk. Add 75g of the cheese and season. Add the macaroni and put in a pie dish. Sprinkle with remaining cheese and brown under the grill or in a hot oven at 220°C/mark 7.

Weight loss: 9.4%

169 Savoury pancakes

112g flour	56g lard (for pan)
300ml whole milk	¼ tsp salt
1 egg	

Method as for sweet pancakes (No. 162).

Weight loss: 20%

170 Cheese and tomato pizza

Dough:	*Topping:*
200g flour	200g tomatoes
1 tsp salt	150g cheese
1 tsp sugar	8 black olives (40g)
150ml warm water	20g oil
15g fresh yeast or 2 tsp dried yeast	

Make the dough, proving once. Knead and roll out shape. Leave for 10 mins. Arrange sliced or pulped tomatoes on top, then cheese and olives. Brush with oil. Bake for 30 mins at 230°C/mark 8.

Weight loss: 14.1%

173 **Risotto**

224g long grain rice	56g margarine
550g stock	1 tsp salt
84g chopped onion	1g pepper

Melt margarine, add onion and fry until soft. Add washed rice and stir over low heat for 10 mins. Pour in stock, bring to the boil and simmer until all is absorbed.

Weight loss: 37%

174 **Meat samosas**

80g flour	10g fresh green chillies
10g butter	15g chopped onion
25ml water	30g chopped potatoes
5g oil (for frying ingredients)	30g peas
70g minced lamb	(190g vegetable oil)

Make a dough with flour, butter and water, divide into two rounds and roll out. Sandwich these together with a little oil and cook in a frying pan without fat for 1 min on each side. Divide into 4 quarters and separate to give 8 samosa cases. Fry minced lamb in oil with chillies, onion, potato and peas. Fill the cases with the mixture, fold over pastry edges to make a triangle, seal with flour and water paste. Deep fry in oil.

Weight loss: 13.5%

175 **Vegetable samosas**

80g flour	260g boiled potatoes
10g butter	45g chopped onion
25ml water	63g pea
25ml oil (for frying ingredients)	4½ g mixed spice
½ tsp salt	15ml water
15ml lemon juice	(190g vegetable oil)

Make cases from flour, butter and water as for meat samosas (No. 353). Fry vegetables, lemon juice and spices in oil for 5 mins. Add water and cook for a further 5 mins. Fill the cases with the mixture, fold over pastry edges to make the triangle, seal with flour and water paste. Deep fry in oil.

Weight loss: 25.9% for filling, 13.5% for pastry

177 Sage and onion stuffing

224g onion
112g white breadcrumbs
4g sage
1g salt

½ g pepper
1 egg
56g margarine

Slice onions and sage, parboil, drain and chop, mix with breadcrumbs. Melt margarine and add to stuffing. Mix thoroughly.

Weight loss: 19.3%

180 Yorkshire pudding

100g flour
1 tsp salt
1 egg

250ml milk
20g dripping

Sieve flour and salt into a basin. Break in the egg and add about 100ml of milk, stirring until smooth. Add the rest of the milk and beat to a smooth batter. Pour into a tin containing the hot dripping. Bake for about 40 mins at 220°C/mark 7.

Weight loss: 16%

220/221 Dream Topping made up with milk

1 packet Dream Topping powder (40g)
150ml whole or semi-skimmed milk

Sprinkle powder onto milk. Whisk briskly.

259 Tzatziki

250g Greek cows milk yogurt
5g fresh garlic
fresh chopped mint

213g cucumber
4g salt

Recipe from yogurt manufacturers.

264 Chocolate nut sundae

115g ice cream
45ml double cream
70g chocolate sauce

6g chopped nuts
wafer (1g)

Cover ice cream with whipped cream and chocolate sauce. Sprinkle with nuts, add wafer.

273 Lemon sorbet

568ml water
224g caster sugar
grated rind of 2 lemons

140ml lemon juice
65g egg whites

Boil water, lemon rind and sugar for 10 mins. Leave to cool and strain. Strain lemon juice and add to syrup. Freeze until nearly firm then beat in stiff egg whites. Continue freezing until firm.

Weight loss: 32.5% for syrup

276/277 Custard made up with milk

500ml whole or skimmed milk
25g custard powder
25g sugar

Blend custard powder with a little of the milk. Add sugar to remainder of milk and bring to the boil. Pour immediately over paste, stirring all the time. Return to pan, bring back to boiling point while stirring.

Weight loss: 20.9%

280/281 Instant dessert powder made up with milk

- 1 packet instant dessert powder (66g)
300ml whole or skimmed milk

Sprinkle powder onto milk. Whisk briskly.

282 Jelly made with water

130g jelly cubes   440ml water

Dissolve jelly cubes in hot water. Add rest of the cold water. Pour into a mould and allow to set.

283/284 Milk puddings

500ml whole or skimmed milk
50g rice, sago, semolina or tapioca
25g sugar

Simmer until cooked or bake in a moderate oven at 180°C/mark 4.

Weight loss: 19.1%

288 Trifle

75g sponge cake   250g custard
25g jam   25ml double cream
50g fruit juice   10g mixed nuts
75g tinned fruit   10g angelica and cherries
25ml sherry

Slit sponge cake, spread with jam and sandwich together. Cut into 4cm cubes. Soak in the fruit juice and sherry. Mix with fruit, cover with cold custard and decorate with whipped cream, nuts and angelica.

296 Scrambled eggs with milk

2 eggs   20g butter
15ml milk   ½ level tsp salt

Melt butter in pan, stir in beaten egg, milk and seasoning. Cook over gentle heat until mixture thickens.

Weight loss: 10.9%

298 Egg fried rice

35g vegetable oil
45g chopped onion

1½ beaten eggs
350g cooked white rice

Heat oil, add egg and remaining ingredients. Cook, turning mixture over, for 3 mins.

Weight loss: 17.3%

299 Meringue

4 egg whites

200g caster sugar

Whisk egg whites until stiff. Fold in the sugar. Pipe onto baking sheet and bake at 130°C/mark ½ for 3 hrs.

Weight loss: 33.3%

300 Meringues filled with cream

40% meringue

60% whipping cream

Proportions derived from a number of shop-bought samples.

301 Omelette

2 eggs
10ml water
10g butter

½ level tsp salt
pepper

Beat eggs with salt and water. Heat butter in an omelette pan. Pour in the mixture and stir until it begins to thicken evenly. While still creamy, fold the omelette and serve.

Weight loss: 5.7%

302 Cheese omelette

115g omelette, cooked
60g Cheddar cheese

Proportions are derived from recipe review.

303/304 Cheese and egg quiche, plain or wholemeal

200g raw plain or wholemeal shortcrust pastry
150g cheese
150g milk

3 eggs

Line a 20cm flan ring with the shortcrust pastry. Fill with grated cheese. Beat eggs in the warmed milk and pour into pastry case. Bake for 10 mins at 200°C/mark 6 and then 30 mins at 180°C/mark 4.

Weight loss: 10%

522 Sausage rolls, flaky

100g raw flaky pastry   40g pork sausage meat

Make the pastry, roll out and cut into 10cm squares. Divide the sausage meat equally and place in the centre of each square. Fold over and seal. Bake for 20-30 minutes at 220°C/mark 7.

Weight loss: 14.3%

523 Sausage rolls, short

100g shortcrust pastry   50g pork sausage meat

Make the pastry, roll out and cut into 10cm squares. Divide the sausage meat equally and place in the centre of each square. Fold over and seal. Bake for 20-30 minutes at 220°C/mark 7.

Weight loss: 8%

535 Steak and Kidney pie

350g raw flaky pastry   100ml water
400g raw stewing meat   2 level teaspoons salt
200g raw kidney   15g flour

Make the pastry. Prepare steak and kidney, cut into pieces and roll in seasoned flour. Place in a pie dish with water. Cover with pastry. Bake pie for 20 minutes at 200°C/mark 6, then lower the heat to 150°C/mark 2 and cover with greaseproof paper. Cook for a further 2-2½ hours.

Weight loss: 21.2%

542 Beef Kheema

125g butter   120g tomatoes
210g cooking fat or vegetable ghee   1360g beef, minced
20g green chillies

Heat butter and fat. Add beef, tomatoes and chillies and fry until brown. Simmer gently until cooked.

Weight loss: 23%

543 Beef Koftas

450g beef topside, minced
10g salt   100g onions
2g mixed spices   30g garlic, chopped
2g chilli powder   10g coriander leaves, chopped
60 g cooking fat or vegetable ghee   1 egg

Mix ingredients except fat and form the mixture into balls. Fry until browned and cooked.

Weight loss: 40%

544 Beef Steak pudding

Suet crust:
200g flour
100g suet
1 ½ level teaspoons baking powder
½ level teaspoons salt
130ml water

Filling:
500g raw stewing steak
130g onion, chopped
50 g flour
25ml water or stock
1 level teaspoon salt
pepper

Make the suet crust pastry, and line a pudding basin, leaving sufficient for a lid. Cut the meat into slices and roll in the seasoned flour. Put into the basin with the onion. Add a little water and cover with the remaining pastry. Steam for about 3 hours.

Weight gain: +1.7%

545 Beef stew

250g raw stewing steak
75g onion
75g carrot
15g dripping

300ml water or stock
15g flour
1 level teaspoon salt
pepper

Melt the dripping in a casserole and brown the pieces of meat. Remove the meat and brown the onion. Add the flour and cook the roux. Gradually blend in the water, add the meat, carrots and seasoning, bring to the boil and finish cooking at 180°C/mark 4, for about 2 hours.

Weight loss: 24.5%

546 Bolognese sauce

25g vegetable oil
75g onion
75g carrot
50g celery
200g minced beef

10g tomato paste
200g canned tomatoes
250ml water or stock
1 level teaspoon salt
pepper, herbs

Brown the onion, carrot and celery in oil. Add the minced beef stirring thoroughly to brown. Add the tomatoes, stock and seasoning and simmer for 45 minutes with the lid on.

Weight loss: 40.9%

547/548 Chicken curry

150g cooking fat or vegetable ghee
225g onions, chopped
20g garlic, chopped
100g tomatoes
35g ginger root

20g curry powder
15g salt
780g chicken
35g liver
425g water

Fry onion and garlic briefly, add remaining ingredients and fry until brown. Then add water and simmer until cooked.

Weight loss: 30%

551 Chilli con carne

450g minced beef
415g canned tomatoes
100g green pepper, chopped
10g chilli powder
5g sugar
115g red kidney beans, dry
160g water in cooked beans
120g water

15g vegetable oil
115g onion, sliced
15g vinegar
30g tomato purée
5g salt

Soak the kidney beans overnight and cook in water. Fry the beef in the oil until lightly browned. Add the onion and pepper and fry until soft. Blend the seasoning and chilli powder with the vinegar, sugar and tomato purée and stir into the meat with the tomatoes and water. Cover and simmer gently for 40 minutes. Add the cooked kidney beans and continue cooking for a further 10 minutes.

Weight loss: 30.6%

552 Curried meat

250g cooked meat
200g onion, peeled and chopped
50g vegetable oil
75g apple, peeled and chopped
50g sultanas

15g desiccated coconut
20g flour
20g curry powder
400ml water
2 level teaspoons salt

Fry the onions in the oil. Add the apple, sultanas and coconut, then the flour and curry powder and fry for a minute or two. Add the water and bring to the boil. Simmer for 5 minutes. Add the cooked meat, cut into pieces and heat thoroughly.

Weight loss: 18.9%

553 Hot pot

250g raw stewing steak
250g potatoes
150g onions

100g carrots
125ml stock
2 level teaspoons salt; pepper

Cut the steak into small pieces and arrange in layers with slices of carrot and onion. Add water and seasoning. Cover with a layer of sliced potatoes. Cover and bake at 180°C/mark 4, for 2½ hours, removing the lid for the last 30 minutes to brown the potatoes.

Weight loss: 29.4%

554 Irish stew

250g neck of mutton (weighed with bone)
250g potato
125g onions

350ml water
1 level teaspoon salt; pepper

Cut up the meat, potato and onion and put into a saucepan. Add water and bring to the boil. Skim well and allow to simmer slowly for 1½ hours.

Weight loss: 34.3%

556 Lamb kheema

200g butter
200g onion
200g tomatoes
5g ginger
2g garlic

5g chillies, fresh, green
4g mixed spices
14g salt
900g lamb, minced

Fry onion in butter until brown. Add tomatoes, ginger, garlic, chillies and mixed spices, and simmer for 5 minutes. Add minced lamb and simmer for another 30 minutes until fat separates out.

Weight loss: 25%

558 Moussaka

250g minced beef
250g aubergines or potatoes
150g onions
30g vegetable oil
100ml water or stock
20g tomato paste
1 level teaspoon salt

Sauce:
150ml milk
15g flour
15g vegetable oil
50g cheese, grated
½ egg

Fry the sliced onions in the oil until soft and remove from pan. Fry the aubergines until transparent then brown the meat. Arrange layers of aubergines, meat and onions in a casserole. Add the tomato paste and the seasoned stock. Pour the cheese sauce over the top, and cook for 1 hour at 190°C/mark 5.

Weight loss: 22.1%

559 Mutton biriani

480g lamb
300g onion, chopped
400g tomatoes
5g ginger
2g garlic
4g chilli powder
4g mixed spices

7g salt
600g water
460g rice
240g butter
30g raisins
15g whole almonds
150g yogurt, natural

Simmer lamb pieces with ⅓ of the onion, tomatoes, ginger, garlic, chilli powder, spices, salt and water for half an hour. Stir in 60g of the butter and raisins, almonds and yogurt. Soak rice in warm water for 30 minutes. Fry ⅓ of the onion in 60g of butter with spices and salt until soft. Add the rice and water to the onion and simmer for 8 minutes. Fry remainder of onion in rings in 120g butter until dark brown. Layer rice and meat in large pan or casserole dish and pour over cooking liquid from meat. Top with the fried onion rings. Cook over low heat or in low oven for 5 to 6 minutes.

Weight loss: 31%

560 Mutton curry

200g butter
300g onion
5g ginger
2g garlic
5g chillies, fresh, green

4g mixed spices
14g salt
100g water
200g tomatoes
900g lamb, leg

Fry onion in butter until brown. Add ginger, garlic, chillies, mixed spices and salt and stir with a little water to prevent burning. Add tomatoes and pieces of lamb and fry for a few minutes. Pour in remainder of water and bring to boil. Simmer on low heat for 60 minutes until meat is tender and fat has separated out.

Weight loss: 42%

562 Shepherd's pie

350g cooked minced beef
100g onion boiled and chopped
150ml water
500g boiled potatoes

50ml milk
20g margarine
2 level teaspoons salt;
pepper

Mix the beef and onion, moisten with water and add seasoning. Place in a pie dish. Mash the potato with the milk and margarine. Pile on top of the meat and bake in the oven for 25 minutes to brown, 190°C/mark 5.

Weight loss: 1.4%

655 Fish pie

200g cooked cod
400g mashed potato

Sauce:
150ml milk
15g margarine
15g flour
½ level teaspoon salt

Flake the fish and mix with the white sauce. Pipe a potato border round a dish, pour in the fish mixture. Brown in the oven, 200°C/mark 6, for 30 minutes.

Weight loss: 10.1%

656 Kedgeree

200g smoked haddock fillet, steamed
2 eggs
½ level teaspoon salt
25g margarine

50g rice
pepper

Boil the rice. Hard boil one egg. Melt the margarine and stir in the flaked fish, rice, seasoning and one beaten egg. Stir in chopped hard boiled egg and heat thoroughly.

Weight gain: +25.4%

983 Marzipan

300g ground almonds
150g caster sugar
150g icing sugar

1 egg
20ml lemon juice

Mix almonds and sugar, add beaten egg and knead all ingredients until smooth.

1030 Candied popcorn

45ml vegetable oil
75g popping corn

Glaze:
45ml water
200g caster sugar
25g butter

Prepare corn as for plain popcorn (No. 20). Heat glaze ingredients until sugar has dissolved, boil to soft ball stage. Add the popped corn and stir until coated.

1031 Plain popcorn

45ml vegetable oil

75g popping corn

Heat oil gently in saucepan until test corn pops. Remove from heat, add corn, cover and return to heat until all corn has popped.

1123 Lentil soup

100g lentils
25g carrot
50g turnip
50g onion
1 ham bone

25g margarine
25g flour
1 litre stock
125ml milk
herbs (bouquet garni)
salt and pepper to taste

Melt the dripping and toss the lentils and sliced vegetables in it over a gentle heat. Add the stock, seasoning, herbs and ham bone and bring to the boil. Simmer for 2-2½ hours stirring at intervals. Remove the bone. Sieve or liquidise and return to the pan with the flour blended to a smooth cream with the milk. Simmer for 5 minutes, and adjust seasoning.

Weight loss: 49.1%

1144/1145 Bread sauce

250ml whole or semi-skimmed milk
50g fresh breadcrumbs
5g margarine
½ tsp salt

2 cloves
mace
1 small onion

Put milk and onion, stuck with cloves, in a saucepan and bring to the boil. Add breadcrumbs, and simmer for about 20 minutes over gentle heat. Remove onion, stir in margarine and season.

Weight loss: 6.8%

1146/1147 Cheese sauce

350ml whole or semi-skimmed milk
75g cheese
½ level tsp salt
cayenne pepper

25g flour
25g margarine

Melt the fat in a pan, add flour and cook gently for a few minutes stirring all the time. Add milk and cook until mixture thickens, stirring continually. Add grated cheese and seasoning. Reheat to soften the cheese, serve immediately.

Weight loss: 15.2%

1148/1149 Cheese sauce, packet mix, made up

1 pkt cheese sauce mix (33g)
284ml whole or semi-skimmed milk

Prepared as packet directions.

Weight loss: 9.1%

1150/1151 Onion sauce

350ml whole or semi-skimmed milk
200g cooked onion
1 level tsp salt
pepper

25g flour
25g margarine

Make the white sauce (as nos. 1152/1153), add the chopped onion and seasoning.

Weight loss: 12.6%

1152/1153 Savoury white sauce

350ml whole or semi-skimmed milk
25g flour
½ level tsp salt

25g margarine

Melt fat in a pan. Add flour and cook for a few minutes, stirring constantly. Add milk and salt, and cook gently until mixture thickens.

Weight loss: 18.1%

1154/1155 Sweet white sauce

350ml whole or semi-skimmed milk
25g flour
25g margarine

30g sugar

As savoury white sauce (Nos 1152/1153) except adding sugar and omitting salt.

Weight loss: 16.7%

1156 Apple chutney

500g cooking apples	1 level teaspoon salt
400g onions	1 level teaspoon curry powder
100g raisins	½ level teaspoon mustard
400ml vinegar	½ level teaspoon pepper
450g sugar	½ level teaspoon ground ginger

Peel and core the apples and peel the onions and chop into small pieces. Mix all the ingredients except the sugar and boil gently till soft. Add the sugar and boil for a further 30 minutes.

Weight loss: 32.1%

1157 French dressing

25ml vinegar	½ level teaspoon salt
75g olive oil	½ level teaspoon pepper

Shake the ingredients together in a screw-topped jar or bottle.

1163 Tomato chutney

1kg tomatoes	500g sugar
125g cooking apples	1 level teaspoon salt
500g onions	½ level teaspoon mustard
100g sultanas	½ level teaspoon pepper
450ml vinegar	2 level teaspoons curry powder

Peel the tomatoes, chop the apples and onions into small pieces. Mix all the ingredients except the sugar and boil gently until soft. Add the sugar and boil for a further 30 minutes.

Weight loss: 38.1%

1173 Tomato sauce

400g tomatoes	250ml stock
25g carrot	25g flour
50g onion	½ level teaspoon salt
25g bacon, streaky	herbs (bouquet garni)
15g margarine	

Fry the chopped vegetables gently with the margarine and bacon. Stir in the flour, blended with some of the stock, then the rest of the stock and the herbs. Simmer for 40 minutes, then sieve or liquidise if desired. Reheat, adjust seasoning and serve.

Weight loss: 44.4%

- Foods are listed below in the same order as in the main tables.
- The alternative names listed in the left-hand column below are those that were most frequently encountered during data collection and are included to help in identifying foods. It is important to recognise that in some cases such names may be used for more than one food and that all such usages may not appear in this list.
- To see if a name is listed, the food index should be consulted first. If the term is included as an alternative name, a cross reference entry indicates the food name to which it refers. This allows all alternatives to be listed together.
- Taxonomic names listed in the right-hand column refer as specifically as possible to the data used. Where two or more taxonomic names are listed, the data are representative of a mixture of these varieties.
- The abbreviation 'var' is used to indicate the specific variety or unspecified variety(ies); 'sp' and 'spp' are used to indicate that one or more than one species of the specified Genus is included.

Alternative names	Food names	Taxonomic names

Cereals

	Oats	*Avena sativa*
	Rye	*Secale cereale*
	Sago	*Metroxylon* spp
	Tapioca	*Manihot esculenta*
	Wheat	*Triticum aestivum*
	Rice	*Oryza sativa*
	Pasta wheat	*Triticum durum*

Meat

	Beef	*Bos taurus*
	Lamb	*Ovis aries*
	Pork	*Sus scrofa*
	Veal	*Bos taurus*

Poultry

	Chicken	*Gallus domesticus*

Alternative names	Food names	Taxonomic names
	Duck	*Anas platyrhynchos*
	Goose	*Anser anser*
	Grouse	*Lagopus scroticus*
	Partridge	*Perdix perdix*
	Pheasant	*Phasianus colchicus*
	Pigeon	*Columba* spp
	Turkey	*Meleagris gallopavo*
Game		
	Hare	*Lepus europaeus*
	Rabbit	*Lepus cuniculus*
	Venison	*Cervus* spp

Fish

White fish

Alternative names	Food names	Taxonomic names
	Cod	*Gadus morhua*
Rock eel Rock salmon	**Dogfish**	Probably *Squalus acanthias*
	Haddock	*Melanogrammus aeglefinus*
	Halibut	*Hippoglossus hippoglossus*
	Lemon sole	*Microstomus kitt*
	Plaice	*Pleuronectes platessa*
Coalfish Coley	**Saithe**	*Pollachius virens*
	Skate	*Raja* spp
	Whiting	*Merlangius merlangus*

Fatty fish

Alternative names	Food names	Taxonomic names
	Anchovies	*Engraulis encrasicholus*
	Herring	*Clupea harengus*

Alternative names	Food names	Taxonomic names
	Kipper	*Clupea harengus*
	Mackerel	*Scomber scombrus*
	Pilchards	*Sardinops sagex ocellata*
	Salmon, Atlantic red	*Salmo salar* *Oncorhynchus nerka*
	Sardines	*Sardina pilchardus*
	Trout, brown	*Salmo trutta*
	Tuna	*Euthynnus* sp *Katsuwonus pelamis*
	Whitebait	Young of *Clupea harengus* and *Sprattus sprattus*
Crustacea		
	Crab	*Cancer pagurus*
	Lobster	*Homarus vulgaris*
	Prawns	*Paleamon serratus*
	Scampi	*Nephrops norvegicus*
	Shrimps	*Crangon crangon* *Pandalus montagui* *Pandalus borealis*
Molluscs		
	Cockles	*Cardium edule*
	Mussels	*Mytilus edulis*
	Squid	*Loligo vulgaris*
	Whelks	*Buccinum undatum*
	Winkles	*Littorina littorea*
Vegetables		
Potatoes Aloo Batata	**Potatoes**	*Solanum tuberosum*

Alternative names	Food names	Taxonomic names
Beans and lentils		
Adzuki beans	**Aduki beans**	*Vigna angularis*
	Baked beans	*Phaseolus vulgaris* (navy beans)
	Beansprouts, mung	*Phaseolus aureus*
Alad Urad	**Black gram**, urad gram	*Vigna mungo*
Blackeye peas Cowpeas Chori Lobia	**Blackeye beans**	*Vigna unguiculata*
	Broad beans	*Vicia faba*
Lima beans	**Butter beans**	*Phaseolus lunatus*
Channa Common gram Garbanzo Yellow gram	**Chick peas**	*Cicer arietinum*
Fansi	**Green beans/French beans**	*Phaseolus vulgaris*
Continental lentils Masur	**Lentils**, green and brown	*Lens esculenta*
Masoor dahl Masur dahl	**Lentils**, red	*Lens esculenta*
Green gram Golden gram Moong beans	**Mung beans**	
	Red kidney beans	*Phaseolus vulgaris*
	Runner beans	*Phaseolus coccineus*
	Soya beans	*Glycine max*
Peas		
Snowpeas	**Mange-tout peas**	*Pisum sativum* var *macrocarpum*

Alternative names	Food names	Taxonomic names
Badla Mattar Vatana	**Peas**	*Pisum sativum*
Other vegetables		
	Asparagus	*Asparagus officinalis* var *altilis*
Baingan Brinjal Eggplant Jew's apple Ringana	**Aubergine**	*Solanum melongena* var ovigerum
	Beetroot	*Beta vulgaris*
Calabrese	**Broccoli**, green	*Brassica oleracea* var *botrytis*
Chote bund gobhi Nhanu kobi	**Brussels sprouts**	*Brassica oleracea* var *gemmifera*
Bund gobhi Kobi	**Cabbage**	*Brassica oleracea*
	Cabbage, January King	*Brassica oleracea* var *capitata*
	Cabbage, white	*Brassica oleracea* var
Gajjar	**Carrots**	*Daucus carota*
Pangoli Phool gobhi	**Cauliflower**	*Brassica oleracea* var *botrytis*
	Celery	*Apium graveolens* var *dulce*
Belgian chicory Witloof	**Chicory**	*Cichorium intybus*
Zucchini	**Courgette**	*Cucurbita pepo*
Kakdi Khira	**Cucumber**	*Cucumis sativus*
Borecole Kale	**Curly kale**	*Brassica oleracea* var acephala
	Fennel, Florence	*Foeniculum vulgare* var *dulce*

Alternative names	Food names	Taxonomic names
Lassan Lehsan	**Garlic**	*Allium sativum*
	Gherkins	*Cucumis sativus*
Bitter gourd Balsam apple	**Gourd**, karela	*Momordica charantia*
	Leeks	*Allium ampeloprasum* var *porrum*
	Lettuce	*Lactuca sativa*
	Marrow	*Cucurbita pepo*
	Mushrooms, common	*Agaricus campestris*
	Mustard and cress	*Brassica* and *Lepidium* spp
Bhendi Bhinda Bhindi Gumbo Lady's fingers	**Okra**	*Hibiscus esculentus*
Dungli Kanda Piyaz	**Onions**	*Allium cepa*
	Parsnip	*Pastinaca sativa*
Pimento	**Peppers**, capsicum, chilli, green	*Capsicum annuum*
Bell peppers Motamircha Simla mirch Sweet peppers	**Peppers**, capsicum (green/red)	*Capsicum annuum* var *grossum*
	Plantain	*Musa paradisiaca*
Kumra Lal kaddu Lal phupala	**Pumpkin**	*Cucurbita* sp

Alternative names	Food names	Taxonomic names
	Quorn, myco-protein	*Fusarium graminearum*
	Radish, red	*Raphanus sativus*
Palak Saag	**Spinach**	*Spinacia oleracea*
	Spring greens	*Brassica oleracea* var
	Spring onions	*Allium cepa*
Neeps (England) Rutabaga Yellow turnip	**Swede**	*Brassica napus* var *napobrassica*
Shakaria Yam (USA)	**Sweet potato**	*Ipomoea batatas*
	Sweetcorn	*Zea mays*
	Tomatoes	*Lycopersicon esculentum*
Neeps (Scotland) Shalgam	**Turnip**	*Brassica rapa* var *rapifera*
	Watercress	*Nasturtium officinale*
	Yam	*Dioscorea* sp

Herbs and spices

	Food names	Taxonomic names
	Cinnamon	*Cinnamomum verum* *Cinnamomum aromaticum*
	Mint	*Mentha spicata*
	Mustard	*Sinapis alba* *Brassica hirta*
	Nutmeg	*Myristica fragrans*
	Paprika	*Capsicum annuum*
	Parsley	*Petroselinum crispum*
	Pepper, black	*Piper nigrum*
	Pepper, white	*Piper nigrum*

Alternative names	Food names	Taxonomic names
	Rosemary	*Rosmarinus officinalis*
	Sage	*Salvia officinalis*
	Thyme	*Thymus vulgaris*

Fruit

Alternative names	Food names	Taxonomic names
Tarel	**Apples**	*Malus pumila*
	Apricots	*Prunus armeniaca*
	Avocado	*Persea americana*
Kula	**Bananas**	*Musa* spp
	Blackberries	*Rubus ulmifolius*
	Blackcurrants	*Ribes nigrum*
	Cherries	*Prunus avium*
	Clementines	*Citrus reticulata* var *Clementine*
	Currants	*Vitis vinifera*
	Damsons	*Prunus domestica* subsp *institia*
	Dates	*Phoenix dactylifera*
Gullar	**Figs**	*Ficus carica*
	Gooseberries	*Ribes grossularia*
	Grapefruit	*Citrus paradisi*
	Grapes	*Vitis vinifera*
	Guava	*Psidium guajava*
Chinese gooseberry	**Kiwi fruit**	*Actinidia chinensis*
	Lemons	*Citrus limon*
Chinese cherry Lichee Lichi Litchee Litchi	**Lychees**	*Litchi chinensis*

Alternative names	Food names	Taxonomic names
	Mandarin oranges	*Citrus reticulata*
	Mangoes	*Mangifera indica*
	Melon, Canteloupe-type	*Cucumis melo* var *cantaloupensis*
	Melon, Galia	*Cucumis melo* var *reticulata*
	Melon, Honeydew	*Cucumis melo* var *indorus*
	Melon, watermelon	*Citrullus lanatus*
	Nectarines	*Prunus persica* var *nectarina*
	Olives	*Olea europaea*
	Oranges	*Citrus sinensis*
Purple grenadillo	**Passion fruit**	*Passiflora edulis f edulis*
Papai Papaya	**Paw-paw**	*Carica papaya*
	Peaches	*Prunus persica*
	Pears	*Pyrus communis*
	Pineapple	*Ananas comosus*
	Plums	*Prunus domestica* subsp *domestica*
	Prunes	*Prunus domestica*
	Raisins	*Vitis vinifera*
	Raspberries	*Rubus idaeus*
	Rhubarb	*Rheum rhaponticum*
	Satsumas	*Citrus reticulata*
	Strawberries	*Fragaria sp*
	Sultanas	*Vitis vinifera*
	Tangerines	*Citrus reticulata*

Alternative names	Food names	Taxonomic names
Nuts and seeds		
Badam	**Almonds**	*Prunus amygdalus*
	Brazil nuts	*Bertholletia excelsa*
Kaju	**Cashew nuts**	*Anacardium occidentale*
	Chestnuts	*Castanea vulgaris*
	Coconut	*Cocos nucifera*
	Hazelnuts	*Corylus avellana* *Corylus maxima*
Queensland nuts	**Macadamia nuts**	*Macadamia integrifolia* *Macadamia tetraphylla*
Groundnuts Monkey nuts	**Peanuts**	*Arachis hypogaea*
Hickory nuts	**Pecan nuts**	*Carya illinoensis*
Indian nuts Pignolias Pine kernels	**Pine nuts**	*Pinus pinea* *Pinus edulis*
Pista	**Pistachio nuts**	*Pistacia vera*
Benniseed Gingelly Til	**Sesame seeds**	*Sesamum indicum*
	Sunflower seeds	*Helianthus annuus*
Akhrot Madeira nuts	**Walnuts**	*Juglans regia*

3.7 REFERENCES

Supplements to 'The Composition of Foods'

Paul, A. A., Southgate, D. A. T. and Russell, J. (1980) *First supplement to McCance and Widdowson's The Composition of Foods, 4th edition: Amino acid composition (mg per 100g food) and fatty acid composition (g per 100g food)*, HMSO, London

Tan, S. P., Wenlock, R. W. and Buss, D. H. (1985) *Second supplement to McCance and Widdowson's The Composition of Foods, 4th edition: Immigrant Foods.* HMSO, London

Holland, B., Unwin, I. D. and Buss, D. H. (1988) *Third supplement to McCance and Widdowson's The Composition of Foods , 4th edition: Cereals and Cereal Products.* Royal Society of Chemistry, Nottingham

Holland, B., Unwin, I. D. and Buss, D. H. (1989) *Fourth supplement to McCance and Widdowson's The Composition of Foods, 4th edition: Milk Products and Eggs.* Royal Society of Chemistry, Cambridge

Holland, B., Unwin, I. D. and Buss, D. H. (1991) *Fifth supplement to McCance and Widdowson's The Composition of Foods, 4th edition: Vegetables, Herbs and Spices.* Royal Society of Chemistry, Cambridge

Holland, B., Unwin, I. D. and Buss, D. H. (In preparation) *First supplement to McCance and Widdowson's The Composition of Foods, 5th edition: Fruit and Nuts.* Royal Society of Chemistry, Cambridge

---

References to the Introduction, Tables and Appendices

Allen, L. H. (1982) Calcium bioavailability and absorption: a review. *Am. J. Clin. Nutr.* **35**, 783–808

AOAC (1975) *Official methods of analysis, 12th edition.* Association of Official Analytical Chemists, Washington DC

Bell, J. G. (1974) Microbiological assay of vitamins of the B group in foodstuffs. *Lab. Pract.* **23**, 235–242, 252

Bender, A. E. (1989) Nutritional significance of bioavailability. In: *Nutrient availability: chemical and biological aspects*, edited by D. A. T. Southgate, I. T. Johnson and G. R. Fenwick. Special publication No. 72. Royal Society of Chemistry, Cambridge. pp 3–9

Bingham, S. A. and Day, K. C. (1987) Average portion weights of foods consumed by a randomly selected British population sample. *Hum. Nutr. : Appl. Nutr.* **41A**, 258–264

Bingham, S. A. (1987) The dietary assessment of individuals; methods, accuracy, new techniques and recommendations. *Nutr. Abs. Rev. (Series A)* **57**, 705–742

The Bread and Flour Regulations (1984) Statutory Instrument No. 1304. HMSO, London

Brubacher, G., Müller-Mulot, W. and Southgate, D. A. T. (1985) *Methods for the determination of vitamins in food.* Elsevier Applied Science Publishers Ltd, London

Brubacher, G. B. and Weiser H. (1985) The vitamin A activity of β-carotene. *Internat. J. Vit. Res.* **55**, 5–15

Cameron, M. E., and van Staveren, W. A. (1988) *Manual on methodology for food consumption studies.* Oxford University Press

Carter, E. G. A. and Carpenter, K. J. (1982) The bioavailability for humans of bound niacin from wheat bran. *Am. J. Clin. Nutr.* **36**, 855–861

Cashel, K., English, R. and Lewis, J. (1989) *Composition of Foods, Australia. Volume 1.* Department of Community Services and Health, Canberra

Christie, A. A., Dean, A. C., and Millburn, B. A. (1973) The determination of vitamin E in food by colorimetry and gas-liquid chromatography. *Analyst* **98**, 161–167

Chughtai, M. I. D. and Waheed Khan, A. (1960) *Nutritive value of food-stuffs and planning of satisfactory diets in Pakistan, Part 1. Composition of raw food-stuffs,* Punjab University Press, Lahore

Crawley, H. (1988) *Food portion sizes.* HMSO, London

Davies, J. and Dickerson J. (1991) *Nutrient content of food portions.* Royal Society of Chemistry, Cambridge

Dean, A. C. (1978) Method for the estimation of available carbohydrate in foods. *Food Chem.* **3**, 241–250

Department of Health (1991) *Dietary reference values for food energy and nutrients for the United Kingdom.* Report on Health and Social Subjects No. 41, HMSO, London

Egan, H., Kirk, R. S. and Sawyer, R. (1981) *Pearson's Chemical Analysis of Foods,* 8th edition. Churchill Livingstone, Edinburgh

Englyst, H. N. and Cummings, J. H. (1984) Simplified method for the measurement of total non-starch polysaccharides by gas-liquid chromatography of constituent sugars as alditol acetates. *Analyst* **109**, 937–942

Englyst, H. N. and Cummings, J. H. (1988) An improved method for the measurement of dietary fibre as the non-starch polysaccharides in plant foods. *J. Assoc. Off. Anal. Chem.* **71**, 808–814

Englyst, H. N., Wiggins, H. S. and Cummings, J. H. (1982) Determination of the non-starch polysaccharides in plant foods by gas-liquid chromatography of constituent sugars as alditol acetates. *Analyst* **107**, 307–318

Fairweather-Tait, S. J. (1991) The metabolism of iron and its bioavailability in foods. *Proc. Nutr. Soc.* (in press)

FAO/WHO (1973) *Energy and protein requirements*. Report of a Joint FAO/WHO *Ad Hoc* Expert Committee. FAO Nutrition Meetings Report Series, No. 52; WHO Technical Report Series, No. 522

FAO(1988) *Requirements of vitamin A, iron, folate and vitamin B_{12}*. Report of a Joint FAO/WHO Expert Consultation. Food and Nutrition Series No. 23. Food and Agriculture Organization of the United Nations, Rome

Finglas, P. M. and Faulks, R. M. (1984) The HPLC analysis of thiamin and riboflavin in potatoes. *Food Chem.* **15,** 37–44

Gopalan, C., Rama Sastri, B. V. and Balasubramanian, S. C. (1980) *Nutritive value of Indian foods*, National Institute of Nutrition, Indian Council of Medical Research, Hyderabad

Hartman, B. G. and Hillig, F. (1934) Acid constituents of food products, with special reference to citric, malic and tartaric acids. *J. Ass. Off. Agric. Chem.* **27**, 522–531

Haytowitz, D. B. and Matthews, R. H. (1984) *Composition of foods: vegetables and vegetable products, raw, processed and prepared*. Agriculture Handbook No. 8-11, US Department of Agriculture, Washington DC

Haytowitz, D. B. and Matthews, R. H. (1986) *Composition of foods: legumes and legume products, raw, processed and prepared*. Agriculture Handbook No. 8-16, US Department of Agriculture, Washington DC

Holland, B., Unwin, I. D. and Buss, D. H. (1991) *Fifth supplement to McCance and Widdowson's The Composition of Foods, 4th edition: Vegetables, Herbs and Spices*. Royal Society of Chemistry, Cambridge

Hulme, A. L. (1971) (Editor) *The biochemistry of fruits and their products, volume 2*. Academic press, London and New York

IUPAC (1976) Standard methods for the analysis of oils, fats and soaps. 4th supplement to the 5th edition. Method II D.19 Preparation of fatty acid methyl esters. Method II D.25 Gas liquid chromatography of fatty acid methyl esters.

Kwiatkowska, C. A., Finglas, P. M. and Faulks, R. M. (1989) The vitamin content of retail vegetables in the UK. *J. Hum. Nut. Diet.* **2**, 159–172

McCance, R. A. and Lawrence, R. D. (1929) *The carbohydrate content of foods*. Medical Research Council Special Report Series No. 135. HMSO, London

McCance. R. A. and Shipp, H. L. (1933) *The chemistry of flesh foods and their losses on cooking*. Medical Research Council Special Report Series, No. 187. HMSO, London

McCance, R. A. and Widdowson, E. M. (1940) *The chemical composition of foods*. Medical Research Council Special Report Series No. 235. HMSO, London

McCance, R. A. and Widdowson, E. M. (1960) *The composition of foods*. HMSO, London

McCance, R. A., Widdowson, E. M. and Shackleton, L. R. B. (1936) *The nutritive value of fruits, vegetables and nuts.* Medical Research Council Special Report Series, No. 213. HMSO, London

McCarthy, M. A. and Matthews, R. H. (1984) *Composition of foods: nut and seed products, raw, processed and prepared.* Agriculture Handbook No. 8-12, US Department of Agriculture, Washington DC

McClaughlin, P. J. and Weihrauch, J. L. (1979) Vitamin E content of foods. *J. Am. Diet. Assoc.* **75,** 647–665

Marsh, A. C. (1980) *Composition of foods: soups, sauces, and gravies, raw, processed and prepared.* Agriculture Handbook No. 8-6, US Department of Agriculture, Washington DC

Marsh, A. C., Moss, M. K. and Murphy, E. W. (1977) *Composition of foods: spices and herbs, raw, processed and prepared.* Agriculture Handbook No. 8-2, US Department of Agriculture, Washington DC

Medical Research Council: Accessory Food Factors Committee (1945) *Nutritive values of wartime foods.* Medical Research Council War Memorandum No. 14. HMSO, London

Michie, N. D., Dixon, E. J. and Bunton, N. G. (1978) Critical review of AOAC fluorimetric method for determining selenium in foods. *J. Assoc. Off. Analyt. Chem.* **61,** 48–51

Miller, D. D. (1989) Calcium in the diet: food sources, recommended intakes and nutritional bioavailibility. *Adv. Food Nutr. Res.* **33,** 103–156

Moxon, R. E. D. (1983) A rapid method for the determination of sodium in butter. *J. Assoc. Publ. Analysts* **21,** 83–87

Moxon, R. E. D. and Dixon, E. J. (1980) Semi-automated method for the determination of total iodine in food. *Analyst* **105,** 344–352

Olson, J. A. (1989) Provitamin A function of carotenoids: The conversion of β-carotene into vitamin A. *J. Nutr.* **119,** 105–108

Paul, A. A. and Southgate, D. A. T. (1977) A study on the composition of retail meat: dissection into lean, separable fat and inedible portion. *J. Hum. Nutr.* **31,** 259–272

Paul, A. A. and Southgate, D. A. T. (1978) *McCance and Widdowson's The Composition of Foods Fourth edition.* HMSO, London

Paul, A. A., Southgate, D. A. T. and Russell, J. (1980) *First supplement to McCance and Widdowson's The Composition of Foods, 4th edition: Amino acid composition (mg per 100g food) and fatty acid composition (g per 100g food),* HMSO, London

Pellet, P. L. and Shadarevian, S. (1970) *Food composition tables for use in the Middle East.* American University of Beirut, Beirut

Phillips, D. I. W., Nelson, M., Barker, D. J. P., Morris, J. A. and Wood, T. J. (1988) Iodine in milk and the incidence of thyrotoxicosis in England. *Clin. Endocrinol.* **28,** 61–66

Phillips, D. R. and Wright A. J. A. (1983) Studies on the response of *Lactobacillus casei* to folate vitamin in foods. *Br. J. Nutr.* **49**, 181–186

Polacchi, W., McHargue, J. S. and Perloff, B. P. (1982) *Food composition tables for the near east.* Food and Agriculture Organization of the United Nations, Rome

Posati, L. P. and Orr, M. L. (1976) *Composition of foods: dairy and egg products, raw, processed and prepared.* Agriculture Handbook No. 8-1, US Department of Agriculture, Washington DC

Royal Society (1972) *Metric units, conversion factors and nomenclature in nutritional and food sciences.* Report of the subcommittee on metrication of the British National Committee for Nutritional Sciences

Sanstead, H. H. (1982) Copper bioavailibility and requirements. *Am. J. Clin. Nutr.* **35**, 809–814

Schwartz, R., Apgar, B. J. and Wien, E. M. (1986) Apparent absorption and retention of calcium, copper, magnesium, manganese and zinc from a diet containing bran. *Am. J. Clin. Nutr.* **43**, 444–455

Sivell, L. M., Bull, N. L., Buss, D. H., Wiggins, R. A., Scuffam, D. and Jackson, P. A. (1984) Vitamin A activity in foods of animal origin. *J. Sci. Food Agric.* **35**, 931–939

Society of Public Analysts and Other Analytical Chemists: Analytical Methods Committee (1951) The chemical assay of aneurine in foodstuffs. *Analyst* **76**, 127–133

Solomons, N. W. (1982) Biological availability of zinc in humans. *Am. J. Clin. Nutr.* **35**, 1048–1075

Southgate, D. A. T. (1969) Determination of carbohydrates in foods II Unavailable carbohydrates. *J. Sci. Food Agric.* **20**, 331–335

Southgate, D. A. T. (1976) *Determination of food carbohydrates.* Applied Science Publishers, London

Southgate, D. A. T. (1974) *Guidelines for the preparation of tables of food composition.* S. Karger, Basel

Southgate, D. A. T. (1987) Minerals, trace elements and potential hazards. *Am. J. Clin. Nutr.* **45**, 1256–1266

Southgate, D. A. T. and Durnin J. V. G. A. (1970) Calorie conversion factors: an experimental reassessment of the factors used in the calculations of the energy value of human diets. *Brit. J. Nutr.* **24**, 517–535

Southgate, D. A. T., Paul, A. A., Dean, A. C. and Christie, A. A. (1978) Free sugars in foods. *J. Hum. Nutr.* **32**, 335–347

Swaminathan, M. (1974) *Essentials of food and nutrition, volume 1. Fundamental aspects.* Garesh & Co., Madras

Watt, B. K. and Merrill, A. L. (1963) *Composition of foods, raw, processed and prepared.* Agriculture Handbook No. 8, US Department of Agriculture, Washington DC

Wenlock, R. W., Sivell, L. M. and Agater, I. B. (1985) Dietary fibre fractions in cereal and cereal-containing products in Britain. *J. Sci. Food Agric.* **36**, 113–121

Wharton, P. A., Eaton, P. M. and Day, K. C. (1983) Sorrento Asian food tables: food tables, recipes and customs of mothers attending Sorrento Maternity Hospital, Birmingham, England. *Hum. Nutr. : Appl. Nutr.,* **37A**, 378–402

Wiles, S. J., Nettleton, P. A., Black. A. E. and Paul, A. A. (1980) The nutrient composition of some cooked dishes eaten in Britain: A supplementary food composition table. *J. Hum. Nutr.* **34**, 189–223

Wu Leung, W. T., Busson, F. and Jardin, C. (1968) *Food composition table for use in Africa.* Food and Agriculture Organization and US Department of Health, Education and Welfare, Bethesda

Wu Leung, W. T., Butrum, R. R., Chang, F. H., Narayana Rao, M. and Polacchi, W. (1972) *Food composition table for use in East Asia.* Food and Agriculture Organization and US Department of Health, Education and Welfare, Bethesda

Young, V. R., Nahapetian, A. and Janghorbani, M. (1982) Selenium bioavailability with reference to human nutrition. *Am. J. Clin. Nutr.* **35**, 1076–1088

- Foods are indexed by their food number and **not** by page number.

- Cross references in this index give access to the individual foods items through this index and to alternative names given in the Alternative and Taxonomic Names list on pages 425–434.

Beans, butter	703
Beans, green	see **Green beans/French beans**
Beans, French	see **Green beans/French beans**
Beans, mung	see **Mung beans**
Beans, red kidney	see **Red kidney beans**
Beans, runner	see **Runner beans**
Beans, soya	see **Soya beans**
Beansprouts, mung, raw	696
Beansprouts, mung, stir-fried in blended oil	697
Beef, brisket, lean and fat, boiled	365
Beef, brisket, lean and fat, raw	364
Beef, fat only, cooked, average	362
Beef, fat only, raw, average	361
Beef, forerib, lean and fat, raw	366
Beef, forerib, lean and fat, roast	367
Beef, forerib, lean only, roast	368
Beef, lean only, raw, average	363
Beef, mince, raw	369
Beef, mince, stewed	370
Beef, rump steak, lean and fat, fried	372
Beef, rump steak, lean and fat, grilled	373
Beef, rump steak, lean and fat, raw	371
Beef, rump steak, lean only, fried	374
Beef, rump steak, lean only, grilled	375
Beef, salted, dried, raw	377
Beef, salted, fat removed, raw	376
Beef, silverside, lean and fat, salted, boiled	378
Beef, silverside, lean only, salted, boiled	379
Beef, sirloin, lean and fat, raw	380
Beef, sirloin, lean and fat, roast	381
Beef, sirloin, lean only, roast	382
Beef, stewing steak, lean and fat, raw	383
Beef, stewing steak, lean and fat, stewed	384
Beef, topside, lean and fat, raw	385
Beef, topside, lean and fat, roast	386
Beef, topside, lean only, roast	387
Beef chow mein	539
Beef curry, retail	540
Beef curry, with rice	541
Beef dripping	317
Beef kheema	542
Beef koftas	543
Beef sausages, fried	525
Beef sausages, grilled	526
Beef sausages, raw	524
Beef steak pudding	544
Beef stew	545
Beefburgers, frozen, fried	504
Beefburgers, frozen, raw	503
Beer, bitter, canned	1094
Beer, bitter, draught	1095
Beer, bitter, keg	1096

Beer, mild, draught	1097
Beetroot, boiled in salted water	742
Beetroot, pickled, drained	743
Beetroot, raw	741
Belgian chicory	see **Chicory**
Bell peppers	see **Peppers**, capsicum, green
Belly rashers pork, lean and fat, grilled	418
Belly rashers pork, lean and fat, raw	417
Benniseed	see **Sesame seeds**
Bhendi	see **Okra**
Bhinda	see **Okra**
Bhindi	see **Okra**
Biriani, mutton	559
Biscuits, chocolate, full coated	93
Biscuits, digestive, chocolate	96
Biscuits, digestive, plain	97
Biscuits, ginger	99
Biscuits, homemade, creaming method	100
Biscuits, sandwich	103
Biscuits, semi-sweet	104
Biscuits, short-sweet	105
Biscuits, wafer, filled	107
Bitter beer, canned	1094
Bitter beer, draught	1095
Bitter beer, keg	1096
Bitter gourd	see **Gourd**, karela
Black gram, urad gram, dried, boiled in unsalted water	699
Black gram, urad gram, dried, raw	698
Black pepper	847
Black pudding, fried	505
Black treacle	1013
Blackberries, raw	869
Blackberries, stewed with sugar	870
Blackberries, stewed without sugar	871
Blackcurrant pie, pastry top and bottom	150
Blackcurrants, canned in juice	874
Blackcurrants, canned in syrup	875
Blackcurrants, raw	872
Blackcurrants, stewed with sugar	873
Blackeye beans, dried, boiled in unsalted water	701
Blackeye beans, dried, raw	700
Blackeye peas	see **Blackeye beans**
Blended vegetable oil, average	333
Boiled eggs, chicken	293
Boiled sweets	1025
Bolognese sauce	546
Bombay Mix	1034
Borecole	see **Curly kale**
Bounty bar	1014
Bournvita powder	1043

Bournvita powder, made up with semi-skimmed milk	1045
Bournvita powder, made up with whole milk	1044
Bovril	1175
Bran Flakes	66
Bran, wheat	1
Brandy	1122
Brandy, cherry	1120
Brawn	506
Brazil nuts	974
Brazil nuts, weighed with shells	975
Bread, brown, average	33
Bread, brown, toasted	34
Bread, currant	37
Beard, currant, toasted	38
Bread, granary	39
Bread, Hovis, average	40
Bread, Hovis, toasted	41
Bread, malt	42
Bread, naan	43
Bread, pitta, white	45
Bread, rye	46
Bread, Vitbe, average	47
Bread, white, average	48
Bread, white, French stick	53
Bread, white, fried in blended oil	50
Bread, white, fried in lard	51
Bread, white, sliced	49
Bread, white, toasted	52
Bread, white, 'with added fibre'	54
Bread, white, 'with added fibre', toasted	55
Bread, wholemeal, average	56
Bread, wholemeal, toasted	57
Bread pudding	151
Bread sauce, made with semi-skimmed milk	1145
Bread sauce, made with whole milk	1144
Breaded chicken, fried in vegetable oil	444
Breast lamb, lean and fat, raw	391
Breast lamb, lean and fat, roast	392
Breast lamb, lean only, roast	393
Brie cheese	226
Brinjal	see **Aubergine**
Brisket beef, lean and fat, boiled	365
Brisket beef, lean and fat, raw	364
Broad beans, frozen, boiled in unsalted water	702
Broccoli, green, boiled in unsalted water	745
Broccoli, green, raw	744
Brown ale, bottled	1098
Brown bread, average	33
Brown bread, toasted	34
Brown chapati flour	2
Brown flour	12

Brown lentils	see **Lentils**, green and brown
Brown rice, boiled	19
Brown rice, raw	18
Brown rolls, crusty	58
Brown rolls, soft	59
Brown sauce, bottled	1165
Brown trout, steamed	629
Brown trout, steamed, weighed with bones	630
Brussels sprouts, boiled in unsalted water	747
Brussels sprouts, frozen, boiled in unsalted water	748
Brussels sprouts, raw	746
Build-up powder	1046
Build-up powder, made up with semi-skimmed milk	1048
Build-up powder, made up with skimmed milk	1049
Build-up powder, made up with whole milk	1047
Bund gobhi	see **Cabbage**
Buns, Chelsea	130
Buns, currant	133
Buns, Hamburger	61
Buns, hot cross	141
Butter	306
Butter beans, canned, drained	703
Butter ghee	335
Butterhead lettuce, raw	778
Cabbage, boiled in unsalted water, average	750
Cabbage, January King, boiled in salted water	752
Cabbage, January King, raw	751
Cabbage, raw, average	749
Cabbage, white, raw	753
Cake mix, made up	110
Cake, Battenburg	109
Cake, Eccles	138
Cake, fruit, plain, retail	113
Cake, fruit, rich	114
Cake, fruit, rich, iced	115
Cake, fruit, wholemeal	116
Cake, Madeira	118
Cake, sponge	119
Cake, sponge, fatless	120
Cake, sponge, jam filled	121
Cake, sponge, with butter icing	122
Cakes, crispie	111
Cakes, fancy iced, individual	112
Calabrese	see **Broccoli**, green
Calf liver, fried	484
Calf liver, raw	483
Camembert cheese	227
Candied popcorn	1030
Canned anchovies, in oil, drained	609

Canned apricots, in juice	864
Canned apricots, in syrup	863
Canned baby sweetcorn, drained	823
Canned baked beans in tomato sauce	694
Canned baked beans in tomato sauce, reduced sugar	695
Canned blackcurrants, in juice	874
Canned blackcurrants, in syrup	875
Canned butter beans, drained	703
Canned carrots, drained	758
Canned cherries, in syrup	878
Canned chick peas, drained	706
Canned cook-in-sauces	1166
Canned crab	636
Canned cream of chicken soup, condensed	1125
Canned cream of chicken soup, condensed, ready to serve	1126
Canned cream of chicken soup, ready to serve	1124
Canned cream of mushroom soup, ready to serve	1127
Canned cream of tomato soup, condensed	1129
Canned cream of tomato soup, condensed ready to serve	1130
Canned cream of tomato soup, ready to serve	1128
Canned curry sauce	1167
Canned custard	278
Canned fruit cocktail, in juice	891
Canned fruit cocktail, in syrup	892
Canned gooseberries, dessert, in syrup	898
Canned grapefruit, in juice	901
Canned grapefruit, in syrup	902
Canned guava, in syrup	907
Canned ham	360
Canned ham and pork, chopped	513
Canned low calorie soup	1131
Canned luncheon meat	515
Canned lychees, in syrup	913
Canned mandarin oranges, in juice	914
Canned mandarin oranges, in syrup	915
Canned mangoes, in syrup	918
Canned mushy peas	728
Canned new potatoes, drained	663
Canned oxtail soup, ready to serve	1132
Canned paw-paw, in juice	937
Canned peaches, in juice	940
Canned peaches, in syrup	941
Canned pears, in juice	945
Canned pears, in syrup	946
Canned peas, drained	733
Canned pilchards, in tomato sauce	621
Canned pineapple, in juice	948
Canned pineapple, in syrup	949

Canned plums, in syrup	954
Canned pork and ham, chopped	513
Canned prunes, in juice	955
Canned prunes, in syrup	956
Canned processed peas, drained	736
Canned raspberries, in syrup	960
Canned red kidney beans, drained	718
Canned rhubarb, in syrup	964
Canned rice pudding	287
Canned salmon	625
Canned sardines, in oil, drained	628
Canned sardines, in tomato sauce	627
Canned shrimps, drained	643
Canned stewed steak, with gravy	536
Canned strawberries, in syrup	968
Canned sweetcorn kernels, drained	824
Canned tomatoes, whole contents	832
Canned tongue	537
Canned tuna, in brine, drained	632
Canned tuna, in oil, drained	631
Canned vegetable soup, ready to serve	1133
Canteloupe	919
Canteloupe, weighed with skin	920
Capsicum peppers, chilli, green, raw	801
Capsicum peppers, green, boiled in salted water	803
Capsicum peppers, green, raw	802
Capsicum peppers, red, boiled in salted water	805
Capsicum peppers, red, raw	804
Carrots, canned, drained	758
Carrots, old, boiled in unsalted water	755
Carrots, old, raw	754
Carrots, young, boiled in unsalted water	757
Carrots, young, raw	756
Cashew nuts, roasted and salted	976
Cauliflower cheese	166
Cauliflower, boiled in unsalted water	760
Cauliflower, raw	759
Celery, boiled in salted water	762
Celery, raw	761
Channa	see **Chick peas**, whole
Channel Island milk, semi-skimmed, UHT	197
Channel Island milk, whole, pasteurised	194
Channel Island milk, whole, pasteurised, summer	195
Channel Island milk, whole, pasteurised, winter	196
Chapati flour, brown	2
Chapati flour, white	3
Chapatis, made with fat	35
Chapatis, made without fat	36
Cheddar cheese, average	228
Cheddar cheese, vegetarian	229
Cheddar-type cheese, reduced fat	230
Cheese, Brie	226

Cheese, Camembert	227
Cheese, Cheddar, average	228
Cheese, Cheddar, vegetarian	229
Cheese, Cheddar-type, reduced fat	230
Cheese, cottage, plain	232
Cheese, cottage, plain, reduced fat	234
Cheese, cottage, plain, with additions	233
Cheese, cream	235
Cheese, Danish blue	236
Cheese, Edam	237
Cheese, Feta	238
Cheese, Gouda	243
Cheese, hard, average	244
Cheese, Lymeswold	245
Cheese, Parmesan	247
Cheese, processed, plain	248
Cheese, soft, full fat	242
Cheese, soft, medium fat	246
Cheese, Stilton, blue	249
Cheese, white, average	250
Cheese and egg quiche	303
Cheese and egg quiche, wholemeal	304
Cheese omelette	302
Cheese sauce packet mix, made up with semi-skimmed milk	1149
Cheese sauce packet mix, made up with whole milk	1148
Cheese sauce, made with semi-skimmed milk	1147
Cheese sauce, made with whole milk	1146
Cheese spread, plain	231
Cheesecake, frozen	274
Chelsea buns	130
Cherries, canned in syrup	878
Cherries, glacé	879
Cherries, raw	876
Cherries, raw, weighed with stones	877
Cherry brandy	1120
Cherry pie filling	880
Chestnuts	977
Chick peas, canned, drained	706
Chick peas, whole, dried, boiled in unsalted water	705
Chick peas, whole, dried, raw	704
Chicken, boiled, dark meat	437
Chicken, boiled, light meat	436
Chicken, boiled, meat only	435
Chicken, breaded, fried in vegetable oil	444
Chicken, dark meat, raw	434
Chicken, leg quarter, roast, meat only, weighed with bone	443
Chicken, light meat, raw	433
Chicken, meat and skin, raw	432
Chicken, meat only, raw	431
Chicken, roast, dark meat	441
Chicken, roast, light meat	440
Chicken, roast, meat and skin	439
Chicken, roast, meat only	438
Chicken, wing quarter, roast, meat only, weighed with bone	442
Chicken curry, retail	549
Chicken curry, with bone	547
Chicken curry, with rice	550
Chicken curry, without bone	548
Chicken liver, fried	486
Chicken liver, raw	485
Chicken noodle soup, dried	1134
Chicken noodle soup, dried, ready to serve	1135
Chicken soup, cream of, canned, ready to serve	1124
Chicken soup, cream of, condensed, canned	1125
Chicken soup, cream of, condensed, canned, ready to serve	1126
Chicory, raw	763
Chilli con carne	551
Chilli peppers, green, raw	801
Chilli powder	838
Chinese cherry	see **Lychees**
Chinese gooseberry	see **Kiwi fruit**
Chips, fine cut, frozen, fried in blended oil	684
Chips, fine cut, frozen, fried in corn oil	685
Chips, fine cut, frozen, fried in dripping	686
Chips, French fries, retail	680
Chips, homemade, fried in blended oil	674
Chips, homemade, fried in corn oil	675
Chips, homemade, fried in dripping	676
Chips, oven, frozen, baked	687
Chips, retail, fried in blended oil	677
Chips, retail, fried in dripping	678
Chips, retail, fried in vegetable oil	679
Chips, straight cut, frozen, fried in blended oil	681
Chips, straight cut, frozen, fried in corn oil	682
Chips, straight cut, frozen, fried in dripping	683
Choc ice	263
Chocolate, milk	1015
Chocolate, plain	1016
Chocolate, white	1017
Chocolate biscuits, full coated	93
Chocolate digestive biscuits	96
Chocolate mousse	285
Chocolate nut spread	999
Chocolate nut sundae	264
Chocolate Swiss rolls, individual	123
Chocolate, drinking, made up with semi-skimmed milk	1066
Chocolate, drinking, made up with whole milk	1065

Cottage cheese, plain, reduced fat	234
Cottage cheese, plain, with additions	233
Cottonseed oil	323
Courgette, boiled in unsalted water	765
Courgette, fried in corn oil	766
Courgette, raw	764
Cowpeas	see **Blackeye beans**
Crab, boiled	634
Crab, boiled, weighed with shell	635
Crab, canned	636
Crackers, cream	94
Crackers, wholemeal	108
Cream, fresh, clotted	216
Cream, fresh, double	215
Cream, fresh, half	211
Cream, fresh, single	212
Cream, fresh, soured	213
Cream, fresh, whipping	214
Cream, sterilised, canned	217
Cream, UHT, canned spray	218
Cream cheese	235
Cream crackers	94
Cream horns	131
Cream of chicken soup, canned, ready to serve	1124
Cream of chicken soup, condensed, canned	1125
Cream of chicken soup, condensed, canned, ready to serve	1126
Cream of mushroom soup, canned, ready to serve	1127
Cream of tomato soup, canned, ready to serve	1128
Cream of tomato soup, condensed, canned	1129
Cream of tomato soup, condensed, canned, ready to serve	1130
Creme caramel	275
Creme eggs	1019
Cress, mustard and	see **Mustard and cress**
Cress, water	see **Watercress**
Crispbread, rye	95
Crispie cakes	111
Crisps, potato	1037
Crisps, potato, low fat	1038
Croissants	60
Croquettes, potato, fried in blended oil	690
Crumble, fruit	154
Crumble, fruit, wholemeal	155
Crumpets, toasted	132
Crunchy Nut Corn Flakes	70
Cucumber, raw	767
Curacao	1121
Curly kale, boiled in salted water	769
Curly kale, raw	768
Currant bread	37

Currant bread, toasted	38
Currant buns	133
Currants	883
Curried meat	552
Curry, beef, retail	540
Curry, beef, with rice	541
Curry, chicken, retail	549
Curry, chicken, with bone	547
Curry, chicken, with rice	550
Curry, chicken, without bone	548
Curry, mutton	560
Curry powder	840
Curry sauce, canned	1167
Custard, canned	278
Custard, made up with skimmed milk	277
Custard, made up with whole milk	276
Custard powder	5
Custard tarts, individual	134
Cutlet veal, fried in vegetable oil	428
Cutlets lamb, lean and fat, grilled	400
Cutlets lamb, lean and fat, grilled, weighed with bone	401
Cutlets lamb, lean and fat, raw	399
Cutlets lamb, lean only, grilled	402
Cutlets lamb, lean only, grilled, weighed with fat and bone	403
Dairy ice cream, flavoured	268
Dairy ice cream, vanilla	267
Dairy spread	307
Damsons, raw, weighed with stones	884
Damsons, stewed with sugar	885
Danish blue cheese	236
Danish pastries	135
Dates, dried, weighed with stones	887
Dates, raw, weighed with stones	886
Demerara sugar	1010
Dessert Top	219
Digestive biscuits, chocolate	96
Digestive biscuits, plain	97
Dogfish, in batter, fried in blended oil	573
Dogfish, in batter, fried in blended oil, weighed with waste	574
Dogfish, in batter, fried in dripping	575
Dogfish, in batter, fried in dripping, weighed with waste	576
Doughnuts, jam	136
Doughnuts, ring	137
Draught bitter beer	1095
Draught mild beer	1097
Dream Topping, made up with semi-skimmed milk	221

Dream Topping, made up with whole milk	220
Dried chicken noodle soup	1134
Dried chicken noodle soup, ready to serve	1135
Dried cod, salted, boiled	572
Dried instant soup powder	1136
Dried instant soup powder, made up with water	1137
Dried minestrone soup	1138
Dried minestrone soup, ready to serve	1139
Dried mixed fruit	888
Dried oxtail soup	1140
Dried oxtail soup, ready to serve	1141
Dried skimmed milk	200
Dried skimmed milk, with vegetable fat	201
Dried tomato soup	1142
Dried tomato soup, ready to serve	1143
Drinking chocolate powder	1064
Drinking chocolate powder, made up with semi-skimmed milk	1066
Drinking chocolate powder, made up with whole milk	1065
Drinking yogurt	251
Dripping, beef	317
Duck, meat only, raw	445
Duck, meat, fat and skin, raw	446
Duck, roast, meat only	447
Duck, roast, meat, fat and skin	448
Duck eggs, whole, raw	297
Dumplings	167
Dungli	see **Onions**
Easy cook white rice, boiled	23
Easy cook white rice, raw	22
Eating apples, average, raw	856
Eating apples, average, raw, peeled	858
Eating apples, average, raw, peeled, weighed with skin and core	859
Eating apples, average, raw, weighed with core	857
Eccles cake	138
Eclairs, frozen	139
Edam cheese	237
Eggs, chicken, boiled	293
Eggs, chicken, fried in vegetable oil	294
Eggs, chicken, poached	295
Eggs, chicken, scrambled, with milk	296
Eggs, chicken, white, raw	291
Eggs, chicken, whole, raw	290
Eggs, chicken, yolk, raw	292
Eggs, duck, whole, raw	297
Egg and cheese quiche	303
Egg and cheese quiche, wholemeal	304
Egg fried rice	298
Egg noodles, boiled	28

Egg noodles, raw	27
Eggplant	see **Aubergine**
Elmlea, double	224
Elmlea, single	222
Elmlea, whipping	223
Evaporated milk, whole	202
Faggots	509
Fancy iced cakes, individual	112
Fansi	see **Green beans/French beans**
Fat, cooking, compound	316
Fat spread	307
Fennel, Florence, boiled in salted water	771
Fennel, Florence, raw	770
Feta cheese	238
Figs, dried	889
Figs, ready-to-eat, semi-dried	890
Filled wafer biscuits	107
Fish cakes, fried	650
Fish fingers, fried in blended oil	651
Fish fingers, fried in lard	652
Fish fingers, grilled	653
Fish paste	654
Fish pie	655
Flaky pastry, cooked	125
Flaky pastry, raw	124
Flapjacks	98
Flavoured low fat yogurt	256
Flavoured milk	203
Flavoured soya milk	210
Florence fennel, boiled in salted water	771
Florence fennel, raw	770
Flour, chapati, brown	2
Flour, chapati, white	3
Flour, rye, whole	7
Flour, soya, full fat	9
Flour, soya, low fat	10
Flour, wheat, brown	12
Flour, wheat, white, breadmaking	13
Flour, wheat, white, plain	14
Flour, wheat, white, self-raising	15
Flour, wheat, wholemeal	16
Forerib beef, lean and fat, raw	366
Forerib beef, lean and fat, roast	367
Forerib beef, lean only, roast	368
Frankfurters	510
French beans	see **Green beans/French beans**
French dressing	1157
French fries, retail	680
French stick, white	53
Fresh cream, clotted	216
Fresh cream, double	215

Fresh cream, half	211
Fresh cream, single	212
Fresh cream, soured	213
Fresh cream, whipping	214
Fried bread, white, fried in blended oil	50
Fried bread, white, fried in lard	51
Fried eggs	294
Fried rice, white , fried in lard/dripping	24
Fromage frais, fruit	239
Fromage frais, plain	240
Fromage frais, very low fat	241
Frosties	71
Fruit 'n Fibre	72
Fruit cake, plain, retail	113
Fruit cake, rich	114
Fruit cake, rich, iced	115
Fruit cake, wholemeal	116
Fruit cocktail, canned in juice	891
Fruit cocktail, canned in syrup	892
Fruit crumble	154
Fruit crumble, wholemeal	155
Fruit fromage frais	239
Fruit gums	1026
Fruit mousse	286
Fruit pie filling	893
Fruit pie, individual	158
Fruit pie, one crust	156
Fruit pie, pastry top and bottom	157
Fruit pie, wholemeal, one crust	159
Fruit pie, wholemeal, pastry top and bottom	160
Fruit salad, homemade	894
Fruit scones	145
Fruit yogurt, low fat	257
Fruit yogurt, whole milk	261
Full fat soft cheese	242
Gajjar	see **Carrots**
Galia melon	921
Galia melon, weighed with skin	922
Gammon joint, lean and fat, boiled	345
Gammon joint, lean and fat, raw	344
Gammon joint, lean only, boiled	346
Gammon rasher, lean and fat, grilled	347
Gammon rasher, lean only, grilled	348
Garam masala	841
Garbanzo	see **Chick peas**
Garden peas, canned, drained	733
Garlic, raw	772
Gateau	117
Gelatin	1176
Ghee, butter	335
Ghee, palm	336

Ghee, vegetable	337
Gherkins, pickled, drained	773
Gin	1122
Gingelly	see **Sesame seeds**
Gingernut biscuits	99
Glacé cherries	879
Glucose liquid, BP	1000
Goats milk, pasteurised	204
Gobhi, bund	see **Cabbage**
Gobhi, chote bund	see **Brussels sprouts**
Gobhi, phool	see **Cauliflower**
Golden gram	see **Mung beans**
Golden syrup	1012
Goose, roast, meat only	449
Gooseberries, raw	895
Gooseberries, stewed with sugar	896
Gooseberries, stewed without sugar	897
Gooseberries, dessert, canned in syrup	898
Gouda cheese	243
Gourd, bitter	see **Gourd**, karela
Gourd, karela, raw	774
Gram, black	see **Black gram,** urad gram
Gram, common	see **Chick peas**
Gram, golden	see **Mung beans**
Gram, green	see **Mung beans**
Gram, yellow	see **Chick peas**
Granary bread	39
Grape juice, unsweetened	1088
Grapefruit, canned in juice	901
Grapefruit, canned in syrup	902
Grapefruit, raw	899
Grapefruit, raw, weighed with peel and pips	900
Grapefruit juice, unsweetened	1089
Grapes, average	903
Grapes, weighed with pips	904
Gravy instant granules	1177
Gravy instant granules, made up with water	1178
Greek pastries	140
Greek yogurt, cows	252
Greek yogurt, sheep	253
Green and brown lentils, whole, dried, boiled in salted water	711
Green and brown lentils, whole, dried, raw	710
Green beans/French beans, frozen, boiled in unsalted water	708
Green beans/French beans, raw	707
Green broccoli, boiled in unsalted water	745
Green broccoli, raw	744
Green chilli peppers, raw	801
Green gram	see **Mung beans**
Green peppers, boiled in salted water	803
Green peppers, raw	802

January King cabbage, raw	751
Jelly, made with water	282
Jew's apple	see **Aubergine**
Juice, apple, unsweetened	1087
Juice, grape, unsweetened	1088
Juice, grapefruit, unsweetened	1089
Juice, lemon	1090
Juice, orange, unsweetened	1091
Juice, pineapple, unsweetened	1092
Juice, tomato	1093
Kaju	see **Cashew nuts**
Kakdi	see **Cucumber**
Kale	see **Curly kale**
Kanda	see **Onions**
Karela gourd, raw	774
Kedgeree	656
Keg bitter beer	1096
Ketchup, tomato	1172
Kheema, beef	542
Kheema, lamb	556
Khira	see **Cucumber**
Kidney, lamb, fried	478
Kidney, lamb, raw	477
Kidney, ox, raw	479
Kidney, ox, stewed	480
Kidney, pig, raw	481
Kidney, pig, stewed	482
Kidney beans	see **Red kidney beans**
Kipper, baked	615
Kipper, baked, weighed with bones	616
Kit Kat	1020
Kiwi fruit	908
Kiwi fruit, weighed with skin	909
Kobi	see **Cabbage**
Kobi, nhanu	see **Brussels sprouts**
Koftas, beef	543
Kula	see **Bananas**
Kumra	see **Pumpkin**
Lady's fingers	see **Okra**
Lager, bottled	1099
Lal kaddu	see **Pumpkin**
Lal phupala	see **Pumpkin**
Lamb, breast, lean and fat, raw	391
Lamb, breast, lean and fat, roast	392
Lamb, breast, lean only, roast	393
Lamb, chops, loin, lean and fat, grilled	395
Lamb, chops, loin, lean and fat, grilled, weighed with bone	396
Lamb, chops, loin, lean and fat, raw	394
Lamb, chops, loin, lean only, grilled	397

Lamb, chops, loin, lean only, grilled, weighed with fat and bone	398
Lamb, cutlets, lean and fat, grilled	400
Lamb, cutlets, lean and fat, grilled, weighed with bone	401
Lamb, cutlets, lean and fat, raw	399
Lamb, cutlets, lean only, grilled	402
Lamb, cutlets, lean only, grilled, weighed with fat and bone	403
Lamb, fat only, cooked, average	389
Lamb, fat only, raw, average	388
Lamb, lean only, raw, average	390
Lamb, leg, lean and fat, raw	404
Lamb, leg, lean and fat, roast	405
Lamb, leg, lean only, roast	406
Lamb, scrag and neck, lean and fat, raw	407
Lamb, scrag and neck, lean and fat, stewed	408
Lamb, scrag and neck, lean only, stewed	409
Lamb, scrag and neck, lean only, stewed, weighed with fat and bone	410
Lamb, shoulder, lean and fat, raw	411
Lamb, shoulder, lean and fat, roast	412
Lamb, shoulder, lean only, roast	413
Lamb heart, raw	472
Lamb kheema	556
Lamb kidney, fried	478
Lamb kidney, raw	477
Lamb liver, fried	488
Lamb liver, raw	487
Lamb sweetbread, fried	496
Lamb sweetbread, raw	495
Lamb tongue, raw	497
Lard	318
Lasagne, frozen, cooked	557
Lassan	see **Garlic**
Leeks, boiled in unsalted water	776
Leeks, raw	775
Leg lamb, lean and fat, raw	404
Leg lamb, lean and fat, roast	405
Leg lamb, lean only, roast	406
Leg pork, lean and fat, raw	424
Leg pork, lean and fat, roast	425
Leg pork, lean only, roast	426
Leg quarter, roast chicken, meat only, weighed with bone	443
Lehsan	see **Garlic**
Lemon curd, starch base	1007
Lemon juice	1090
Lemon meringue pie	161
Lemon sole, in crumbs, fried	592
Lemon sole, in crumbs, fried, weighed with bones	593

Lemon sole, raw	589
Lemon sole, steamed	590
Lemon sole, steamed, weighed with bones and skin	591
Lemon sorbet	273
Lemonade, bottled	1081
Lemons, whole, without pips	910
Lentil soup	1123
Lentils, continental see **Lentils**, green and brown	
Lentils, green and brown, whole, dried, boiled in salted water	711
Lentils, green and brown, whole, dried, raw	710
Lentils, red, split, dried, boiled in unsalted water	713
Lentils, red, split, dried, raw	712
Lettuce, average, raw	777
Lettuce, butterhead, raw	778
Lettuce, Iceberg, raw	779
Lichee	see **Lychees**
Lichi	see **Lychees**
Lima beans	see **Butter beans**
Lime juice cordial, undiluted	1083
Liquorice allsorts	1027
Litchee	see **Lychees**
Litchi	see **Lychees**
Liver, calf, fried	484
Liver, calf, raw	483
Liver, chicken, fried	486
Liver, chicken, raw	485
Liver, lamb, fried	488
Liver, lamb, raw	487
Liver, ox, raw	489
Liver, ox, stewed	490
Liver, pig, raw	491
Liver, pig, stewed	492
Liver pate	517
Liver pate, low fat	518
Liver sausage	514
Lobia	see **Blackeye beans**
Lobster, boiled	637
Lobster, boiled, weighed with shell	638
Loin chops, lamb, lean and fat, grilled	395
Loin chops, lamb, lean and fat, grilled, weighed with bone	396
Loin chops, lamb, lean and fat, raw	394
Loin chops, lamb, lean only, grilled	397
Loin chops, lamb, lean only, grilled, weighed with fat and bone	398
Loin chops, pork, lean and fat, grilled	420
Loin chops, pork, lean and fat, grilled, weighed with bone	421
Loin chops, pork, lean and fat, raw	419

Loin chops, pork, lean only, grilled	422
Loin chops, pork, lean only, grilled, weighed with fat and bone	423
Low calorie soup, canned	1131
Low calorie yogurt	254
Low fat potato crisps	1038
Low fat sausages, fried	531
Low fat sausages, grilled	532
Low fat sausages, raw	530
Low fat yogurt, flavoured	256
Low fat yogurt, fruit	257
Low fat yogurt, plain	255
Low-fat spread	308
Lucozade	1082
Luncheon meat, canned	515
Lychees, canned in syrup	913
Lychees, raw	911
Lychees, raw, weighed with skin and stone	912
Lymeswold cheese	245
Macadamia nuts, salted	982
Macaroni cheese	168
Macaroni, boiled	26
Macaroni, raw	25
Mackerel, fried	618
Mackerel, fried, weighed with bones	619
Mackerel, raw	617
Mackerel, smoked	620
Madeira cake	118
Madeira nuts	see **Walnuts**
Malt bread	42
Mandarin oranges, canned in juice	914
Mandarin oranges, canned in syrup	915
Mange-tout peas, boiled in salted water	726
Mange-tout peas, raw	725
Mange-tout peas, stir-fried in blended oil	727
Mango chutney, oily	1158
Mangoes, canned in syrup	918
Mangoes, raw	916
Mangoes, raw, weighed with skin and stone	917
Margarine	309
Margarine, hard, animal and vegetable fat	310
Margarine, hard, vegetable fat only	311
Margarine, polyunsaturated	314
Margarine, soft, animal and vegetable fat	312
Margarine, soft, vegetable fat only	313
Marmalade	1008
Marmite	1179
Marrow, boiled in unsalted water	781
Marrow, raw	780
Mars bar	1021
Marzipan, homemade	983

Marzipan, retail	984
Masoor dahl	see **Lentils**, red
Masur	see **Lentils**, green and brown
Masur dahl	see **Lentils**, red
Mattar	see **Peas**
Mature human milk	207
Mayonnaise, retail	1159
Meat, curried	552
Meat paste	516
Meat samosas	174
Medium fat soft cheese	246
Melon, Canteloupe	919
Melon, Canteloupe, weighed with skin	920
Melon, Galia	921
Melon, Galia, weighed with skin	922
Melon, Honeydew	923
Melon, Honeydew, weighed with skin	924
Meringue	299
Meringue, with cream	300
Mild beer, draught	1097
Milk, Channel Island, semi-skimmed, UHT	197
Milk, Channel Island, whole, pasteurised	194
Milk, Channel Island, whole, pasteurised, summer	195
Milk, Channel Island, whole, pasteurised, winter	196
Milk, condensed, skimmed, sweetened	198
Milk, condensed, whole, sweetened	199
Milk, dried skimmed	200
Milk, dried skimmed, with vegetable fat	201
Milk, evaporated, whole	202
Milk, flavoured	203
Milk, goats, pasteurised	204
Milk, human, colostrum	205
Milk, human, mature	207
Milk, human, transitional	206
Milk, semi-skimmed, average	185
Milk, semi-skimmed, pasteurised	186
Milk, semi-skimmed, pasteurised,fortified plus Skimmed Milk Powder	187
Milk, semi-skimmed, UHT	188
Milk, sheeps, raw	208
Milk, skimmed, average	181
Milk, skimmed, pasteurised	182
Milk, skimmed, pasteurised, fortified plus Skimmed Milk Powder	183
Milk, skimmed, UHT, fortified	184
Milk, soya, flavoured	210
Milk, soya, plain	209
Milk, whole, average	189
Milk, whole, pasteurised	190
Milk, whole, pasteurised, summer	191
Milk, whole, pasteurised, winter	192
Milk, whole, sterilised	193
Milk chocolate	1015
Milk pudding, made with skimmed milk	284
Milk pudding, made with whole milk	283
Milk shake powder	1073
Milk shake powder, made up with semi-skimmed milk	1075
Milk shake powder, made up with whole milk	1074
Milk shake, purchased	1072
Milky Way	1022
Mince beef, raw	369
Mince beef, stewed	370
Mince pies, individual	144
Mincemeat	1009
Minestrone soup, dried	1138
Minestrone soup, dried, ready to serve	1139
Minestrone soup, low calorie, canned	1131
Mint sauce	1169
Mint, fresh	842
Mixed fruit, dried	888
Mixed nuts	985
Mixed peel	926
Mixed vegetables, frozen, boiled in salted water	782
Monkey nuts	see **Peanuts**
Moong beans	see **Mung beans**
Motamircha	see **Peppers**, capsicum, green
Moussaka	558
Mousse, chocolate	285
Mousse, fruit	286
Muesli, Swiss style	73
Muesli, with no added sugar	74
Mung beans, whole, dried, boiled in unsalted water	715
Mung beans, whole, dried, raw	714
Mung beansprouts, raw	696
Mung beansprouts, stir-fried in blended oil	697
Mushroom soup, cream of, canned, ready to serve	1127
Mushrooms, boiled in salted water	784
Mushrooms, fried in blended oil	785
Mushrooms, fried in butter	786
Mushrooms, fried in corn oil	787
Mushrooms, raw	783
Mushy peas, canned, re-heated	728
Mussels, boiled	645
Mussels, boiled, weighed with shell	646
Mustard and cress, raw	788
Mustard powder	843
Mustard, smooth	1180
Mustard, wholegrain	1181
Mutton biriani	559
Mutton curry	560
Myco-protein Quorn	811

Naan bread	43
Nectarines	927
Nectarines, weighed with stones	928
Neeps (England)	see **Swede**
Neeps (Scotland)	see **Turnip**
New potatoes, average, raw	660
New potatoes, boiled in unsalted water	661
New potatoes, canned, re-heated, drained	663
New potatoes, in skins, boiled in unsalted water	662
Nhanu kobi	see **Brussels sprouts**
Non-dairy ice cream, flavoured	270
Non-dairy ice cream, mixes	271
Non-dairy ice cream, vanilla	269
Noodles, egg, boiled	28
Noodles, egg, raw	27
Nutmeg, ground	844
Nuts, Brazil	974
Nuts, Brazil, weighed with shells	975
Nuts, cashew, roasted and salted	976
Nuts, mixed	985
Oat and Wheat Bran	75
Oat Bran Flakes, Common Sense	68
Oatcakes, retail	102
Oatmeal, quick cook, raw	6
Oil, coconut	320
Oil, cod liver	321
Oil, corn	322
Oil, cottonseed	323
Oil, olive	324
Oil, palm	325
Oil, peanut	326
Oil, rapeseed, high erucic acid	327
Oil, rapeseed, low erucic acid	328
Oil, safflower	329
Oil, sesame	330
Oil, soya	331
Oil, sunflowerseed	332
Oil, vegetable, blended, average	333
Oil, wheatgerm	334
Okra, boiled in unsalted water	790
Okra, raw	789
Okra, stir-fried in corn oil	791
Old potatoes, average, raw	664
Old potatoes, baked, flesh and skin	665
Old potatoes, baked, flesh only	666
Old potatoes, baked, flesh only, weighed with skin	667
Old potatoes, boiled in unsalted water	668
Old potatoes, mashed with butter	669
Old potatoes, mashed with margarine	670
Old potatoes, roast in blended oil	671

Old potatoes, roast in corn oil	672
Old potatoes, roast in lard	673
Olive oil	324
Olives, in brine	929
Olives, in brine, weighed with stones	930
Omelette, cheese	302
Omelette, plain	301
Onion sauce, made with semi-skimmed milk	1151
Onion sauce, made with whole milk	1150
Onions, boiled in unsalted water	793
Onions, fried in blended oil	794
Onions, fried in corn oil	795
Onions, fried in lard	796
Onions, pickled, cocktail/silverskin, drained	798
Onions, pickled, drained	797
Onions, raw	792
Onions, spring	see **Spring onions**
Orange drink, undiluted	1084
Orange juice, unsweetened	1091
Oranges	931
Oranges, weighed with peel and pips	932
Ovaltine powder	1076
Ovaltine powder, made up with semi-skimmed milk	1078
Ovaltine powder, made up with whole milk	1077
Oven chips, frozen, baked	687
Ox heart, raw	473
Ox heart, stewed	474
Ox kidney, raw	479
Ox kidney, stewed	480
Ox liver, raw	489
Ox liver, stewed	490
Ox tongue, boiled	499
Ox tongue, pickled, raw	498
Oxo cubes	1182
Oxtail soup, canned, ready to serve	1132
Oxtail soup, dried	1140
Oxtail soup, dried, ready to serve	1141
Oxtail, stewed	493
Oxtail, stewed, weighed with fat and bones	494
Palak	see **Spinach**
Pale ale, bottled	1100
Palm ghee	336
Palm oil	325
Pancake roll	561
Pancakes, savoury, made with whole milk	169
Pancakes, Scotch	148
Pancakes, sweet, made with whole milk	162
Pangoli	see **Cauliflower**
Papadums, fried in vegetable oil	44
Papai	see **Paw-paw**

Papaya	see **Paw-paw**
Paprika	845
Parmesan cheese	247
Parsley, fresh	846
Parsnip, boiled in unsalted water	800
Parsnip, raw	799
Partridge, roast, meat only	452
Partridge, roast, weighed with bone	453
Passion fruit	933
Passion fruit, weighed with skin	934
Pasta sauce, tomato based	1170
Paste, fish	654
Paste, meat	516
Pasteurised goats milk	204
Pasteurised semi-skimmed milk	186
Pasteurised semi-skimmed milk, fortified plus Skimmed Milk Powder	187
Pasteurised skimmed milk	182
Pasteurised skimmed milk, fortified plus Skimmed Milk Powder	183
Pasteurised whole milk	190
Pasteurised whole milk, summer	191
Pasteurised whole milk, winter	192
Pastie, Cornish	508
Pastilles	1028
Pastries, Danish	135
Pastries, Greek	140
Pastry, flaky, cooked	125
Pastry, flaky, raw	124
Pastry, shortcrust, cooked	127
Pastry, shortcrust, raw	126
Pastry, wholemeal, cooked	129
Pastry, wholemeal, raw	128
Pate, liver	517
Pate, liver, low fat	518
Paw-paw, canned in juice	937
Paw-paw, raw	935
Paw-paw, raw, weighed with skin and pips	936
Peaches, canned in juice	940
Peaches, canned in syrup	941
Peaches, raw	938
Peaches, raw, weighed with stone	939
Peanut butter, smooth	986
Peanut oil	326
Peanuts and raisins	1036
Peanuts, dry roasted	989
Peanuts, plain	987
Peanuts, plain, weighed with shells	988
Peanuts, roasted and salted	990
Pears, average, raw	942
Pears, average, raw, peeled	944
Pears, average, raw, weighed with core	943

Pears, canned in juice	945
Pears, canned in syrup	946
Peas, blackeye	see **Blackeye beans**
Peas, boiled in unsalted water	730
Peas, canned, drained	733
Peas, chick, canned, drained	706
Peas, chick, whole, dried, boiled in unsalted water	705
Peas, chick, whole, dried, raw	704
Peas, frozen, boiled in salted water	731
Peas, frozen, boiled in unsalted water	732
Peas, mange-tout	see **Mange-tout peas**
Peas, mushy	see **Mushy peas**
Peas, petit pois	see **Petit pois**
Peas, processed	see **Processed peas**
Peas, raw	729
Pecan nuts	991
Pepper, black	847
Pepper, white	848
Peppermints	1029
Peppers, capsicum, chilli, green, raw	801
Peppers, capsicum, green, boiled in salted water	803
Peppers, capsicum, green, raw	802
Peppers, capsicum, red, boiled in salted water	805
Peppers, capsicum, red, raw	804
Petit pois, frozen, boiled in salted water	734
Petit pois, frozen, boiled in unsalted water	735
Pheasant, roast, meat only	454
Pheasant, roast, weighed with bone	455
Phool gobhi	see **Cauliflower**
Pickle, sweet	1160
Pickled beetroot, drained	743
Pickled gherkins, drained	773
Pickled onions, cocktail/silverskin, drained	798
Pickled onions, drained	797
Pie, blackcurrant, pastry top and bottom	150
Pie, fish	655
Pie, fruit, individual	158
Pie, fruit, one crust	156
Pie, fruit, pastry top and bottom	157
Pie, fruit, wholemeal, one crust	159
Pie, fruit, wholemeal, pastry top and bottom	160
Pie, lemon meringue	161
Pie, pork, individual	520
Pie, steak and kidney, individual	534
Pie, steak and kidney, pastry top only	535
Pie, with pie filling	163
Pie filling, cherry	880
Pie filling, fruit	893
Pies, mince, individual	144
Pig heart, raw	475

Pig kidney, raw	481
Pig kidney, stewed	482
Pig liver, raw	491
Pig liver, stewed	492
Pigeon, roast, meat only	456
Pigeon, roast, weighed with bone	457
Pignolias	see **Pine nuts**
Pilchards, canned in tomato sauce	621
Pimento	see **Peppers**, capsicum, chilli, green
Pine kernels	see **Pine nuts**
Pine nuts	992
Pineapple juice, unsweetened	1092
Pineapple, canned in juice	948
Pineapple, canned in syrup	949
Pineapple, raw	947
Pista	see **Pistachio nuts**
Pistachio nuts, weighed with shells	993
Pitta bread, white	45
Piyaz	see **Onions**
Pizza	170
Pizza, frozen	171
Plaice, in batter, fried in blended oil	597
Plaice, in batter, fried in dripping	598
Plaice, in crumbs, fried, fillets	599
Plaice, raw	594
Plaice, steamed	595
Plaice, steamed, weighed with bones and skin	596
Plain scones	146
Plantain, boiled in unsalted water	807
Plantain, raw	806
Plantain, ripe, fried in vegetable oil	808
Plums, average, raw	950
Plums, average, raw, weighed with stones	951
Plums, average, stewed with sugar, weighed with stones	952
Plums, average, stewed without sugar, weighed with stones	953
Plums, canned in syrup	954
Poached eggs, chicken	295
Polony	519
Polyunsaturated margarine	314
Popcorn, candied	1030
Popcorn, plain	1031
Porage	see **Porridge**
Pork, belly rashers, lean and fat, grilled	418
Pork, belly rashers, lean and fat, raw	417
Pork, chops, loin, lean and fat, grilled	420
Pork, chops, loin, lean and fat, grilled, weighed with bone	421
Pork, chops, loin, lean and fat, raw	419
Pork, chops, loin, lean only, grilled	422
Pork, chops, loin, lean only, grilled, weighed with fat and bone	423
Pork, fat only, cooked, average	415
Pork, fat only, raw, average	414
Pork, lean only, raw, average	416
Pork, leg, lean and fat, raw	424
Pork, leg, lean and fat, roast	425
Pork, leg, lean only, roast	426
Pork, trotters and tails, salted, boiled	427
Pork and ham, chopped, canned	513
Pork pie, individual	520
Pork sausages, fried	528
Pork sausages, grilled	529
Pork sausages, raw	527
Porridge, made with water	76
Porridge, made with whole milk	77
Port	1113
Potato crisps	1037
Potato crisps, low fat	1038
Potato croquettes, fried in blended oil	690
Potato hoops	1039
Potato powder, instant, made up with water	688
Potato powder, instant, made up with whole milk	689
Potato waffles, frozen, cooked	691
Potato, sweet	see **Sweet potato**
Potatoes, new, average, raw	660
Potatoes, new, boiled in unsalted water	661
Potatoes, new, canned, drained	663
Potatoes, new, in skins, boiled in unsalted water	662
Potatoes, old, average, raw	664
Potatoes, old, baked, flesh and skin	665
Potatoes, old, baked, flesh only	666
Potatoes, old, baked, flesh only, weighed with skin	667
Potatoes, old, boiled in unsalted water	668
Potatoes, old, mashed with butter	669
Potatoes, old, mashed with margarine	670
Potatoes, old, roast in blended oil	671
Potatoes, old, roast in corn oil	672
Potatoes, old, roast in lard	673
Prawns, boiled	639
Prawns, boiled, weighed with shell	640
Processed cheese, plain	248
Processed peas, canned, drained	736
Prunes, canned in juice	955
Prunes, canned in syrup	956
Prunes, ready-to-eat	957
Pudding, bread	151
Pudding, Christmas	152
Pudding, Christmas, retail	153
Pudding, milk, made with skimmed milk	284

Pudding, milk, made with whole milk	283
Pudding, rice, canned	287
Pudding, sponge	164
Puffed Wheat	78
Pumpkin, boiled in salted water	810
Pumpkin, raw	809
Purple grenadillo	see **Passion fruit**
Queensland nuts	see **Macadamia nuts**, salted
Quiche, cheese and egg	303
Quiche, cheese and egg, wholemeal	304
Quick cook oatmeal, raw	6
Quorn, myco-protein	811
Rabbit, raw, meat only	468
Rabbit, stewed, meat only	469
Rabbit, stewed, weighed with bone	470
Radish, red, raw	812
Raisin Splitz	79
Raisins	958
Raisins and peanuts	1036
Rapeseed oil, high erucic acid	327
Rapeseed oil, low erucic acid	328
Raspberries, canned in syrup	960
Raspberries, raw	959
Ravioli, canned in tomato sauce	172
Ready Brek	80
Red kidney beans, canned, drained	718
Red kidney beans, dried, boiled in unsalted water	717
Red kidney beans, dried, raw	716
Red lentils, split, dried, boiled in unsalted water	713
Red lentils, split, dried, raw	712
Red peppers, boiled in salted water	805
Red peppers, raw	804
Red wine	1107
Reduced calorie salad cream	1162
Reduced sugar jam	1006
Rhubarb, canned in syrup	964
Rhubarb, raw	961
Rhubarb, stewed with sugar	962
Rhubarb, stewed without sugar	963
Ribena, undiluted	1085
Rice, brown, boiled	19
Rice, brown, raw	18
Rice, egg fried	298
Rice, savoury, cooked	21
Rice, savoury, raw	20
Rice, white, easy cook, boiled	23
Rice, white, easy cook, raw	22
Rice, white, fried	24
Rice Krispies	81
Rice pudding, canned	287
Rich fruit cake	114
Rich fruit cake, iced	115
Rich iced fruit cake	115
Ricicles	82
Ring doughnuts	137
Ringana	see **Aubergine**
Risotto, plain	173
Roast chicken, dark meat	441
Roast chicken, light meat	440
Roast chicken, meat and skin	439
Roast chicken, meat only	438
Roast turkey, dark meat	465
Roast turkey, light meat	464
Roast turkey, meat and skin	463
Roast turkey, meat only	462
Rock eel	see **Dogfish**
Rock salmon	see **Dogfish**
Roe, cod, hard, fried	657
Roe, herring, soft, fried	658
Rolls, brown, crusty	58
Rolls, brown, soft	59
Rolls, white, crusty	62
Rolls, white, soft	63
Rolls, wholemeal	64
Rosé wine, medium	1108
Rosehip syrup, undiluted	1086
Rosemary, dried	849
Rum	1122
Rump steak beef, lean and fat, fried	372
Rump steak beef, lean and fat, grilled	373
Rump steak beef, lean and fat, raw	371
Rump steak beef, lean only, fried	374
Rump steak beef, lean only, grilled	375
Runner beans, boiled in unsalted water	720
Runner beans, raw	719
Rutabaga	see **Swede**
Rye bread	46
Rye crispbread	95
Rye flour, whole	7
Saag	see **Spinach**
Safflower oil	329
Sage and onion stuffing	177
Sage, dried, ground	850
Sago, raw	8
Saithe, raw	600
Saithe, steamed	601
Saithe, steamed, weighed with bones and skin	602
Salad cream	1161
Salad cream, reduced calorie	1162
Salami	521

Salmon, canned	625
Salmon, raw	622
Salmon, smoked	626
Salmon, steamed	623
Salmon, steamed, weighed with bones and skin	624
Salt, block	1183
Salt, table	1184
Salted beef, dried, raw	377
Salted beef, fat removed, raw	376
Samosas, meat	174
Samosas, vegetable	175
Sandwich biscuits	103
Sardines, canned in oil, drained	628
Sardines, canned in tomato sauce	627
Satsumas	965
Satsumas, weighed with peel	966
Sauce, barbecue	1164
Sauce, Bolognese	546
Sauce, bread, made with semi-skimmed milk	1145
Sauce, bread, made with whole milk	1144
Sauce, brown, bottled	1165
Sauce, cheese, made with semi-skimmed milk	1147
Sauce, cheese, made with whole milk	1146
Sauce, cook-in-sauces, canned	1166
Sauce, curry, canned	1167
Sauce, horseradish	1168
Sauce, mint	1169
Sauce, onion, made with semi-skimmed milk	1151
Sauce, onion, made with whole milk	1150
Sauce, pasta, tomato based	1170
Sauce, soy, dark, thick	1171
Sauce, tomato	1173
Sauce, white, savoury, made with semi-skimmed milk	1153
Sauce, white, savoury, made with whole milk	1152
Sauce, white, sweet, made with semi-skimmed milk	1155
Sauce, white, sweet, made with whole milk	1154
Sauce packet mix, cheese, made up with semi-skimmed milk	1149
Sauce packet mix, cheese, made up with whole milk	1148
Sausages, beef, fried	525
Sausages, beef, grilled	526
Sausages, beef, raw	524
Sausages, low fat, fried	531
Sausages, low fat, grilled	532
Sausages, low fat, raw	530
Sausages, pork, fried	528
Sausages, pork, grilled	529
Sausages, pork, raw	527
Sausage roll, flaky pastry	522
Sausage roll, short pastry	523
Saveloy	533
Savoury complan powder	1057
Savoury complan powder, made up with water	1058
Savoury pancakes, made with whole milk	169
Savoury rice, cooked	21
Savoury rice, raw	20
Savoury white sauce, made with semi-skimmed milk	1153
Savoury white sauce, made with whole milk	1152
Scampi, in breadcrumbs, frozen, fried	641
Scones, fruit	145
Scones, plain	146
Scones, wholemeal	147
Scotch eggs, retail	305
Scotch pancakes	148
Scrag and neck lamb, lean and fat, raw	407
Scrag and neck lamb, lean and fat, stewed	408
Scrag and neck lamb, lean only, stewed	409
Scrag and neck lamb, lean only, stewed, weighed with fat and bone	410
Scrambled eggs, with milk	296
Semi-skimmed Channel Island milk, UHT	197
Semi-skimmed milk, average	185
Semi-skimmed milk, pasteurised	186
Semi-skimmed milk, pasteurised, fortified plus Skimmed Milk Powder	187
Semi-skimmed milk, UHT	188
Semi-sweet biscuits	104
Sesame oil	330
Sesame seeds	994
Shakaria	see **Sweet potato**
Shalgam	see **Turnip**
Sheep heart, roast	476
Sheep tongue, stewed	500
Sheeps milk, raw	208
Shepherd's pie	562
Sherry, dry	1114
Sherry, medium	1115
Sherry, sweet	1116
Short-sweet biscuits	105
Shortbread	106
Shortcrust pastry, cooked	127
Shortcrust pastry, raw	126
Shoulder lamb, lean and fat, raw	411
Shoulder lamb, lean and fat, roast	412
Shoulder lamb, lean only, roast	413
Shredded Wheat	83
Shreddies	84
Shrimps, canned, drained	643
Shrimps, frozen, shell removed	642
Silverside beef, lean and fat, salted, boiled	378

Silverside beef, lean only, salted, boiled	379
Silverskin onions, drained	798
Simla mirch see **Peppers**, capsicum, green	
Sirloin beef, lean and fat, raw	380
Sirloin beef, lean and fat, roast	381
Sirloin beef, lean only, roast	382
Skate, in batter, fried	603
Skate, in batter, fried, weighed with waste	604
Skimmed condensed milk, sweetened	198
Skimmed milk, average	181
Skimmed milk, pasteurised	182
Skimmed milk, pasteurised, fortified plus Skimmed Milk Powder	183
Skimmed milk, UHT, fortified	184
Smacks	85
Smartie-type sweets	1023
Smoked haddock, steamed	584
Smoked haddock, steamed, weighed with bones and skin	585
Smoked mackerel	620
Smoked salmon	626
Snowpeas see **Mange-tout peas**	
Soft margarine, animal and vegetable fat	312
Soft margarine, vegetable fat only	313
Sole, lemon, in crumbs, fried	592
Sole, lemon, in crumbs, fried, weighed with bones	593
Sole, lemon, raw	589
Sole, lemon, steamed	590
Sole, lemon, steamed, weighed with bones and skin	591
Sorbet, lemon	273
Soup, instant, dried	1136
Soup, instant, dried, ready to serve	1137
Soup, chicken noodle, dried	1134
Soup, chicken noodle, dried, ready to serve	1135
Soup, lentil	1123
Soup, low calorie, canned	1131
Soup, minestrone, dried, ready to serve	1139
Soup, minestrone, dried	1138
Soup, oxtail, canned, ready to serve	1132
Soup, oxtail, dried	1140
Soup, oxtail, dried, ready to serve	1141
Soup, tomato, dried	1142
Soup, tomato, dried, ready to serve	1143
Soup, vegetable, canned, ready to serve	1133
Soy sauce	1171
Soya bean tofu, steamed	723
Soya bean tofu, steamed, fried	724
Soya beans, dried, boiled in unsalted water	722
Soya beans, dried, raw	721
Soya flour, full fat	9

Soya flour, low fat	10
Soya milk, flavoured	210
Soya milk, plain	209
Soya oil	331
Soya yogurt	258
Spaghetti, canned in tomato sauce	176
Spaghetti, white, boiled	30
Spaghetti, white, raw	29
Spaghetti, wholemeal, boiled	32
Spaghetti, wholemeal, raw	31
Sparkling white wine	1111
Special K	86
Spinach, boiled in unsalted water	814
Spinach, frozen, boiled in unsalted water	815
Spinach, raw	813
Spirits, 40% volume	1122
Sponge cake	119
Sponge cake, fatless	120
Sponge cake, jam filled	121
Sponge cake, with butter icing	122
Sponge pudding	164
Spread, chocolate nut	999
Spread, dairy/fat	307
Spread, low-fat	308
Spread, very low fat	315
Spring greens, boiled in unsalted water	817
Spring greens, raw	816
Spring onions, bulbs and tops, raw	818
Sprouts, Brussels see **Brussels sprouts**	
Squid, frozen, raw	647
Start	87
Steak, rump, lean and fat, fried	372
Steak, rump, lean and fat, grilled	373
Steak, rump, lean and fat, raw	371
Steak, rump, lean only, fried	374
Steak, rump, lean only, grilled	375
Steak, stewed, canned, with gravy	536
Steak, stewing, lean and fat, raw	383
Steak, stewing, lean and fat, stewed	384
Steak and kidney pie, individual	534
Steak and kidney pie, pastry top only	535
Sterilised cream, canned	217
Sterilised whole milk	193
Stew, beef	545
Stew, Irish	554
Stew, Irish, weighed with bones	555
Stewed apples, cooking, stewed with sugar	854
Stewed apples, cooking, stewed without sugar	855
Stewed blackberries, stewed with sugar	870
Stewed blackberries, stewed without sugar	871
Stewed blackcurrants, stewed with sugar	873
Stewed damsons, stewed with sugar	885

Treacle, golden syrup	1012
Trifle	288
Trifle, with fresh cream	289
Tripe, dressed	501
Tripe, dressed, stewed	502
Trotters and tails pork, salted, boiled	427
Trout, brown, steamed	629
Trout, brown, steamed, weighed with bones	630
Tuna, canned in brine, drained	632
Tuna, canned in oil, drained	631
Turkey, dark meat, raw	461
Turkey, light meat, raw	460
Turkey, meat and skin, raw	459
Turkey, meat only, raw	458
Turkey, roast, dark meat	465
Turkey, roast, light meat	464
Turkey, roast, meat and skin	463
Turkey, roast, meat only	462
Turkish delight, without nuts	1033
Turnip, boiled in unsalted water	834
Turnip, raw	833
Turnip, yellow	see **Swede**
Twiglets	1042
Twix	1024
Tzatziki	259
UHT Channel Island milk, semi-skimmed	197
UHT cream, canned spray	218
UHT semi-skimmed milk	188
UHT skimmed milk, fortified	184
Urad	see **Black gram**, urad gram
Urad gram black gram, dried, boiled in unsalted water	699
Urad gram black gram, dried, raw	698
Vatana	see **Peas**
Veal, cutlet, fried in vegetable oil	428
Veal, fillet, raw	429
Veal, fillet, roast	430
Vegetable ghee	337
Vegetable oil, blended, average	333
Vegetable samosas	175
Vegetable soup, canned, ready to serve	1133
Vegetable soup, low calorie, canned	1131
Venison, roast	471
Vermouth, dry	1117
Vermouth, sweet	1118
Very low fat fromage frais	241
Very low fat spread	315
Vinegar	1185
Vintage cider	1106
Vitbe, average	47

Wafer biscuits, filled	107
Wafers, ice cream	272
Waffles, potato, frozen, cooked	691
Walnuts	997
Walnuts, weighed with shells	998
Water	1186
Watercress	835
Watermelon	925
Weetabix	90
Weetaflake	91
Weetos	92
Wheat and Oat Bran	75
Wheat bran	1
Wheat flour, brown	12
Wheat flour, white, breadmaking	13
Wheat flour, white, plain	14
Wheat flour, white, self-raising	15
Wheat flour, wholemeal	16
Wheatgerm	17
Wheatgerm oil	334
Whelks, boiled, weighed with shell	648
Whisky	1122
White bread, average	48
White bread, French stick	53
White bread, fried in blended oil	50
White bread, fried in lard	51
White bread, sliced	49
White bread, toasted	52
White bread, 'with added fibre'	54
White bread, 'with added fibre', toasted	55
White cabbage, raw	753
White chapati flour	3
White cheese, average	250
White chocolate	1017
White flour, breadmaking	13
White flour, plain	14
White flour, self-raising	15
White pepper	848
White pitta bread	45
White pudding	538
White rice, easy cook, boiled	23
White rice, easy cook, raw	22
White rice, fried in lard/dripping	24
White rolls, crusty	62
White rolls, soft	63
White sauce, savoury, made with semi-skimmed milk	1153
White sauce, savoury, made with whole milk	1152
White sauce, sweet, made with semi-skimmed milk	1155
White sauce, sweet, made with whole milk	1154

White sugar	1011
White wine, dry	1109
White wine, medium	1110
White wine, sparkling	1111
White wine, sweet	1112
Whitebait, fried	633
Whiting, in crumbs, fried	607
Whiting, in crumbs, fried, weighed with bones	608
Whiting, steamed	605
Whiting, steamed, weighed with bones	606
Whole Channel Island milk, pasteurised	194
Whole Channel Island milk, pasteurised, summer	195
Whole Channel Island milk, pasteurised, winter	196
Whole condensed milk, sweetened	199
Whole evaporated milk	202
Whole milk yogurt, fruit	261
Whole milk yogurt, plain	260
Whole milk, average	189
Whole milk, pasteurised	190
Whole milk, pasteurised, summer	191
Whole milk, pasteurised, winter	192
Whole milk, sterilised	193
Wholegrain mustard	1181
Wholemeal bread, average	56
Wholemeal bread, toasted	57
Wholemeal crackers	108
Wholemeal flour	16
Wholemeal fruit cake	116
Wholemeal fruit pie, one crust	159
Wholemeal fruit pie, pastry top and bottom	160
Wholemeal pastry, cooked	129
Wholemeal pastry, raw	128
Wholemeal rolls	64
Wholemeal scones	147

Wholemeal spaghetti, boiled	32
Wholemeal spaghetti, raw	31
Wholemeal wheat flour	16
Wine, red	1107
Wine, rosé, medium	1108
Wine, white, dry	1109
Wine, white, medium	1110
Wine, white, sparkling	1111
Wine, white, sweet	1112
Wing quarter, roast chicken, meat only, weighed with bone	442
Winkles, boiled, weighed with shell	649
Witloof	see **Chicory**
Yam (USA)	see **Sweet potato**
Yam, boiled in unsalted water	837
Yam, raw	836
Yeast, bakers, compressed	1187
Yeast, bakers, dried	1188
Yellow gram	see **Chick peas**
Yellow turnip	see **Swede**
Yogurt, drinking	251
Yogurt, Greek, cows	252
Yogurt, Greek, sheep	253
Yogurt, low calorie	254
Yogurt, low fat, flavoured	256
Yogurt, low fat, fruit	257
Yogurt, low fat, plain	255
Yogurt, soya	258
Yogurt, whole milk, fruit	261
Yogurt, whole milk, plain	260
Yorkshire pudding	180
Zucchini	see **Courgette**